Advances in Experimental Medicine and Biology

Volume 895

More information about this series at http://www.springer.com/series/5584

Toni Cathomen • Matthew Hirsch
Matthew Porteus
Editors

Genome Editing

The Next Step in Gene Therapy

Springer

Editors
Toni Cathomen
Medical Center - University of Freiburg
Freiburg, Germany

Matthew Hirsch
University of North Carolina
Chapel Hill, NC, USA

Matthew Porteus
Stanford Medical School
Palo Alto, CA, USA

ISSN 0065-2598 ISSN 2214-8019 (electronic)
Advances in Experimental Medicine and Biology
ISBN 978-1-4939-3507-9 ISBN 978-1-4939-3509-3 (eBook)
DOI 10.1007/978-1-4939-3509-3

Library of Congress Control Number: 2015958888

Springer New York Heidelberg Dordrecht London

Printed on acid-free paper

Springer Science+Business Media LLC New York is part of Springer Science+Business Media (www.springer.com)

This book is dedicated to the memory of Carlos F. Barbas III. Among his incredibly long list of accomplishments, he was best known in the field of genome engineering for his invaluable contributions in designing zinc-finger arrays that have been used in artificial transcription factors, zinc-finger nucleases, and custom recombinases.
"It has been a dream of mine to develop drugs that make a difference"
(Carlos Barbas, 2008)

Preface

In the early 1900s, human genetic engineering remained the fanciful speculation of science fiction writers, scientists, and philosophers. The surfaced ideas perceived consequences that, already more than 100 years ago, invoked ethical dilemmas: Would cavalier applications of genetic engineering interrupt "natural" evolution? Would engineered creatures be capable of destroying us? Would an altered DNA blueprint make us susceptible to presently harmless "pathogens"? The concerns appeared limitless. In our day, most of our genome editing efforts are driven by the allure of potentially curing genetic diseases. In fact, some newly developed genome editing technologies are so effective and easy to use that the genes of nonhuman primate embryos could be altered. In view of that, it is conceivable that the same technique, which is used to cure genetic disorders, could also be employed to introduce genetic changes in the human germline in order to enhance human qualities, like intelligence or good looks. Where do we draw the line? To give us more time for public discussion and to better understand how safe the current genome engineering tools are, scientist and ethicist have called for a moratorium on human germline editing.

Human genetic engineering is at the forefront of disease therapy research based on seminal observations that have collectively increased the frequency of the ability of a cell to "process" its DNA. Depending on the cellular decision, both chromosomal disruption and the precise tailoring of a native locus with endogenous or exogenous DNA remain possible. To complement our ability to induce DNA alterations, great strides have been made to deliver nucleic acids efficiently throughout the human body. As we are on the crest of this genetic tsunami, it appears timely to coalesce the current understanding of the early twentieth century.

We feel happy that some of the world's most prominent geneticists, biologists, and bioinformaticians, each having an expertise we felt will continue to shape our understanding and our ability of human genome engineering, have contributed to this book. In attempts to make this volume relevant to a broad target audience, the authors of the individual chapters and the editors have made an effort to provide sufficient background for the respective genre. The final product represents the most comprehensive work on the many facets of human genetic engineering and our stepwise progression toward a dream of a disease-free existence. Please enjoy.

Freiburg, Germany
Chapel Hill, NC
Palo Alto, CA

Toni Cathomen
Matthew Louis Hirsch
Matthew Porteus

Contents

Contributors

Raman Bahal, Ph.D. Department of Therapeutic Radiology, Yale University School of Medicine, New Haven, CT, USA

Gang Bao, Ph.D. Department of Biomedical Engineering, Georgia Institute of Technology and Emory University, Atlanta, GA, USA

Michele P. Calos, Ph.D. Department of Genetics, Stanford University School of Medicine, Stanford, CA, USA

Paula M. Cannon, Ph.D. Department of Molecular Microbiology and Immunology, Keck School of Medicine, University of Southern California, Los Angeles, CA, USA

Dana Carroll, Ph.D. Department of Biochemistry, University of Utah School of Medicine, Salt Lake City, UT, USA

Toni Cathomen, Ph.D. Institute for Cell and Gene Therapy, Medical Center - University of Freiburg, Freiburg, Germany

Thomas James Cradick Department of Biomedical Engineering, Georgia Institute of Technology and Emory University, Atlanta, GA, USA

Philippe Duchateau, Ph.D. Research Department, Paris, France

Colin M. Exline, Ph.D. Department of Molecular Microbiology and Immunology, Keck School of Medicine, University of Southern California, Los Angeles, CA, USA

Eli J. Fine Department of Biomedical Engineering, Georgia Institute of Technology and Emory University, Atlanta, GA, USA

Charles A. Gersbach, Ph.D. Department of Biomedical Engineering, Duke University, Durham, NC, USA

Peter M. Glazer, M.D., Ph.D. Department of Therapeutic Radiology, Yale University, New Haven, CT, USA

Department of Genetics, Yale University, New Haven, CT, USA

Anisha Gupta, Ph.D. Department of Therapeutic Radiology, Yale University School of Medicine, New Haven, CT, USA

Matthew Louis Hirsch, Ph.D. Gene Therapy Center, University of North Carolina at Chapel Hill, Chapel Hill, NC, USA

Department of Ophthalmology, University of North Carolina at Chapel Hill, Chapel Hill, NC, USA

Nathalia Holt, Ph.D. Department of Molecular Microbiology and Immunology, Keck School of Medicine, University of Southern California, Los Angeles, CA, USA

Maria Jasin, Ph.D. Developmental Biology Program, Memorial Sloan Kettering Cancer Center, New York, NY, USA

Alexandre Juillerat, Ph.D. Research Department, Paris, France

George Nicholas Llewellyn, Ph.D. Department of Molecular Microbiology and Immunology, Keck School of Medicine, University of Southern California, Los Angeles, CA, USA

Angela M. Mitchell, Ph.D. Gene Therapy Center, University of North Carolina at Chapel Hill, Chapel Hill, NC, USA

David Mittelman Department of Biological Sciences, Virginia Tech, Blacksburg, VA, USA

Virginia Bioinformatics Institute, Blacksburg, VA, USA

Rachel Moser, B.A. Department of Ophthalmology, University of North Carolina at Chapel Hill, Chapel Hill, NC, USA

Christopher Moye, Ph.D. Verna and Marrs McLean Biochemistry and Molecular Biology, Baylor College of Medicine, Houston, TX, USA

Claudio Mussolino, Ph.D. Center for Chronic Immunodeficiency, Medical Center - University of Freiburg, Freiburg, Germany

Mark J. Osborn, Ph.D. Department of Pediatrics, Division of Blood and Marrow Transplantation, Masonic Cancer Center, Stem Cell Institute, and Center for Genome Engineering, University of Minnesota, Minneapolis, MN, USA

David G. Ousterout, Ph.D. Department of Biomedical Engineering, Duke University, Durham, NC, USA

Patrick Ruff, Ph.D. Department of Microbiology and Immunology, Columbia University Medical Center, New York, NY, USA

Richard Jude Samulski, Ph.D. Department of Pharmacology, Gene Therapy Center, University of North Carolina at Chapel Hill, Chapel Hill, NC, USA

Francesca Storici, Ph.D. School of Biology, Georgia Institute of Technology, Atlanta, GA, USA

Jakub Tolar, M.D., Ph.D. Department of Pediatrics, Division of Blood and Marrow Transplantation, Stem Cell Institute, University of Minnesota, Minneapolis, MN, USA

John H. Wilson, Ph.D. Verna and Marrs McLean Biochemistry and Molecular Biology, Baylor College of Medicine, Houston, TX, USA

About the Editors

Toni Cathomen, Ph.D., is professor and director of the Institute for Cell and Gene Therapy at the University Medical Center Freiburg, Germany. The institute provides the Medical Center with blood and cell products as well as all transfusion and transplantation-related diagnostic services. Toni Cathomen received his Ph.D. from the University of Zurich, Switzerland. Before his appointment in Freiburg, he was a postdoctoral fellow at the Salk Institute in San Diego, USA, assistant professor of molecular virology at Charité Medical School in Berlin, and associate professor of experimental hematology at Hannover Medical School. Toni Cathomen's main research goals are (1) to further improve safe genome editing tools (incl. TALENs, CRISPR/Cas9) for therapeutic applications in human stem cells, (2) to develop disease models and cell therapies based on induced pluripotent stem cells (iPSCs), and (3) to translate cell and gene therapy efforts into the clinic.

Matthew Hirsch is an assistant professor of ophthalmology at the University of North Carolina at Chapel Hill. He also holds appointments in microbiology and immunology, genetics and molecular biology, and in the Gene Therapy Center at UNC. Dr. Hirsch obtained his Ph.D. from West Virginia University working on *E. coli* and *Salmonella* genetics in Morgantown, WV. He completed his postdoc with Jude Samulski at UNC studying both episomal and chromosomal genetic engineering using adeno-associated virus (AAV). Dr. Hirsch continues these basic AAV studies and has several reagents under preclinical evaluation for the treatment of blindness and muscular dystrophies.

Dr. Matthew Porteus was raised in California and was a local graduate of Gunn High School before completing an A.B. degree in history and science at Harvard University where he graduated magna cum laude and wrote a thesis entitled "Safe or Dangerous Chimeras: The Recombinant DNA Controversy as a Conflict Between Differing Socially Constructed Interpretations of Recombinant DNA Technology." He then returned to the area and completed his combined M.D. and Ph.D. at Stanford Medical School with his Ph.D. focused on understanding the molecular basis of mammalian forebrain development with his Ph.D. thesis entitled "Isolation and

Characterization of TES-1/DLX-2: A Novel Homeobox Gene Expressed During Mammalian Forebrain Development." After completion of his dual degree program, he was an intern and resident in pediatrics at Boston Children's Hospital and then completed his pediatric hematology/oncology fellowship in the combined Boston Children's Hospital/Dana Farber Cancer Institute program. For his fellowship and postdoctoral research, he worked with Dr. David Baltimore at MIT and Caltech where he began his studies in developing homologous recombination as a strategy to correct disease-causing mutations in stem cells as definitive and curative therapy for children with genetic diseases of the blood, particularly sickle cell disease. Following his training with Dr. Baltimore, he took an independent faculty position at UT Southwestern in the Departments of Pediatrics and Biochemistry before again returning to Stanford in 2010 as an associate professor. During this time his work has been the first to demonstrate that gene correction could be achieved in human cells at frequencies that were high enough to potentially cure patients and is considered one of the pioneers and founders of the field of genome editing—a field that now encompasses thousands of labs and several new companies throughout the world. His research program continues to focus on developing genome editing by homologous recombination as curative therapy for children with genetic diseases but also has interests in the clonal dynamics of heterogeneous populations and the use of genome editing to better understand diseases that affect children including infant leukemias and genetic diseases that affect the muscle. Clinically, Dr. Porteus attends at the Lucille Packard Children's Hospital where he takes care of pediatric patients undergoing hematopoietic stem cell transplantation.

Gene Editing 20 Years Later

Maria Jasin

Abstract Directed modification of the genome is critical for interrogating gene function and can also be applied for gene therapy. Two decades ago a double-strand break (DSB) in the genome was discovered to induce efficient gene modification, either by homologous recombination with introduced DNA, i.e., gene targeting, or imprecise joining of DNA ends leading to mutagenesis. The accelerating development of technologies—meganucleases, ZFNs, TALENs, and CRISPR/Cas9— to introduce DSBs at specific sites in the genome for the purposes of modification is revolutionizing the biological and biomedical sciences. This chapter provides an overview of the research that led to these advances in gene editing and also summarizes DSB repair mechanisms in mammalian cells.

Keywords Gene targeting • Gene editing • Genome engineering • Homologous recombination • Nonhomologous end-joining • NHEJ • Cas9 • Double-strand break • Homology-directed repair • Genome rearrangement

Introduction

The ability to make directed modifications of the genome is critical to understanding the function of genes and it also holds great promise for gene therapy [1]. The discovery that DNA double-strand breaks (DSBs) induce efficient genome modification—termed gene editing—and the subsequent development of technologies to introduce DSBs at specific sites in the genome are revolutionizing the biological and biomedical sciences. The discoveries that led to these advances will be summarized in this chapter. A summary of DSB repair mechanisms in mammalian cells will also be provided.

M. Jasin, Ph.D. (✉)
Developmental Biology Program, Memorial Sloan Kettering Cancer Center,
1275 York Avenue, New York, NY 10065, USA
e-mail: m-jasin@ski.mskcc.org

T. Cathomen et al. (eds.), *Genome Editing*, Advances in Experimental
Medicine and Biology 895, DOI 10.1007/978-1-4939-3509-3_1

Facile Gene Modification in Yeast but Impediments in Mammalian Cells

Geneticists discovered in 1978 that DNA introduced into budding yeast integrates into the genome at homologous sequences by homologous recombination (HR) [2]. Subsequently, a variety of approaches based on HR were developed to exploit this observation [3, 4]. As a result, the yeast genome could be efficiently modified to study gene function, propelling budding yeast as the model organism of choice for molecular genetic studies of basic cellular processes in eukaryotes.

By contrast, DNA introduced into mammalian cells was found to integrate at nonhomologous sequences in the genome [5]. Homologous integration—also called gene targeting—was observed at low frequency, requiring extensive screening to be detected [6]. The use of drug markers for selection/counterselection [7] and promoterless selection markers [8] facilitated the recovery of gene-targeted cells, such that gene targeting has become commonplace in some systems, e.g., mouse embryonic stem cells to make mutant mice [9]. Despite improvements in gene targeting approaches, however, applications have been limited due to several factors, including the requirement for selection due to its inefficiency, inability to target both alleles in diploid cells in a single round of screening, and the need for embryonic stem cell lines for creating modified animals.

DSBs as Initiators of Homologous Recombination

The preponderance of nonhomologous integration of plasmid DNA into the genome of mammalian cells had dissuaded investigators from considering that HR played a vital role in mammalian cells. However, continued discoveries in yeast provided an entry point to tackle the problem. The introduction of a DSB in a plasmid within a sequence homologous to the chromosome was found to substantially increase the efficiency of the gene targeting events [10]. Two types of events were observed: The DSB could be repaired without integration of the plasmid (Fig. 1a); thus, if a deletion occurred at the DSB site in the plasmid (i.e., a gap), the plasmid could be repaired leading to restoration of the sequence. Alternatively, the plasmid could integrate into the genome during the repair process (Fig. 1b).

These studies, foreshadowed by studies of γ-irradiation-induced DSBs in yeast (see [11]), and studies in bacteria and their phages [12] demonstrated that DNA ends are recombinogenic in yeast and prokaryotic systems and led to the proposal of the DSB repair model for HR [13]. In this model, DSBs are initiators of HR and the DNA that is broken (recipient) is converted to the sequence of the unbroken, homologous DNA (donor). A central intermediate of this model is a double Holliday junction which can be resolved to give rise to either a crossover or a noncrossover. Considering plasmid DSB repair events involving a chromosome, a crossover leads to integration of the plasmid (Fig. 1b) whereas a noncrossover restores plasmid sequences without integration (Fig. 1a).

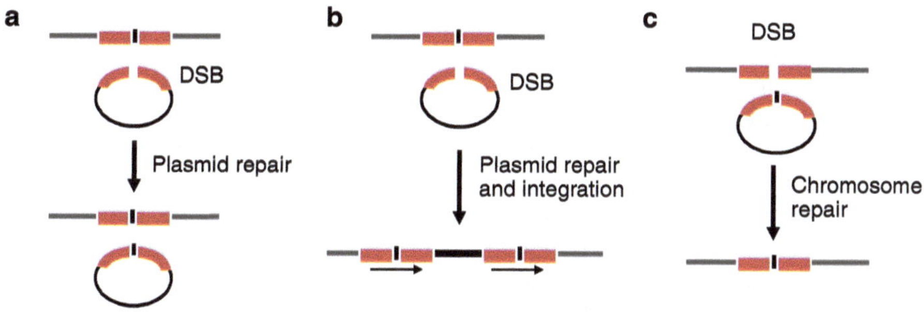

Fig. 1 DSB-induced homologous recombination (HR) between chromosomal and plasmid DNA. Homology between the chromosome and plasmid is represented by the *red bar*. (**a**) A DSB or gap in plasmid DNA can be repaired using the chromosome as a template by a simple gene conversion. Such noncrossovers can be detected if the plasmid contains an origin of replication. (**b**) A DSB or gap in plasmid DNA can be repaired using the chromosome as a template by gene conversion with a crossover, leading to plasmid integration during gene targeting. Crossovers are suppressed in mammalian cells. (**c**) A DSB in the chromosome can be repaired using the plasmid as a template, resulting in efficient genome modification

Publication of DSB repair studies in yeast and the proposal of the DSB repair model led us to directly consider whether mammalian DSB repair could also occur by HR [14]. Initial studies involved a DSB (gap) on a plasmid that could be repaired from the chromosome (Fig. 1a), leading to the production of intact virus. These experiments demonstrated efficient DSB repair by HR, such that ~10% of the linearized plasmids were estimated to have undergone HR leading to a noncrossover. Although genome modification was not achieved with this experimental design, it provided clear proof that HR could be a prominent DSB repair mechanism in mammalian cells, which had previously been unsuspected.

To detect genome modification, the experimental design was revised to target the same locus with a promoterless selectable marker gene [8]. DSB repair leading to plasmid integration was observed (Fig. 1b), modifying the genome and further demonstrating the recombinogenic nature of DSBs in mammalian cells. The enrichment of homologous integration events with a DSB was estimated to be 100-fold [8]. Despite this large effect, these integration events (crossovers) were noted to be significantly less frequent than non-integrative events (noncrossovers) [14]. This bias suggested that specific mechanisms exist that suppress crossing-over in mammalian cells, which is now well supported (e.g., [15]).

Groundbreaking Experiments: DSBs in the Genome Induce Gene Targeting Orders of Magnitude

The finding that DSBs in plasmids induced recombination with the chromosome suggested that homologous DSB repair could be co-opted in mammalian cells for genome engineering. However, they also pointed to a limitation: Because the DNA with the DSB is the recipient of genetic information [13], the model implied that a DSB should be in the genome, not the plasmid, for direct genome modification (Fig. 1c). While cleaving the genome was not done in yeast because homologous integration of plasmid DNA is readily achieved, the model suggested that mammalian events could be greatly enhanced by the introduction of a site-specific DSB into the genome. Clues as to how to achieve this came from the specialized mating-type switching system in yeast, in which a site-specific DSB induces HR between two chromosomal sequences [16]. By moving the DSB site to another genomic location and expressing the mating-type HO endonuclease, HR between this other location and homologous chromosomal sequences was stimulated [17]. Studies using transposons in *Drosophila* also showed that HR between chromosomal sequences could be induced by DSBs [18].

To determine if a DSB would stimulate gene targeting (Fig. 1c), we induced a DSB in the genome while also introducing a homologous fragment that could be used to repair the DSB [19]. In particular, we expressed I-SceI endonuclease [20], which is related to HO endonuclease and which induces another type of site-specific HR event in yeast [21]. I-SceI endonuclease was used for this purpose because its cleavage site was well defined and long—18 bp [22]—such that its expression was not expected to be lethal to cells with complex genomes [20].

The I-SceI recognition site was integrated into the mammalian genome and a 700-bp fragment homologous to sequences flanking the DSB was provided when I-SceI was expressed (Fig. 2a). Gene targeting was elevated several orders of magnitude [19], indicating that the introduction of a chromosomal DSB is a viable way to increase gene targeting in organisms refractory to spontaneous gene targeting [23]. Subsequent studies confirmed the recombinogenicity of DSBs for gene targeting in mammalian cells and demonstrated similar results in embryonic stem cells, other cell types, and with circular homologous DNA [24–28]. These experiments performed in 1994 are the biological basis of current genome editing approaches using HR, as will be discussed below.

DSBs in the Genome Induce Mutagenesis

The 1994 experiments that demonstrated that DSBs induce gene targeting also demonstrated that DSBs induce mutagenesis (Fig. 2b). In this case, the DNA ends generated by I-SceI were rejoined without homology or with just a few bp of homology, termed microhomology [19]. Estimates from these experiments were that these imprecise nonhomologous end-joining (NHEJ) events occurred twice as frequently as gene targeting (HR). Small deletions and insertions were seen at the breakpoint junctions, characteristic of imprecise NHEJ [29–31]. Larger deletions induced by a single DSB were also identified using selection [32]. These experiments form the biological basis of current genome editing approaches using NHEJ, as will be discussed below.

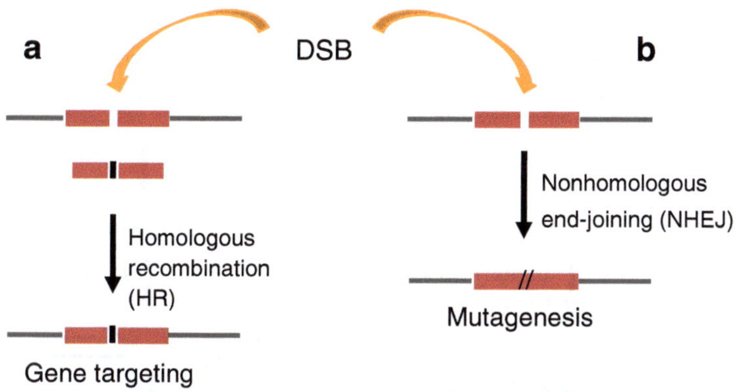

Fig. 2 Gene editing induced by a DSB in the chromosome. A DSB in the chromosome can be repaired by HR with introduced homologous DNA, i.e., gene targeting, (**a**) or by imprecise NHEJ leading to mutagenesis (**b**). Initial studies in mammalian cells with I-SceI endonuclease showed that both kinds of events were efficiently induced, with imprecise NHEJ events somewhat more abundant [19]. Experiments with ZFNs, TALENS, and CRISPR/Cas9 have also shown that both types of gene editing are induced by these nucleases as well

Two DSBs Induce Genomic Rearrangements

These experiments also demonstrated that two DSBs on the same chromosome could lead to deletions [19]. Moreover, other types of chromosomal aberrations involving two DSBs were generated in other experiments. Chromosomal rearrangements are commonly observed in cancer cells, including recurrent, reciprocal translocations [33]. I-SceI endonuclease has been used to induce chromosomal translocations by the placement of an I-SceI site on each participating chromosome [34, 35]. Translocations were not observed when only one chromosome incurred a DSB [27]. Two DSBs were found to induce translocations through NHEJ [36], but not by HR, likely due to crossover suppression [34, 35]. Because cancer translocation breakpoint junctions do not typically show homology between the two chromosomal sequences that are joined, NHEJ-based translocation systems are relevant to understanding the joining mechanisms by which oncogenic translocations arise [37].

HR Studies Using I-SceI Endonuclease

Repair of a DSB by HR (sometimes called homology-direct repair, or HDR) initiates with DNA end resection to generate single-stranded DNA overhangs (for a review, see [38] and references therein) (Fig. 3a). The single strands provide a substrate for a strand exchange protein—RAD51 in eukaryotes—to form a nucleoprotein filament which can then invade the unbroken homologous sequence. Repair

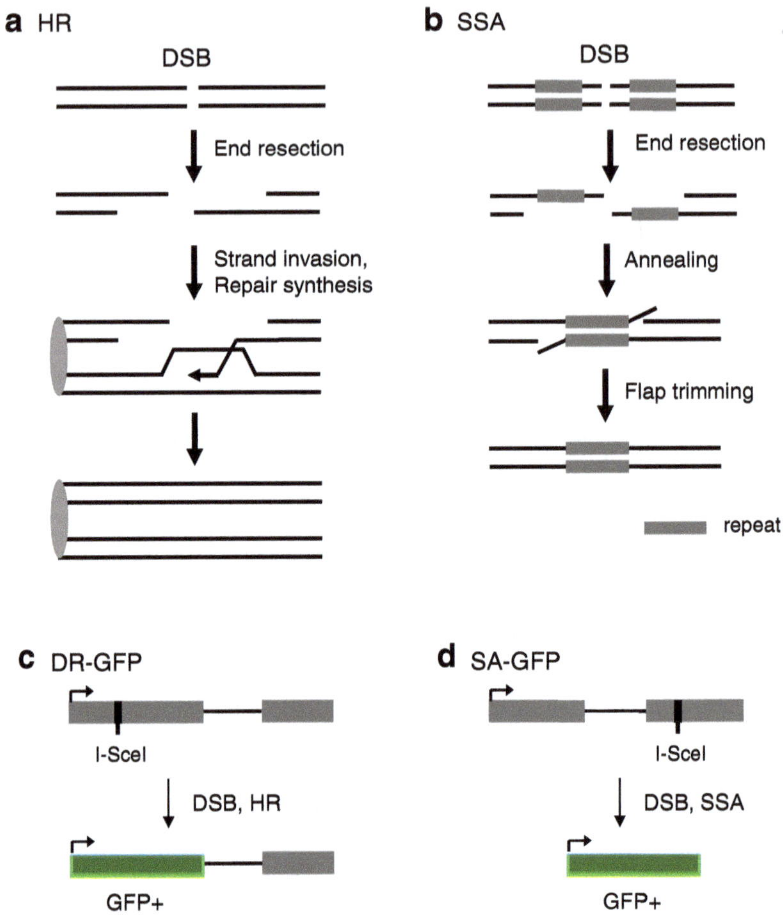

Fig. 3 HR and single-strand annealing (SSA). (**a**) DSB repair by HR, simplified. DNA ends are resected leading to single-stranded DNA tails onto which the RAD51 protein can form a filament to promote strand invasion of an unbroken homologous DNA, typically the sister chromatid, as shown. The 3′ end of the invading strand primes DNA synthesis; use of the sister chromatid can precisely restore the original sequence prior to damage. To complete HR, the newly synthesized strand can dissociate to anneal to the other end, although other outcomes are possible [38]. (**b**) DSB repair by SSA. SSA can occur at a DSB flanked by sequence repeats. As with HR, SSA begins with end resection, but the complementary single-stranded DNA generated on either side of the DSB can anneal. Flaps that are generated can be trimmed by nucleases. (**c**) DR-GFP reporter to measure HR. In the DR-GFP reporter assay, a DSB repaired through HR between the two nonfunctional GFP genes restores a functional GFP gene, as detected by flow cytometry. The *left* GFP gene is nonfunctional due to the presence of the I-SceI site; the *right* GFP gene is nonfunctional because it is truncated at both ends. (**d**) SA-GFP reporter to measure SSA. In the SA-GFP reporter assay, a DSB repaired through SSA between the two nonfunctional GFP genes restores a functional GFP gene. In this case, the *left* GFP gene is truncated at its 3′ end; the *right* GFP gene is truncated at its 5′ end and it also contains an I-SceI site

DNA synthesis is primed by the invading end and uses the homologous sequence as a template. A variety of outcomes are possible after this point, but one of the simplest is the annealing of the newly synthesized strand to the resected DNA end which was not involved in the strand invasion (see [39] and references therein). HR in mammalian cells appears to occur most frequently between sister chromatids, although homologous sequences on homologs and on other chromosomes can be used at lower frequencies [40].

An alternative pathway that uses homology, termed single-strand annealing (SSA) can occur when homologous sequences flank the DSB (Fig. 3b). This pathway also initiates with DNA end resection but the complementary single-strands formed by resection anneal to each other rather than initiating strand invasion. After annealing, DNA flaps can be trimmed prior to ligation. The frequency of SSA relative to HR will be determined in part by the distance between repeats and their sequence identity. Thus far, the role of SSA is unclear, although it has been used in conjunction with HR for distinguishing whether proteins are involved in end resection or later steps of HR.

Understanding HR in mammalian cells and the factors involved has been facilitated by the use of I-SceI endonuclease. Because gene targeting is not thought to reflect physiological HR events, intrachromosomal HR reporters have been developed, the most common of which, DR-GFP, is based on GFP fluorescence [41] (Fig. 3c). In this reporter, a simple conversion of sequences at the DSB site by the unbroken, homologous sequence on the same chromatid or sister chromatid results in GFP positive cells. The DR-GFP reporter has now been introduced into mice to study HR in primary somatic cells [42]. The related SA-GFP reporter is used to assay SSA events [43] (Fig. 3d).

With DR-GFP and related reporters, mammalian HR mutants could be conclusively identified. Thus, proteins related to RAD51 [41, 44] and the BRCA1 and BRCA2 tumor suppressors [45–47] were clearly identified as promoting HR. Comparison of HR and SSA in mutant cells led to the discovery that BRCA1 and BRCA2 work at different steps in the HR pathway: BRCA1 mutant cells were found to be deficient in both HR and SSA, suggesting this protein works at a common early step, i.e., end resection, whereas BRCA2 mutant cells were found to be deficient in HR but to have increased SSA [43]. A role for BRCA1 in end resection has been supported by subsequent studies [48], whereas a role for BRCA2 in the later step of RAD51-mediated strand exchange is clear from biochemical studies [49]. The increase in SSA in BRCA2 mutant cells occurs because end resected molecules that would normally be channeled into HR are free to undergo strand annealing [43, 50].

In addition to the simple HR events detected by the DR-GFP reporter, events involving longer lengths of repair synthesis were also observed, resulting in a duplication of sequences [51] (Fig. 4a). These events have a somewhat different genetic control; for example, they are overrepresented in the residual HR events found in BRCA1 and other mutant cells (see [52] and references therein).

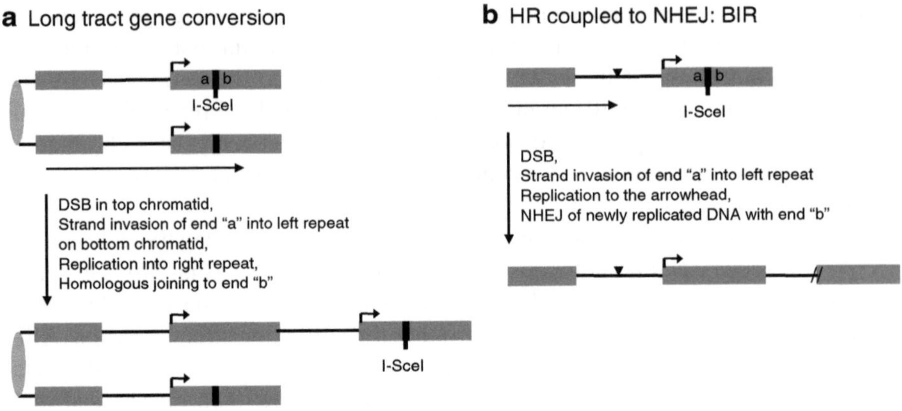

Fig. 4 Other types of HR events detected after I-SceI cleavage. (**a**) Long tract gene conversion is an HR event that involves extensive replication, such that a duplication of sequences occurs. The event is completed at homologous sequences, as depicted. (**b**) HR coupled to NHEJ, also termed BIR. Duplication of sequences also occurs in this case, but the event is completed by NHEJ

Collaboration and Competition Between HR and NHEJ

Several NHEJ factors were identified because they are required for antigen receptor rearrangement, a specialized pathway where DSBs are generated by the RAG proteins [29–31]. The imprecise joining that occurs during NHEJ generates diversity, which is important for the immune response. NHEJ mutants have increased HR and SSA when challenged with an I-SceI DSB [43, 53], indicating competition between DSB repair pathways for repair of a single DSB [54]. This finding is emphasized by recent studies showing that BRCA1 mutant phenotypes can be suppressed by loss of a protein implicated in NHEJ [48].

HR and NHEJ can sometimes be coupled to repair the same DSB. In this case, one DNA end invades a homologous sequence to prime repair synthesis (HR) but the newly synthesized DNA joins to the other DNA end by NHEJ [51, 55] (Fig. 4b). This type of event has recently been termed break-induced replication (BIR) [56], but it differs from similarly termed events in yeast in which replication extends to the end of the chromosome [57], although both have a reliance on a common replication factor [56]. BIR, like the events that terminate in homology (Fig. 4a), duplicate existing chromosomal sequences (Fig. 4b); it has been suggested that large genomic duplications arise in this manner in cancer cells [56].

Genome Editing with an Emerging Range of Nucleases

The initial gene editing studies with I-SceI endonuclease made it clear that genetic manipulation of genomes was possible using rare-cutting endonucleases [23]. Critical for the success of this approach is an endonuclease directed to cleave the locus to be modified. Oligonucleotide reagents seemed to hold promise, because they could in principle target any site in the genome through Watson–Crick base pairing [23], e.g., oligonucleotides with an incorporated chemical cleaving moiety coated with the bacterial strand exchange protein RecA for in vitro cleavage [58]. Homing endonucleases like I-SceI—often called meganucleases—were expected to be difficult to modify to recognize other sequences, which is borne out by the crystal structure showing the complex interactions of I-SceI with DNA [59]. On the other hand, progress has been made in meganuclease redesign (see, for example, [60] or [61] for I-SceI). Although unlikely to be developed as generalized cleavage reagents, meganucleases have characteristics that may make their use desirable in some circumstances. For example, they are small—I-SceI is only 235 amino acids—and a single chain, which facilitates their delivery to target cells, especially in gene therapy applications.

The modularity of zinc finger DNA binding domains presented an alternative route to engineer DNA binding domains with novel recognition specificities [62–64]. The bipartite nature of the restriction enzyme FokI, which has distinct DNA binding and cleavage domains [65, 66], provided a modular cleavage domain that could be fused to zinc finger DNA recognition domains to create nucleases with novel cleavage specificities [67], now called zinc finger nucleases (ZFNs). In principle, each zinc finger would interact with three base pairs, such that several "fingers" could be assembled and fused to the FokI cleavage domain to cleave a unique site in a complex genome [68]. Mutagenesis and gene targeting using ZFNs were first reported in *Drosophila* at the scoreable *yellow* gene, targeting a site containing GNN triplets which are well defined for zinc finger binding [69, 70]. In the multi-component gene targeting system, DNA was excised from the genome using I-SceI endonuclease to provide a donor for HR [70].

The first DSB-induced gene targeting of a human gene was performed at the disease-relevant *IL-2Rγ* gene [71]. The efficiency of gene targeting with ZFN expression using a plasmid donor was extremely high, including in stem cells. What was even more remarkable was the high efficiency of bi-allelic targeting, something that is not achieved by traditional gene targeting approaches. A number of studies in different systems have now used ZFNs for a variety of gene editing purposes [68]. A particularly powerful application of DSB-induced mutagenesis to human disease is the HIV co-receptor *CCR5* gene [72], an approach that is currently being used for gene therapy [1].

Despite these successes, the assembly of zinc finger modules that recognize DNA with high specificity has been difficult to generalize for researchers; for example, each zinc finger unit is not completely independent of each other and zinc fingers to each of the 64 triplets are not readily available. Nuclease design was greatly facilitated by the discovery of the simple DNA recognition code of TAL effector proteins from plant pathogens, in which two amino acids within each module recognize a single base pair [73–76]. As in ZFNs, fusions of the DNA binding domain repeats are made to the FokI cleavage domain to generate TAL effector nucleases (TALENs) [77].

The development of TALENs essentially solved the problem of the ability to make generalized cleavage reagents for the purpose of genome modification. However, the discovery of an RNA-guided nuclease in bacterial adaptive immunity, termed CRISPR/Cas9 [78], has made the approach even easier, given that a single nuclease (Cas9) is used together with an RNA which directs cleavage specificity based on Watson-Crick base pairing. Researchers were quick to adapt CRISPR/Cas9 to editing the genome [79, 80], such that it has rapidly become the approach of choice for researchers [81]. In addition to the simpler construction relative to TALENs, CRISPR/Cas9 is readily adaptable to multiplexing, because only the RNA component needs to be introduced to target cleavage of multiple sites.

As with gene targeting and mutagenesis, designer nucleases have been used to generate chromosomal rearrangements. For example, cancer-relevant chromosomal translocations have been induced by ZFNs, TALENs, and CRISPR/Cas9 in a number of human cell lines [37, 82, 83]. A similar approach has been used to induce an oncogenic chromosomal inversion in mice, giving rise to lung tumors [84].

Conclusions

The last 20 years have seen an acceleration of gene editing approaches. A decade elapsed between our initial gene editing experiment and the use of ZFNs to modify endogenous genes. Five years after the application of ZFNs, the development of TALENs allowed basic researchers to essentially edit any gene, and 3 years later CRISPR/Cas9 was applied to further facilitate gene editing through a simple nucleic acid target design. Investigators now working with almost any organism can consider adapting these technologies to address a variety of biological questions. The expectation is that further improvements will be made to these technologies, although it is difficult to imagine that advances will occur that will be as great those described here which have already revolutionized biomedical science. One clear lesson is the continued importance of basic research, given that the development of genome editing relied on discoveries from diverse systems that could not have been anticipated to have such a revolutionary impact on biomedical science.

Acknowledgments We thank Rohit Prakash and Matt Krawczyk for discussions. This work is supported by NIH CA185660 and GM054668.

References

1. Tebas P, Stein D, Tang WW, Frank I, Wang SQ, Lee G, et al. Gene editing of CCR5 in autologous CD4 T cells of persons infected with HIV. N Engl J Med. 2014;370(10):901–10.
2. Hinnen A, Hicks JB, Fink GR. Transformation of yeast. Proc Natl Acad Sci U S A. 1978;75(4):1929–33.
3. Scherer S, Davis RW. Replacement of chromosome segments with altered DNA sequences constructed in vitro. Proc Natl Acad Sci U S A. 1979;76(10):4951–5.

4. Rothstein RJ. One-step gene disruption in yeast. Methods Enzymol. 1983;101:202–11.
5. Lacy E, Roberts S, Evans EP, Burtenshaw MD, Costantini FD. A foreign beta-globin gene in transgenic mice: integration at abnormal chromosomal positions and expression in inappropriate tissues. Cell. 1983;34(2):343–58.
6. Smithies O, Gregg RG, Boggs SS, Koralewski MA, Kucherlapati RS. Insertion of DNA sequences into the human chromosomal ß-globin locus by homologous recombination. Nature. 1985;317:230–4.
7. Mansour SL, Thomas KR, Capecchi MR. Disruption of the proto-oncogene int-2 in mouse embryo-derived stem cells: a general strategy for targeting mutations to non-selectable genes. Nature. 1988;336(6197):348–52.
8. Jasin M, Berg P. Homologous integration in mammalian cells without target gene selction. Genes Dev. 1988;2:1353–63.
9. Capecchi MR. Gene targeting in mice: functional analysis of the mammalian genome for the twenty-first century. Nat Rev Genet. 2005;6(6):507–12.
10. Orr-Weaver TL, Szostak JW, Rothstein RJ. Yeast transformation: a model system for the study of recombination. Proc Natl Acad Sci U S A. 1981;78(10):6354–8.
11. Resnick MA. The repair of double-strand breaks in DNA; a model involving recombination. J Theor Biol. 1976;59(1):97–106.
12. Thaler DS, Stahl FW. DNA double-chain breaks in recombination of phage lambda and of yeast. Annu Rev Genet. 1988;22:169–97.
13. Szostak JW, Orr-Weaver TL, Rothstein RJ, Stahl FW. The double-strand-break repair model for recombination. Cell. 1983;33(1):25–35.
14. Jasin M, deVilliers J, Weber F, Schaffner W. High frequency of homologous recombination in mammalian cells between endogenous and introduced SV40 genomes. Cell. 1985;43: 695–703.
15. LaRocque JR, Stark JM, Oh J, Bojilova E, Yusa K, Horie K, et al. Interhomolog recombination and loss of heterozygosity in wild-type and Bloom syndrome helicase (BLM)-deficient mammalian cells. Proc Natl Acad Sci U S A. 2011;108(29):11971–6.
16. Strathern JN, Klar AJ, Hicks JB, Abraham JA, Ivy JM, Nasmyth KA, et al. Homothallic switching of yeast mating type cassettes is initiated by a double-stranded cut in the MAT locus. Cell. 1982;31(1):183–92.
17. Nickoloff JA, Chen EY, Heffron F. A 24-base-pair DNA sequence from the MAT locus stimulates intergenic recombination in yeast. Proc Natl Acad Sci U S A. 1986;83(20):7831–5.
18. Gloor GB, Nassif NA, Johnson-Schlitz DM, Preston CR, Engels WR. Targeted gene replacement in Drosophila via P element-induced gap repair. Science. 1991;253(5024):1110–7.
19. Rouet P, Smih F, Jasin M. Introduction of double-strand breaks into the genome of mouse cells by expression of a rare-cutting endonuclease. Mol Cell Biol. 1994;14:8096–106.
20. Rouet P, Smih F, Jasin M. Expression of a site-specific endonuclease stimulates homologous recombination in mammalian cells. Proc Natl Acad Sci U S A. 1994;91:6064–8.
21. Belfort M, Roberts RJ. Homing endonucleases: keeping the house in order. Nucleic Acids Res. 1997;25(17):3379–88.
22. Colleaux L, d'Auriol L, Gailbert F, Dujon B. Recognition and cleavage site of the intron-encoded omega transposase. Proc Natl Acad Sci U S A. 1988;85:6022–6.
23. Jasin M. Genetic manipulation of genomes with rare-cutting endonucleases. Trends Genet. 1996;12(6):224–8.
24. Smih F, Rouet P, Romanienko PJ, Jasin M. Double-strand breaks at the target locus stimulate gene targeting in embryonic stem cells. Nucleic Acids Res. 1995;23(24):5012–9.
25. Choulika A, Perrin A, Dujon B, Nicolas JF. Induction of homologous recombination in mammalian chromosomes by using the I-SceI system of Saccharomyces cerevisiae. Mol Cell Biol. 1995;15(4):1968–73.
26. Liang F, Romanienko PJ, Weaver DT, Jeggo PA, Jasin M. Chromosomal double-strand break repair in Ku80-deficient cells. Proc Natl Acad Sci U S A. 1996;93(17):8929–33.
27. Richardson C, Moynahan ME, Jasin M. Double-strand break repair by interchromosomal recombination: suppression of chromosomal translocations. Genes Dev. 1998;12(24):3831–42.

28. Donoho G, Jasin M, Berg P. Analysis of gene targeting and intrachromosomal homologous recombination stimulated by genomic double-strand breaks in mouse embryonic stem cells. Mol Cell Biol. 1998;18(7):4070–8.
29. Goodarzi AA, Jeggo PA. The repair and signaling responses to DNA double-strand breaks. Adv Genet. 2013;82:1–45.
30. Deriano L, Roth DB. Modernizing the nonhomologous end-joining repertoire: alternative and classical NHEJ share the stage. Annu Rev Genet. 2013;47:433–55.
31. Pannunzio NR, Li S, Watanabe G, Lieber MR. Non-homologous end joining often uses microhomology: Implications for alternative end joining. DNA Repair (Amst). 2014;17:74–80.
32. Sargent RG, Brenneman MA, Wilson JH. Repair of site-specific double-strand breaks in a mammalian chromosome by homologous and illegitimate recombination. Mol Cell Biol. 1997;17(1):267–77.
33. Mitelman F, Johansson B, Mertens F. The impact of translocations and gene fusions on cancer causation. Nat Rev Cancer. 2007;7(4):233–45.
34. Richardson C, Jasin M. Frequent chromosomal translocations induced by DNA double-strand breaks. Nature. 2000;405(6787):697–700.
35. Elliott B, Richardson C, Jasin M. Chromosomal translocation mechanisms at intronic alu elements in mammalian cells. Mol Cell. 2005;17(6):885–94.
36. Weinstock DM, Elliott B, Jasin M. A model of oncogenic rearrangements: differences between chromosomal translocation mechanisms and simple double-strand break repair. Blood. 2006;107(2):777–80.
37. Ghezraoui H, Piganeau M, Renouf B, Renaud JB, Sallmyr A, Ruis B, et al. Chromosomal translocations in human cells are generated by canonical nonhomologous end-joining. Mol Cell. 2014;55(6):829–42.
38. Jasin M, Rothstein R. Repair of strand breaks by homologous recombination. Cold Spring Harb Perspect Biol. 2013;5(11).
39. Ferguson DO, Holloman WK. Recombinational repair of gaps in DNA is asymmetric in Ustilago maydis and can be explained by a migrating D-loop model. Proc Natl Acad Sci U S A. 1996;93(11):5419–24.
40. Johnson RD, Jasin M. Double-strand-break-induced homologous recombination in mammalian cells. Biochem Soc Trans. 2001;29(Pt 2):196–201.
41. Pierce AJ, Johnson RD, Thompson LH, Jasin M. XRCC3 promotes homology-directed repair of DNA damage in mammalian cells. Genes Dev. 1999;13(20):2633–8.
42. Kass EM, Helgadottir HR, Chen CC, Barbera M, Wang R, Westermark UK, et al. Double-strand break repair by homologous recombination in primary mouse somatic cells requires BRCA1 but not the ATM kinase. Proc Natl Acad Sci U S A. 2013;110(14):5564–9.
43. Stark JM, Pierce AJ, Oh J, Pastink A, Jasin M. Genetic steps of mammalian homologous repair with distinct mutagenic consequences. Mol Cell Biol. 2004;24(21):9305–16.
44. Johnson RD, Liu N, Jasin M. Mammalian XRCC2 promotes the repair of DNA double-strand breaks by homologous recombination. Nature. 1999;401(6751):397–9.
45. Moynahan ME, Chiu JW, Koller BH, Jasin M. Brca1 controls homology-directed DNA repair. Mol Cell. 1999;4(4):511–8.
46. Moynahan ME, Cui TY, Jasin M. Homology-directed DNA repair, mitomycin-c resistance, and chromosome stability is restored with correction of a Brca1 mutation. Cancer Res. 2001;61(12):4842–50.
47. Moynahan ME, Pierce AJ, Jasin M. BRCA2 is required for homology-directed repair of chromosomal breaks. Mol Cell. 2001;7(2):263–72.
48. Bunting SF, Callen E, Wong N, Chen HT, Polato F, Gunn A, et al. 53BP1 inhibits homologous recombination in Brca1-deficient cells by blocking resection of DNA breaks. Cell. 2010;141(2):243–54.
49. Jensen RB, Carreira A, Kowalczykowski SC. Purified human BRCA2 stimulates RAD51-mediated recombination. Nature. 2010;467(7316):678–83.

50. Tutt A, Bertwistle D, Valentine J, Gabriel A, Swift S, Ross G, et al. Mutation in Brca2 stimulates error-prone homology-directed repair of DNA double-strand breaks occurring between repeated sequences. EMBO J. 2001;20(17):4704–16.
51. Johnson RD, Jasin M. Sister chromatid gene conversion is a prominent double-strand break repair pathway in mammalian cells. EMBO J. 2000;19(13):3398–407.
52. Chandramouly G, Kwok A, Huang B, Willis NA, Xie A, Scully R. BRCA1 and CtIP suppress long-tract gene conversion between sister chromatids. Nat Commun. 2013;4:2404.
53. Pierce AJ, Hu P, Han M, Ellis N, Jasin M. Ku DNA end-binding protein modulates homologous repair of double-strand breaks in mammalian cells. Genes Dev. 2001;15(24):3237–42.
54. Kass EM, Jasin M. Collaboration and competition between DNA double-strand break repair pathways. FEBS Lett. 2010;584(17):3703–8.
55. Richardson C, Jasin M. Coupled homologous and nonhomologous repair of a double-strand break preserves genomic integrity in mammalian cells. Mol Cell Biol. 2000;20(23):9068–75.
56. Costantino L, Sotiriou SK, Rantala JK, Magin S, Mladenov E, Helleday T, et al. Break-induced replication repair of damaged forks induces genomic duplications in human cells. Science. 2014;343(6166):88–91.
57. Malkova A, Ira G. Break-induced replication: functions and molecular mechanism. Curr Opin Genet Dev. 2013;23(3):271–9.
58. Baliga R, Singleton JW, Dervan PB. RecA.oligonucleotide filaments bind in the minor groove of double-stranded DNA. Proc Natl Acad Sci U S A. 1995;92(22):10393–7.
59. Moure CM, Gimble FS, Quiocho FA. The crystal structure of the gene targeting homing endonuclease I-SceI reveals the origins of its target site specificity. J Mol Biol. 2003;334(4):685–95.
60. Takeuchi R, Choi M, Stoddard BL. Redesign of extensive protein-DNA interfaces of meganucleases using iterative cycles of in vitro compartmentalization. Proc Natl Acad Sci U S A. 2014;111(11):4061–6.
61. Doyon JB, Pattanayak V, Meyer CB, Liu DR. Directed evolution and substrate specificity profile of homing endonuclease I-SceI. J Am Chem Soc. 2006;128(7):2477–84.
62. Miller J, McLachlan AD, Klug A. Repetitive zinc-binding domains in the protein transcription factor IIIA from Xenopus oocytes. EMBO J. 1985;4(6):1609–14.
63. Pavletich NP, Pabo CO. Zinc finger-DNA recognition: crystal structure of a Zif268-DNA complex at 2.1 A. Science. 1991;252(5007):809–17.
64. Nardelli J, Gibson TJ, Vesque C, Charnay P. Base sequence discrimination by zinc-finger DNA-binding domains. Nature. 1991;349(6305):175–8.
65. Li L, Wu LP, Chandrasegaran S. Functional domains in Fok I restriction endonuclease. Proc Natl Acad Sci U S A. 1992;89(10):4275–9.
66. Wah DA, Hirsch JA, Dorner LF, Schildkraut I, Aggarwal AK. Structure of the multimodular endonuclease FokI bound to DNA. Nature. 1997;388(6637):97–100.
67. Kim YG, Cha J, Chandrasegaran S. Hybrid restriction enzymes: zinc finger fusions to Fok I cleavage domain. Proc Natl Acad Sci U S A. 1996;93(3):1156–60.
68. Urnov FD, Rebar EJ, Holmes MC, Zhang HS, Gregory PD. Genome editing with engineered zinc finger nucleases. Nat Rev Genet. 2010;11(9):636–46.
69. Bibikova M, Golic M, Golic KG, Carroll D. Targeted chromosomal cleavage and mutagenesis in Drosophila using zinc-finger nucleases. Genetics. 2002;161(3):1169–75.
70. Bibikova M, Beumer K, Trautman JK, Carroll D. Enhancing gene targeting with designed zinc finger nucleases. Science. 2003;300(5620):764.
71. Urnov FD, Miller JC, Lee YL, Beausejour CM, Rock JM, Augustus S, et al. Highly efficient endogenous human gene correction using designed zinc-finger nucleases. Nature. 2005;435(7042):646–51.
72. Varela-Rohena A, Carpenito C, Perez EE, Richardson M, Parry RV, Milone M, et al. Genetic engineering of T cells for adoptive immunotherapy. Immunol Res. 2008;42(1-3):166–81.
73. Moscou MJ, Bogdanove AJ. A simple cipher governs DNA recognition by TAL effectors. Science. 2009;326(5959):1501.

74. Boch J, Scholze H, Schornack S, Landgraf A, Hahn S, Kay S, et al. Breaking the code of DNA binding specificity of TAL-type III effectors. Science. 2009;326(5959):1509–12.
75. Deng D, Yan C, Pan X, Mahfouz M, Wang J, Zhu JK, et al. Structural basis for sequence-specific recognition of DNA by TAL effectors. Science. 2012;335(6069):720–3.
76. Mak AN, Bradley P, Cernadas RA, Bogdanove AJ, Stoddard BL. The crystal structure of TAL effector PthXo1 bound to its DNA target. Science. 2012;335(6069):716–9.
77. Doyle EL, Stoddard BL, Voytas DF, Bogdanove AJ. TAL effectors: highly adaptable phytobacterial virulence factors and readily engineered DNA-targeting proteins. Trends Cell Biol. 2013;23(8):390–8.
78. Jinek M, Chylinski K, Fonfara I, Hauer M, Doudna JA, Charpentier E. A programmable dual-RNA-guided DNA endonuclease in adaptive bacterial immunity. Science. 2012;337(6096): 816–21.
79. Cong L, Ran FA, Cox D, Lin S, Barretto R, Habib N, et al. Multiplex genome engineering using CRISPR/Cas systems. Science. 2013;339(6121):819–23.
80. Mali P, Yang L, Esvelt KM, Aach J, Guell M, DiCarlo JE, et al. RNA-guided human genome engineering via Cas9. Science. 2013;339(6121):823–6.
81. Hsu PD, Lander ES, Zhang F. Development and applications of CRISPR-Cas9 for genome engineering. Cell. 2014;157(6):1262–78.
82. Piganeau M, Ghezraoui H, De Cian A, Guittat L, Tomishima M, Perrouault L, et al. Cancer translocations in human cells induced by zinc finger and TALE nucleases. Genome Res. 2013;23(7):1182–93.
83. Choi PS, Meyerson M. Targeted genomic rearrangements using CRISPR/Cas technology. Nat Commun. 2014;5:3728.
84. Maddalo D, Manchado E, Concepcion CP, Bonetti C, Vidigal JA, Han YC, et al. In vivo engineering of oncogenic chromosomal rearrangements with the CRISPR/Cas9 system. Nature. 2014;516(7531):423–7.

The Development and Use of Zinc-Finger Nucleases

Dana Carroll

Abstract Zinc-finger nucleases (ZFNs) were the first of the targetable nucleases to be developed and exploited for genome engineering. They have proved remarkably effective, enhancing the frequency of gene targeting several orders of magnitude. The modularity of DNA recognition by zinc fingers has made it possible to design ZFNs for a wide range of genomic targets in a remarkable assortment of organisms and cell types. Use of this platform helped define the parameters and approaches for nuclease-stimulated genome manipulation. Although much of the territory has been ceded in the last few years to the more easily designed TALENs and CRISPR/Cas nucleases, successful ZFNs are still in wide use in a number of applications, including current clinical trials.

Keywords Zinc-finger nucleases (ZFNs) • Nonhomologous end joining (NHEJ) • Homologous recombination (HR) • Gene targeting

Introduction

If you are a geneticist, you have two ways to proceed to get a mutation in your favorite gene—forward or backward. Classically (forward genetics) you would generate random mutations, identify an interesting phenotype, and then endeavor to characterize the gene that harbored the responsible mutation, which might or might not be your gene. With the advent of methods for gene isolation and DNA sequencing, it became plausible to go the other direction (reverse genetics), first identifying the gene of interest and then attacking it specifically to generate mutations and test for phenotypes. Until relatively recently, however, the tools to do this were quite limited.

D. Carroll, Ph.D. (✉)
Department of Biochemistry, University of Utah School of Medicine,
15 N. Medical Drive East, Room 4100, Salt Lake City, UT 84112-5650, USA
e-mail: dana@biochem.utah.edu

T. Cathomen et al. (eds.), *Genome Editing*, Advances in Experimental Medicine and Biology 895, DOI 10.1007/978-1-4939-3509-3_2

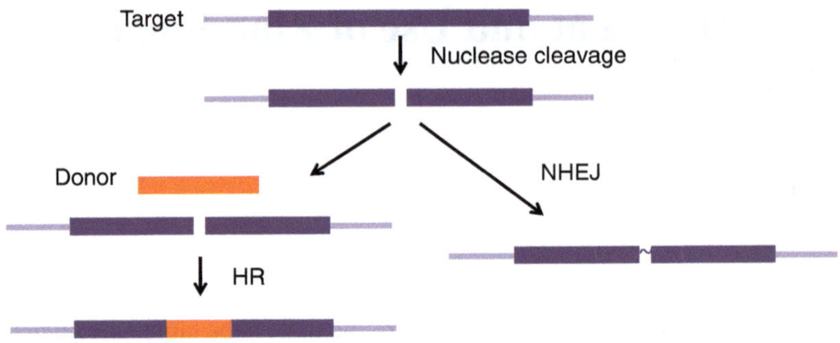

Fig. 1 Pathways of gene modification after a targetable nuclease-induced break. The target is shown as *a purple rectangle*. After nuclease cleavage, in the presence of a homologous donor DNA sequence (*orange*), the break can be repaired by homologous recombination (HR), incorporating sequences from the donor. An alternative repair pathway is nonhomologous end joining (NHEJ), which often leaves small sequence alterations at the site of the break, as indicated by the squiggle

In the 1970s and 1980s, investigators developed methods for gene targeting in yeast [1, 2] and in mice [3, 4], based on homologous recombination (HR) between an introduced DNA molecule and an endogenous target. The absolute frequency of recombination was quite low, even in yeast, but strong selection allowed recovery of the desired cells. For reasons both technical and biological, it proved difficult to extend these methods to other organisms. Beginning in the 1990s, whole genome sequences were being determined, and the desire for a facile approach to manipulate specific genes was growing. Researchers could identify sequences that they would like to alter, but had no reliable way to do so.

An important insight was the recognition that gene targeting relies on cellular DNA repair activities, and that, in normal circumstances, the intended genomic target is intact and not in need of repair. Double-strand breaks (DSBs) in chromosomal DNA constitute potentially lethal damage and must be repaired [5]. Furthermore, DSBs stimulate HR in a range of circumstances, including natural meiotic crossing over. It seemed, therefore, that the key to expanding the range and efficiency of gene targeting was to make the target more susceptible to homologous repair by breaking or otherwise damaging it. In addition to HR, DSBs are repaired in essentially all organisms by an error-prone process, called nonhomologous end joining (NHEJ). In practice, targeted DSBs lead both to local mutagenesis via NHEJ and, in the presence of an appropriate donor DNA, to targeted gene replacement (Fig. 1).

Quite a number of approaches have been taken to addressing specific genomic target sequences, and most of them are reviewed in this volume. The methods that have taken hold involve the use of nuclease proteins with separable recognition and cleavage modules. This class includes zinc-finger nucleases (ZFNs), transcription activator-like effector nucleases (TALENs), and CRISPR/Cas RNA-guided nucleases (CRISPRs). My assignment is to review the development and applications of the pioneers among these, the ZFNs. More extensive reviews are available elsewhere [6–9].

Origins of ZFNs

ZFNs are not natural proteins, but they originated from natural components. In the early 1990s, Chandrasegaran and colleagues discovered that the Type II restriction enzyme, *Fok*I, has separable DNA-recognition and DNA-cleavage domains [10]. This observation stimulated the conjecture that novel specificities could be produced by linking the nonspecific cleavage domain to alternative DNA-binding modules. This was demonstrated first by fusion with the homeo-box from the Drosophila transcription factor, Ubx [11].

Meanwhile, repetitive structural modules, called zinc fingers, were identified in a number of eukaryotic, DNA-binding transcription factors [12, 13]. The structure, determined by Pavletich and Pabo [14], of a set of three fingers bound to their cognate site confirmed the modularity of recognition and the coordination of a single zinc atom by two histidine and two cysteine residues in each finger. The principal contacts made by each finger were to three consecutive base pairs in the DNA [15] (Fig. 2). In his second chimeric restriction enzyme, Chandrasegaran fused two different sets of zinc fingers provided by his colleague, Jeremy Berg, to the *Fok*I cleavage domain and again demonstrated redirected cutting [16]. These fusions were the first ZFNs.

Information available in 1996 suggested that a wide range of DNA sequences could be specified by zinc fingers and that there might even be a code of recognition [17, 18]. The latter prospect has not been borne out [19], but the former certainly has. By design, by selection, and by characterizing natural fingers, researchers have established an extensive catalog of individual fingers and combinations that recognize many different sequences. The establishment of a code has been foiled by the fact that fingers that perform well in one context do not routinely function well in others [20, 21].

Characterization of ZFNs

Experiments by Smith et al. [22] and Bibikova et al. [23] determined the requirements for ZFN cleavage, both in solution and in cells. The *Fok*I cleavage domain must dimerize to be a functional nuclease [22, 24]. Apparently some aspect of dimer formation in the natural restriction enzyme is lost in the zinc-finger fusion. As a consequence, the weak dimer interface alone promotes association only at very high protein concentrations [24]. To achieve efficient cleavage by ZFNs, two sets of zinc fingers are required, each linked to a cleavage domain monomer and directed to sequences in close proximity on the target DNA [22, 23] (Fig. 2). At high local concentration, dimerization is favored and cleavage occurs.

ZFNs were able to cleave a chromatin substrate in intact cells and to stimulate homologous recombination [23]. This was demonstrated using synthetic substrates injected into *Xenopus* oocytes, and it was important because the bacterial *Fok*I

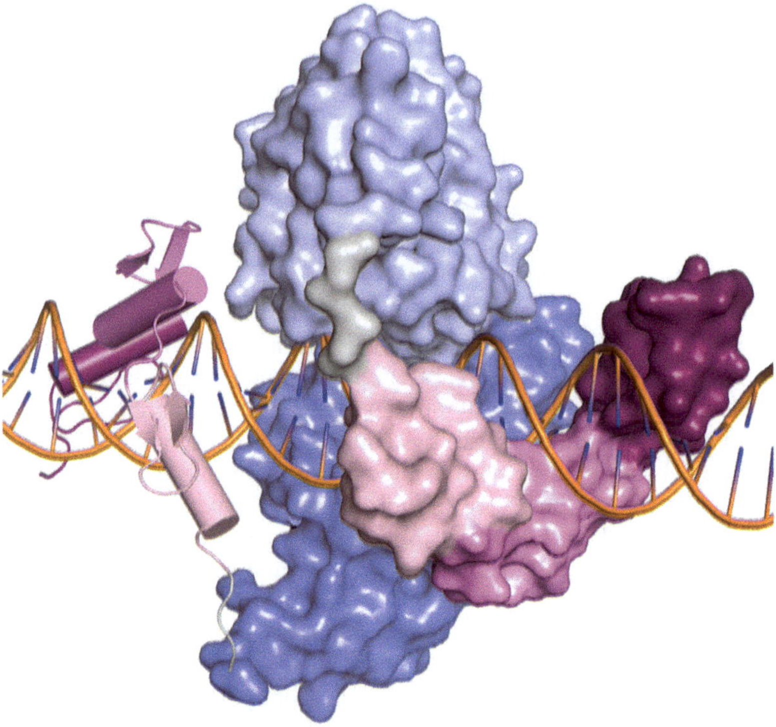

Fig. 2 Model of a pair of three-finger ZFNs bound to DNA. Each zinc finger is in a shade of *pink*, in *ribbon* representation on the *left* and space-filling representation on the *right*. The *Fok*I nuclease domains are in shades of *blue*, and the linker between these and the finger sets are shown in *gray*. The DNA axis runs horizontally through the figure, and the backbone is in *orange*. The zinc finger binding sites are 6 bp apart. This composite model was assembled using Protein Database submissions 1MEY and 2FOK [22]

nuclease would not normally see sequences in this context. How the ZFNs recognize sequences in chromatin is still not known, but there is no indication that this presents a limitation to their effectiveness. Using the oocyte system, the optimum spacer between zinc finger binding sites was shown to be 6 bp, when the linker between the binding and cleavage domains is reduced to four amino acids [23]. This linker is now in common use, and the preference for a 6-bp spacer has been validated in multiple studies [25, 26], although a spacer of 5 bp, and even 4 bp, works in some situations.

Designing ZFNs

As noted above, the modularity of zinc finger recognition does not translate into a simple code. Many groups have derived new fingers and finger combinations, both by design and by selection. Simply making new combinations of well-characterized fingers sometimes works, but not reliably [21, 27–29]. A number of researchers have established selection schemes [30–33], have identified fingers that work well together [34, 35], and have endeavored to understand what governs successful recognition [36]. Many active and specific ZFN pairs have been derived (e.g., see [8, 36]), but still no simple method for producing new ones has emerged.

The Klug group demonstrated a number of years ago that constructing DNA-binding arrays with units of two fingers enhanced their specificity [9, 37]. Six-finger arrays constructed with extended linkers between finger pairs are more sensitive to mismatches to the DNA, perhaps because one mismatch destabilizes the entire two-finger unit [37]. The most extensive library of zinc fingers and finger pairs is maintained by Sangamo Biosciences, Inc., and ZFNs based on this collection are marketed by Sigma-Aldrich (http://www.sigmaaldrich.com/life-science/zinc-finger-nuclease-technology.html). The price of these reagents has decreased substantially in recent years, but they are still rather expensive. The advantages are that Sigma does all the design and testing and ultimately provides validated ZFNs.

The first ZFNs for a genomic target displayed significant toxicity when expressed at high levels [38], due to promiscuous binding and cleavage [39], and this issue has continued to trouble many new designs. Increasing the number of fingers in each monomer is one approach to ameliorating this effect, but that is not always sufficient [40, 41]. The toxicity is frequently a property of one ZFN within a pair, and it appears to be due to inadequate specificity, leading to homodimerization and cleavage at unintended, off-target sites [39]. A major step forward was the introduction of substitutions in the dimer interface of the cleavage domain that prevent formation of homodimers, while allowing the necessary heterodimerization [42, 43]. The second generation of these obligate heterodimer modifications [44] also corrects a deficit in cleavage efficiency seen with the first generation, and these are now in common use with ZFNs and TALENs [45].

In many situations with experimental organisms, cleavage and mutagenesis (by NHEJ) at off-target sites is a tolerable nuisance, since the effects can be minimized by backcrossing, by complementation, or by use of independently derived alleles (e.g. [46],). In applications to humans, however, the issue is more concerning. Methods to detect off-target mutagenesis have been developed, based on determination of in vitro binding and cleavage specificity [9, 47–49]. Ultimately, procedures are needed that detect where secondary mutations are actually made in cells and organisms. [50] Whole genome sequencing would have to be very deep, since rare mutations can be selected from a cell population when introduced into patients, and such mutations can have severe effects [51].

The Utility of ZFNs in Gene Targeting

The first experiments in an intact organism showed that ZFNs directed to a genomic sequence in *Drosophila melanogaster* could stimulate local mutagenesis by NHEJ [38] and sequence replacement by HR. [52] This was followed by experiments with human cells in culture, using synthetic [53] or natural [41] targets, and by studies in several model organisms. By now, ZFNs have been used successfully in more than 25 different species, from yeast to butterflies to humans [7, 8].

Each new organism, cell type or end use requires careful consideration of how the nucleases and, in cases where HR is desired, the donor DNA, will be delivered. This concern applies to the other targeted nucleases as well, and the lessons learned from ZFN studies have contributed to the accelerated progress with TALENs and CRISPRs.

Because the genomic changes induced by ZFNs are permanent, only transient expression is required. In cultured mammalian cells, investigators have delivered ZFNs by plasmid transfection [41, 53], viral vectors [54–56], mRNA transfection, and even direct protein addition to the culture medium [57]. The latter approach seems to work because of the high intrinsic positive charge on the ZFNs, and it is applicable to a range of cell types, albeit with variable efficiencies. Long, double-stranded donor DNAs can be introduced on plasmid or viral vectors, while short, single-stranded donors are usually simply added to the medium [54, 55, 58, 59].

For situations in which manipulation of whole organisms is the goal, genome alterations in the germ line must be achieved. The cells in which the germ line is most accessible are typically in the very early embryo. Injection of ZFN mRNAs at this stage has proved successful in a wide range of organisms, including insects [60–63], fish [64–68], frogs [69], sea urchins [70, 71], mice [72–74], rats [75–77], and rabbits [78]. In the cases where HR products were reported [60, 71, 73–75], the donor DNA was simply included in the injection mix. In pigs and cows, genome modified animals were produced by in vitro mutagenesis of cultured somatic cells followed by nuclear transfer to enucleated eggs [79, 80].

Plants present particular challenges to delivery of genome engineering reagents. In some favorable cases, whole plants can be regenerated from cells or callus cultures, and the manipulations can be done in those contexts. This has worked, for example, in tobacco [81] and maize [82]. ZFN delivery in other plants has been accomplished with T-DNA transfer from Agrobacterium [83–86]. Viral vectors are also being developed [87], but no current approach is applicable to all plants.

The case of genome editing in plants nicely illustrates the fact that the biology of each system will dictate the best experimental approach and the range of outcomes. Experience with two popular experimental organisms emphasizes this potential limitation. The first applications of ZFNs to the nematode, *Caenorhabditis elegans*, achieved very good frequencies of somatic mutagenesis, but nothing in the germ line [88]. The nucleases were delivered in this study by forming extrachromosomal arrays of the transgenes, which were likely subjected to potent RNA interference in the germ line. When researchers instead used mRNA injection directly into the developing gonad, ZFN mutagenesis was observed, albeit at rather low frequency [89]. With the more efficient TALENs [89, 90], and particularly with CRISPRs [90–98], injection of DNA, mRNA and protein have all proven effective.

The other case is the zebrafish. It was among the early success stories for ZFN mutagenesis [66, 67, 99], but HR products were not readily obtained. With more efficient cleavage by TALENs, HR with both DNA oligonucleotides (oligos) and long, double-stranded donors was achieved [100, 101]. Interestingly, many of the oligo HR products appear to be only half-homologous, half-end joined [100]. This presumably reflects the strong preference in zebrafish embryos for DSB repair by NHEJ.

ZFN Contributions

Research with ZFNs and homing endonucleases (also called meganucleases) [102] has provided critical information on how to optimize the results of targeted genome cleavage. As noted above, this includes guidance on nuclease delivery in a variety of organisms and cell types. The balance between repair by NHEJ and by HR has been addressed [60, 103], including the idea of introducing single-strand breaks (nicks), rather than DSBs, to favor HR. [104–107] Design of the donor DNA was investigated [108], and the efficacy of synthetic, single-stranded oligo donors was demonstrated [59, 109]. Homology requirements and the extent of target sequence incorporation at the target (conversion tracts) have been defined [108, 110]. A method for making insertions with only limited terminal homology was demonstrated [111]. Approaches to making a variety of more complex genomic alterations have been made, including precise deletions and inversions [59, 112, 113], gene correction by cDNA insertion [56, 114], and chromosomal translocations [115, 116]. In addition, methods for detecting and quantitating nuclease-induced mutagenesis in the absence of selection have been developed [42, 117, 118].

For many research applications, the ease of design makes TALENs and CRISPR/Cas nucleases very attractive. The CRISPRs have the added advantage that a single, constant protein is involved, and specificity is determined by guide RNAs that can be easily multiplexed—for genome-wide libraries in cell populations [119, 120], or to attack multiple genes in a single cell [121]. TALENs appear to have inherently high specificity that can be enhanced by obligate heterodimer modifications, as noted above [45, 122]. The specificity of CRISPR/Cas nucleases has been questioned [123–127], but some effective solutions have been developed. These include shortening the guide sequence to exacerbate the influence of mismatches [128], using a variant Cas9 that cuts only one strand in conjunction with a pair of guide RNAs that direct nicks to closely spaced sites [126, 129, 130], and fusing fully inactivated Cas9 to the *Fok*I cleavage domain along with two guide RNAs to promote dimerization [131, 132].

Before ceding the playing field entirely to TALENs and CRISPRs, we should note that when a single target is being attacked repeatedly, it doesn't matter what platform is being used. For applications to human therapy, to livestock and to crop plants, the development of the cleavage reagent represents a small part of the cost and effort devoted to the project. Considerations like specificity and ease of delivery then become paramount. In this regard, the smaller size of the ZFNs will offer an advantage in some circumstances. Finally, some existing ZFNs are among the most effective and specific of the nucleases currently in use [49, 114, 133–135]. ZFNs targeted to the human CCR5 gene [49, 136] have been in clinical trials for several years [9, 137] and are proving safe and, to the extent allowed in a Phase I analysis, effective. Additional ZFN pairs have been targeted to other human disease genes [7, 8], and ones that have proved useful in specific applications like these are likely to continue to be exploited in the foreseeable future.

Acknowledgments I am grateful to the people in my lab, who have worked on genome engineering projects over many years, to my collaborators, and to all the researchers in the field of targetable nucleases, who have accelerated progress to such a dramatic extent and made this arena so much fun to watch and to work in. Research in my lab has been supported by grants from the NIH, most recently R01 GM078571.

References

1. Rothstein RJ. One-step gene disruption in yeast. Methods Enzymol. 1983;101:202–11.
2. Scherer S, Davis RW. Replacement of chromosome segments with altered DNA sequences constructed in vitro. Proc Natl Acad Sci U S A. 1979;76:4951–5.
3. Capecchi MR. Gene targeting in mice: functional analysis of the mammalian genome for the twenty-first century. Nat Rev Genet. 2005;6:507–12.
4. Thomas KR, Folger KR, Capecchi MR. High frequency targeting of genes to specific sites in the mammalian genome. Cell. 1986;44(3):419–28.
5. Chapman JR, Taylor MR, Boulton SJ. Playing the end game: DNA double-strand break repair pathway choice. Mol Cell. 2012;47(4):497–510.
6. Carroll D. Genome engineering with zinc-finger nucleases. Genetics. 2011;188(4):773–82.
7. Carroll D. Genome Engineering with Targetable Nucleases. Ann Rev Biochem. 2014;83:409–39.
8. Segal DJ, Meckler JF. Genome engineering at the dawn of the golden age. Annu Rev Genomics Hum Genet. 2013;14:135–58.
9. Urnov FD, Rebar EJ, Holmes MC, Zhang HS, Gregory PD. Genome editing with engineered zinc finger nucleases. Nat Rev Genet. 2010;11(9):636–46.
10. Li L, Wu LP, Chandrasegaran S. Functional domains in FokI restriction endonuclease. Proc Natl Acad Sci U S A. 1992;89:4275–9.
11. Kim Y-G, Chandrasegaran S. Chimeric restriction endonuclease. Proc Natl Acad Sci U S A. 1994;91:883–7.
12. Klug A. The discovery of zinc fingers and their applications in gene regulation and genome manipulation. Annu Rev Biochem. 2010;79:213–31.
13. Miller J, McLachlan AD, Klug A. Repetitive zinc-binding domains in the protein transcription factor IIIA from Xenopus oocytes. EMBO J. 1985;4(6):1609–14.
14. Pavletich NP, Pabo CO. Zinc finger-DNA recognition: crystal structure of a Zif268-DNA complex at 2.1 A resolution. Science. 1991;252:809–17.
15. Pabo CO, Peisach E, Grant RA. Design and selection of novel Cys2His2 zinc finger proteins. Annu Rev Biochem. 2001;70:313–40.
16. Kim Y-G, Cha J, Chandrasegaran S. Hybrid restriction enzymes: zinc finger fusions to FokI cleavage domain. Proc Natl Acad Sci U S A. 1996;93:1156–60.
17. Choo Y, Klug A. Toward a code for the interactions of zinc fingers with DNA: selection of randomized fingers displayed on phage. Proc Natl Acad Sci U S A. 1994;91:11163–7.
18. Desjarlais JR, Berg JM. Toward rules relating zinc finger protein sequences and DNA binding site preferences. Proc Natl Acad Sci U S A. 1992;89:7345–9.
19. Wolfe SA, Nekludova L, Pabo CO. DNA recognition by Cys2His2 zinc finger proteins. Annu Rev Biophys Biomol Struct. 2000;29:183–212.

20. Lam KN, van Bakel H, Cote AG, van der Ven A, Hughes TR. Sequence specificity is obtained from the majority of modular C2H2 zinc-finger arrays. Nucleic Acids Res. 2011;39:4680–90.
21. Ramirez CL, Foley JE, Wright DA, Muller-Lerch F, Rahman SH, Cornu TI, et al. Unexpected failure rates for modular assembly of engineered zinc fingers. Nat Methods. 2008;5(5):374–5.
22. Smith J, Bibikova M, Whitby FG, Reddy AR, Chandrasegaran S, Carroll D. Requirements for double-strand cleavage by chimeric restriction enzymes with zinc finger DNA-recognition domains. Nucleic Acids Res. 2000;28:3361–9.
23. Bibikova M, Carroll D, Segal DJ, Trautman JK, Smith J, Kim Y-G, et al. Stimulation of homologous recombination through targeted cleavage by chimeric nucleases. Mol Cell Biol. 2001;21:289–97.
24. Bitinaite J, Wah DA, Aggarwal AK, Schildkraut I. FokI dimerization is required for DNA cleavage. Proc Natl Acad Sci U S A. 1998;95:10570–5.
25. Handel EM, Alwin S, Cathomen T. Expanding or restricting the target site repertoire of zinc-finger nucleases: the inter-domain linker as a major determinant of target site selectivity. Mol Ther. 2009;17(1):104–11.
26. Shimizu Y, Bhakta MS, Segal DJ. Restricted spacer tolerance of a zinc finger nuclease with a six amino acid linker. Bioorg Med Chem Lett. 2009;19:3970–2.
27. Carroll D, Morton JJ, Beumer KJ, Segal DJ. Design, construction and in vitro testing of zinc finger nucleases. Nat Protoc. 2006;1(3):1329–41.
28. Kim JS, Lee HJ, Carroll D. Genome editing with modularly assembled zinc-finger nucleases. Nat Methods. 2010;7(2):91.
29. Segal DJ, Beerli RR, Blancafort P, Dreier B, Effertz K, Huber A, et al. Evaluation of a modular strategy for the construction of novel polydactyl zinc finger DNA-binding proteins. Biochemistry. 2003;42:2137–48.
30. Christensen RG, Gupta A, Zuo Z, Schriefer LA, Wolfe SA, Stormo GD. A modified bacterial one-hybrid system yields improved quantitative models of transcription factor specificity. Nucleic Acids Res. 2011;39(12), e83.
31. Maeder ML, Thibodeau-Beganny S, Osiak A, Wright DA, Anthony RM, Eichtinger M, et al. Rapid "Open-Source" engineering of customized zinc-finger nucleases for highly efficient gene modification. Mol Cell. 2008;31:294–301.
32. Persikov AV, Rowland EF, Oakes BL, Singh M, Noyes MB. Deep sequencing of large library selections allows computational discovery of diverse sets of zinc fingers that bind common targets. Nucleic Acids Res. 2014;42:1497–508.
33. Segal DJ, Dreier B, Beerli RR, Barbas III CF. Toward controlling gene expression at will: selection and design of zinc finger domains recognizing each of the 5′-GNN-3′ DNA target sequences. Proc Natl Acad Sci U S A. 1999;96:2758–63.
34. Gupta A, Christensen RG, Rayla AL, Lakshmanan A, Stormo GD, Wolfe SA. An optimized two-finger archive for ZFN-mediated gene targeting. Nat Methods. 2012;9(6):588–90.
35. Sander JD, Dahlborg EJ, Goodwin MJ, Cade L, Zhang F, Cifuentes D, et al. Selection-free zinc-finger-nuclease engineering by context-dependent assembly (CoDA). Nat Methods. 2011;8(1):67–9.
36. Zhu C, Gupta A, Hall VL, Rayla AL, Christensen RG, Dake B, et al. Using defined finger-finger interfaces as units of assembly for constructing zinc-finger nucleases. Nucleic Acids Res. 2013;41(4):2455–65.
37. Moore M, Klug A, Choo Y. Improved DNA binding specificity from polyzinc finger peptides by using strings of two-finger units. Proc Natl Acad Sci U S A. 2001;98:1437–41.
38. Bibikova M, Golic M, Golic KG, Carroll D. Targeted chromosomal cleavage and mutagenesis in Drosophila using zinc-finger nucleases. Genetics. 2002;161:1169–75.
39. Beumer K, Bhattacharyya G, Bibikova M, Trautman JK, Carroll D. Efficient gene targeting in Drosophila with zinc-finger nucleases. Genetics. 2006;172(4):2391–403.
40. Shimizu Y, Sollu C, Meckler JF, Adriaenssens A, Zykovich A, Cathomen T, et al. Adding fingers to an engineered zinc finger nuclease can reduce activity. Biochemistry. 2011;50(22):5033–41.

41. Urnov FD, Miller JC, Lee Y-L, Beausejour CM, Rock JM, Augustus S, et al. Highly efficient endogenous gene correction using designed zinc-finger nucleases. Nature. 2005;435:646–51.
42. Miller JC, Holmes MC, Wang J, Guschin DY, Lee Y-L, Rupniewski I, et al. An improved zinc-finger nuclease architecture for highly specific genome cleavage. Nat Biotechnol. 2007;25:778–85.
43. Szczepek M, Brondani V, Buchel J, Serrano L, Segal DJ, Cathomen T. Structure-based redesign of the dimerization interface reduces the toxicity of zinc-finger nucleases. Nat Biotechnol. 2007;25:786–93.
44. Doyon Y, Vo TD, Mendel MC, Greenberg SG, Wang J, Xia DF, et al. Enhancing zinc-finger-nuclease activity with improved obligate heterodimer architectures. Nat Methods. 2011;8:74–9.
45. Joung JK, Sander JD. TALENs: a widely applicable technology for targeted genome editing. Nat Rev Mol Cell Biol. 2013;14(1):49–55.
46. Liu JL, Wu Z, Deryusheva S, Rajendra TK, Beumer KJ, Gao H, et al. Coilin is essential for Cajal body organization in Drosophila melanogaster. Mol Biol Cell. 2009;20:1661–70.
47. Gupta A, Meng X, Zhu LJ, Lawson ND, Wolfe SA. Zinc finger protein-dependent and -independent contributions to the in vivo off-target activity of zinc finger nucleases. Nucleic Acids Res. 2011;39(1):381–92.
48. Pattanayak V, Ramirez CL, Joung JK, Liu DR. Revealing off-target cleavage specificities of zinc-finger nucleases by in vitro selection. Nat Methods. 2011;8(9):765–70.
49. Perez EE, Wang J, Miller JC, Jouvenot Y, Kim KA, Liu O, et al. Establishment of HIV-1 resistance in CD4+ T cells by genome editing using zinc-finger nucleases. Nat Biotechnol. 2008;26(7):808–16.
50. Gabriel R, Lombardo A, Arens A, Miller JC, Genovese P, Kaeppel C, et al. An unbiased genome-wide analysis of zinc-finger nuclease specificity. Nat Biotechnol. 2011;29(9):816–23.
51. Hacein-Bey-Abina S, Von Kalle C, Schmidt M, McCormack MP, Wulfffraat N, Leboulch P, et al. LMO2-associated clonal T cell proliferation in two patients after gene therapy for SCID-X1. Science. 2003;302:415–9.
52. Bibikova M, Beumer K, Trautman JK, Carroll D. Enhancing gene targeting with designed zinc finger nucleases. Science. 2003;300(5620):764.
53. Porteus MH, Baltimore D. Chimeric nucleases stimulate gene targeting in human cells. Science. 2003;300:763.
54. Ellis BL, Hirsch ML, Porter SN, Samulski RJ, Porteus MH. Zinc-finger nuclease-mediated gene correction using single AAV vector transduction and enhancement by Food and Drug Administration-approved drugs. Gene Ther. 2013;20(1):35–42.
55. Handel EM, Gellhaus K, Khan K, Bednarski C, Cornu TI, Muller-Lerch F, et al. Versatile and efficient genome editing in human cells by combining zinc-finger nucleases with adeno-associated viral vectors. Hum Gene Ther. 2012;23(3):321–9.
56. Lombardo A, Genovese P, Beausejour CM, Colleoni S, Lee YL, Kim KA, et al. Gene editing in human stem cells using zinc finger nucleases and integrase-defective lentiviral vector delivery. Nat Biotechnol. 2007;25(11):1298–306.
57. Gaj T, Guo J, Kato Y, Sirk SJ, Barbas 3rd CF. Targeted gene knockout by direct delivery of zinc-finger nuclease proteins. Nat Methods. 2012;9(8):805–7.
58. Asuri P, Bartel MA, Vazin T, Jang JH, Wong TB, Schaffer DV. Directed evolution of adeno-associated virus for enhanced gene delivery and gene targeting in human pluripotent stem cells. Mol Ther. 2012;20(2):329–38.
59. Chen F, Pruett-Miller SM, Huang Y, Gjoka M, Duda K, Taunton J, et al. High-frequency genome editing using ssDNA oligonucleotides with zinc-finger nucleases. Nat Methods. 2011;8(9):753–5.
60. Beumer KJ, Trautman JK, Bozas A, Liu JL, Rutter J, Gall JG, et al. Efficient gene targeting in Drosophila by direct embryo injection with zinc-finger nucleases. Proc Natl Acad Sci U S A. 2008;105(50):19821–6.

61. Merlin C, Beaver LE, Taylor OR, Wolfe SA, Reppert SM. Efficient targeted mutagenesis in the monarch butterfly using zinc-finger nucleases. Genome Res. 2013;23(1):159–68.
62. Takasu Y, Kobayashi I, Beumer K, Uchino K, Sezutsu H, Carroll D, et al. Targeted mutagenesis in the silkworm Bombyx mori using zinc-finger nuclease mRNA injections. Insect Biochem Mol Biol. 2010;40:759–65.
63. Watanabe T, Ochiai H, Sakuma T, Horch HW, Hamaguchi N, Nakamura T, et al. Non-transgenic genome modifications in a hemimetabolous insect using zinc-finger and TAL effector nucleases. Nat Commun. 2012;3:1017.
64. Ansai S, Ochiai H, Kanie Y, Kamei Y, Gou Y, Kitano T, et al. Targeted disruption of exogenous EGFP gene in medaka using zinc-finger nucleases. Dev Growth Differ. 2012;54(5):546–56.
65. Dong Z, Ge J, Li K, Xu Z, Liang D, Li J, et al. Heritable targeted inactivation of myostatin gene in yellow catfish (Pelteobagrus fulvidraco) using engineered zinc finger nucleases. PLoS One. 2011;6(12), e28897.
66. Doyon Y, MaCammon JM, Miller JC, Faraji F, Ngo C, Katibah GE, et al. Heritable targeted gene disruption in zebrafish using designed zinc-finger nucleases. Nat Biotechnol. 2008;26:702–8.
67. Meng X, Noyes MB, Zhu LJ, Lawson ND, Wolfe SA. Targeted gene inactivation in zebrafish using engineered zinc-finger nucleases. Nat Biotechnol. 2008;26:695–701.
68. Yano A, Guyomard R, Nicol B, Jouanno E, Quillet E, Klopp C, et al. An immune-related gene evolved into the master sex-determining gene in rainbow trout, Oncorhynchus mykiss. Curr Biol. 2012;22(15):1423–8.
69. Young JJ, Cherone JM, Doyon Y, Ankoudinova I, Faraji FM, Lee AH, et al. Efficient targeted gene disruption in the soma and germ line of the frog Xenopus tropicalis using engineered zinc-finger nucleases. Proc Natl Acad Sci U S A. 2011;108(17):7052–7.
70. Ochiai H, Fujita K, Suzuki K, Nishikawa M, Shibata T, Sakamoto N, et al. Targeted mutagenesis in the sea urchin embryo using zinc-finger nucleases. Genes Cells. 2010;15(8):875–85.
71. Ochiai H, Sakamoto N, Fujita K, Nishikawa M, Suzuki K, Matsuura S, et al. Zinc-finger nuclease-mediated targeted insertion of reporter genes for quantitative imaging of gene expression in sea urchin embryos. Proc Natl Acad Sci U S A. 2012;109(27):10915–20.
72. Carbery ID, Ji D, Harrington A, Brown V, Weinstein EJ, Liaw L, et al. Targeted genome modification in mice using zinc-finger nucleases. Genetics. 2010;186(2):451–9.
73. Meyer M. Hrabe de Angelis M, Wurst W, Kuhn R. Gene targeting by homologous recombination in mouse zygotes mediated by zinc-finger nucleases. Proc Natl Acad Sci U S A. 2010;107:15022–6.
74. Meyer M, Ortiz O. Hrabe de Angelis M, Wurst W, Kuhn R. Modeling disease mutations by gene targeting in one-cell mouse embryos. Proc Natl Acad Sci U S A. 2012;109:9354–9.
75. Cui X, Ji D, Fisher DA, Wu Y, Briner DM, Weinstein EJ. Targeted integration in rat and mouse embryos with zinc-finger nucleases. Nat Biotechnol. 2011;29(1):64–7.
76. Geurts AM, Cost GJ, Freyvert Y, Zeitler B, Miller JC, Choi VM, et al. Knockout rats via embryo microinjection of zinc-finger nucleases. Science. 2009;325:433.
77. Mashimo T, Takizawa A, Voigt B, Yoshimi K, Hiai H, Kuramoto T, et al. Generation of knockout rats with X-linked severe combined immunodeficiency (X-SCID) using zinc-finger nucleases. PLoS One. 2010;5(1), e8870.
78. Flisikowska T, Thorey IS, Offner S, Ros F, Lifke V, Zeitler B, et al. Efficient immunoglobulin gene disruption and targeted replacement in rabbit using zinc finger nucleases. PLoS One. 2011;6(6), e21045.
79. Hauschild J, Petersen B, Santiago Y, Queisser AL, Carnwath JW, Lucas-Hahn A, et al. Efficient generation of a biallelic knockout in pigs using zinc-finger nucleases. Proc Natl Acad Sci U S A. 2011;108(29):12013–7.
80. Yu S, Luo J, Song Z, Ding F, Dai Y, Li N. Highly efficient modification of beta-lactoglobulin (BLG) gene via zinc-finger nucleases in cattle. Cell Res. 2011;21(11):1638–40.
81. Townsend JA, Wright DA, Winfrey RJ, Fu F, Maeder ML, Joung JK, et al. High-frequency modification of plant genes using engineered zinc-finger nucleases. Nature. 2009;459:442–5.

82. Shukla VK, Doyon Y, Miller JC, DeKelver RC, Moehle EA, Worden SE, et al. Precise genome modification in the crop species Zea mays using zinc-finger nucleases. Nature. 2009;459(7245):437–41.
83. Curtin SJ, Zhang F, Sander JD, Haun WJ, Starker C, Baltes NJ, et al. Targeted mutagenesis of duplicated genes in soybean with zinc-finger nucleases. Plant Physiol. 2011;156(2):466–73.
84. Lloyd A, Plaisier CL, Carroll D, Drews GN. Targeted mutagenesis using zinc-finger nucleases in Arabidopsis. Proc Natl Acad Sci U S A. 2005;102:2232–7.
85. Osakabe K, Osakabe Y, Toki S. Site-directed mutagenesis in Arabidopsis using custom-designed zinc finger nucleases. Proc Natl Acad Sci U S A. 2010;107(26):12034–9.
86. Zhang F, Maeder ML, Unger-Wallace E, Hoshaw JP, Reyon D, Christian M, et al. High frequency targeted mutagenesis in Arabidopsis thaliana using zinc finger nucleases. Proc Natl Acad Sci U S A. 2010;107(26):12028–33.
87. Baltes NJ, Gil-Humanes J, Cermak T, Atkins PA, Voytas DF. DNA replicons for plant genome engineering. Plant Cell. 2014;26(1):151–63.
88. Morton J, Davis MW, Jorgensen EM, Carroll D. Induction and repair of zinc-finger nuclease-targeted double-strand breaks in Caenorhabditis elegans somatic cells. Proc Natl Acad Sci U S A. 2006;103(44):16370–5.
89. Wood AJ, Lo TW, Zeitler B, Pickle CS, Ralston EJ, Lee AH, et al. Targeted genome editing across species using ZFNs and TALENs. Science. 2011;333(6040):307.
90. Lo TW, Pickle CS, Lin S, Ralston EJ, Gurling M, Schartner CM, et al. Heritable Genome Editing Using TALENs and CRISPR/Cas9 to Engineer Precise Insertions and Deletions in Evolutionarily Diverse Nematode Species. Genetics. 2013;195:331–48.
91. Chen C, Fenk LA, de Bono M. Efficient genome editing in Caenorhabditis elegans by CRISPR-targeted homologous recombination. Nucleic Acids Res. 2013;41, e193.
92. Chiu H, Schwartz HT, Antoshechkin I, Sternberg PW. Transgene-Free Genome Editing in Caenorhabditis elegans Using CRISPR-Cas. Genetics. 2013;195:1167–71.
93. Cho SW, Lee J, Carroll D, Kim JS, Lee J. Heritable Gene Knockout in Caenorhabditis elegans by Direct Injection of Cas9-sgRNA Ribonucleoproteins. Genetics. 2013;195:1177–80.
94. Friedland AE, Tzur YB, Esvelt KM, Colaiacovo MP, Church GM, Calarco JA. Heritable genome editing in C. elegans via a CRISPR-Cas9 system. Nat Methods. 2013;10:741–3.
95. Frokjaer-Jensen C. Exciting Prospects for Precise Engineering of Caenorhabditis elegans Genomes with CRISPR/Cas9. Genetics. 2013;195(3):635–42.
96. Katic I, Grosshans H. Targeted Heritable Mutation and Gene Conversion by Cas9-CRISPR in Caenorhabditis elegans. Genetics. 2013;195:1173–6.
97. Tzur YB, Friedland AE, Nadarajan S, Church GM, Calarco JA, Colaiacovo MP. Heritable custom genomic modifications in Caenorhabditis elegans via a CRISPR-Cas9 System. Genetics. 2013;195:1181–5.
98. Waaijers S, Portegijs V, Kerver J, Lemmens BB, Tijsterman M, van den Heuvel S, et al. CRISPR/Cas9-Targeted Mutagenesis in Caenorhabditis elegans. Genetics. 2013;195:1187–91.
99. Foley JE, Yeh JR, Maeder ML, Reyon D, Sander JD, Peterson RT, et al. Rapid mutation of endogenous zebrafish genes using zinc finger nucleases made by Oligomerized Pool ENgineering (OPEN). PLoS One. 2009;4(2), e4348.
100. Bedell VM, Wang Y, Campbell JM, Poshusta TL, Starker CG, Krug 2nd RG, et al. In vivo genome editing using a high-efficiency TALEN system. Nature. 2012;491(7422):114–8.
101. Zu Y, Tong X, Wang Z, Liu D, Pan R, Li Z, et al. TALEN-mediated precise genome modification by homologous recombination in zebrafish. Nat Methods. 2013;10(4):329–31.
102. Stoddard BL. Homing endonucleases: from microbial genetic invaders to reagents for targeted DNA modifications. Structure. 2011;19:7–15.
103. Bozas A, Beumer KJ, Trautman JK, Carroll D. Genetic analysis of zinc-finger nuclease-induced gene targeting in Drosophila. Genetics. 2009;182(3):641–51.
104. Kim E, Kim S, Kim DH, Choi BS, Choi IY, Kim JS. Precision genome engineering with programmable DNA-nicking enzymes. Genome Res. 2012;22(7):1327–33.

105. McConnell Smith A, Takeuchi R, Pellenz S, Davis L, Maizels N, Monnat Jr RJ, et al. Generation of a nicking enzyme that stimulates site-specific gene conversion from the I-AniI LAGLIDADG homing endonuclease. Proc Natl Acad Sci U S A. 2009;106(13):5099–104.
106. Ramirez CL, Certo MT, Mussolino C, Goodwin MJ, Cradick TJ, McCaffrey AP, et al. Engineered zinc finger nickases induce homology-directed repair with reduced mutagenic effects. Nucleic Acids Res. 2012;40:5560–8.
107. Wang J, Friedman G, Doyon Y, Wang NS, Li CJ, Miller JC, et al. Targeted gene addition to a predetermined site in the human genome using a ZFN-based nicking enzyme. Genome Res. 2012;22:1316–26.
108. Beumer KJ, Trautman JK, Mukherjee K, Donor CD, DNA. Utilization during gene targeting with zinc-finger nucleases. G3 (Bethesda). 2013;3:657–64.
109. Radecke S, Radecke F, Cathomen T, Schwarz K. Zinc-finger nuclease-induced gene repair with oligodeoxynucleotides: wanted and unwanted target locus modifications. Mol Ther. 2010;18:743–53.
110. Elliott B, Richardson C, Winderbaum J, Nickoloff JA, Jasin M. Gene conversion tracts from double-strand break repair in mammalian cells. Mol Cell Biol. 1998;18(1):93–101.
111. Orlando SJ, Santiago Y, DeKelver RC, Freyvert Y, Boydston EA, Moehle EA, et al. Zinc-finger nuclease-driven targeted integration into mammalian genomes using donors with limited chromosomal homology. Nucleic Acids Res. 2010;38(15), e152.
112. Lee HJ, Kim E, Kim JS. Targeted chromosomal deletions in human cells using zinc finger nucleases. Genome Res. 2010;20(1):81–9.
113. Lee HJ, Kweon J, Kim E, Kim S, Kim JS. Targeted chromosomal duplications and inversions in the human genome using zinc finger nucleases. Genome Res. 2012;22(3):539–48.
114. Li H, Haurigot V, Doyon Y, Li T, Wong SY, Bhagwat AS, et al. In vivo genome editing restores haemostasis in a mouse model of haemophilia. Nature. 2011;475(7355):217–21.
115. Brunet E, Simsek D, Tomishima M, DeKelver RC, Choi VM, Gregory P, et al. Chromosomal translocations induced as specified loci in human stem cells. Proc Natl Acad Sci U S A. 2009;106:10620–5.
116. Piganeau M, Ghezraoui H, De Cian A, Guittat L, Tomishima M, Perrouault L, et al. Cancer translocations in human cells induced by zinc finger and TALE nucleases. Genome Res. 2013;23(7):1182–93.
117. Dahlem TJ, Hoshijima K, Jurynec MJ, Gunther D, Starker CG, Locke AS, et al. Simple methods for generating and detecting locus-specific mutations induced with TALENs in the zebrafish genome. PLoS Genet. 2012;8(8), e1002861.
118. Kim HJ, Lee HJ, Kim H, Cho SW, Kim JS. Targeted genome editing in human cells with zinc finger nucleases constructed via modular assembly. Genome Res. 2009;19(7):1279–88.
119. Shalem O, Sanjana NE, Hartenian E, Shi X, Scott DA, Mikkelson T, et al. Genome-Scale CRISPR-Cas9 Knockout Screening in Human Cells. Science. 2014;343:84–7.
120. Wang T, Wei JJ, Sabatini DM, Lander ES. Genetic Screens in Human Cells Using the CRISPR/Cas9 System. Science. 2014;343:80–4.
121. Wang H, Yang H, Shivalila CS, Dawlaty MM, Cheng AW, Zhang F, et al. One-step generation of mice carrying mutations in multiple genes by CRISPR/Cas-mediated genome engineering. Cell. 2013;153:910–8.
122. Bogdanove AJ, Voytas DF. TAL effectors: customizable proteins for DNA targeting. Science. 2011;333(6051):1843–6.
123. Cradick TJ, Fine EJ, Antico CJ, Bao G. CRISPR/Cas9 systems targeting beta-globin and CCR5 genes have substantial off-target activity. Nucleic Acids Res. 2013;41:9584–92.
124. Fu Y, Foden JA, Khayter C, Maeder ML, Reyon D, Joung JK, et al. High-frequency off-target mutagenesis induced by CRISPR-Cas nucleases in human cells. Nat Biotechnol. 2013;31:822–6.
125. Hsu PD, Scott DA, Weinstein JA, Ran FA, Konermann S, Agarwala V, et al. DNA targeting specificity of RNA-guided Cas9 nucleases. Nat Biotechnol. 2013;31:827–32.
126. Mali P, Aach J, Stranges PB, Esvelt KM, Moosburner M, Kosuri S, et al. CAS9 transcriptional activators for target specificity screening and paired nickases for cooperative genome engineering. Nat Biotechnol. 2013;31:833–8.

127. Pattanayak V, Lin S, Guilinger JP, Ma E, Doudna JA, Liu DR. High-throughput profiling of off-target DNA cleavage reveals RNA-programmed Cas9 nuclease specificity. Nat Biotechnol. 2013;31(9):839–43.
128. Fu Y, Sander JD, Reyon D, Cascio VM, Joung JK. Improving CRISPR-Cas nuclease specificity using truncated guide RNAs. Nat Biotechnol. 2014;32:279–84.
129. Cho SW, Kim S, Kim Y, Kweon J, Kim HS, Bae S, et al. Analysis of off-target effects of CRISPR/Cas-derived RNA-guided endonucleases and nickases. Genome Res. 2014;24:132–41.
130. Ran FA, Hsu PD, Lin CY, Gootenberg JS, Konermann S, Trevino AE, et al. Double Nicking by RNA-Guided CRISPR Cas9 for Enhanced Genome Editing Specificity. Cell. 2013;154:1380–9.
131. Guilinger JP, Thompson DB, Liu DR. Fusion of catalytically inactive Cas9 to FokI nuclease improves the specificity of genome modification. Nat Biotechnol. 2014;32:577–82.
132. Tsai SQ, Wyvekens N, Khayter C, Foden JA, Thapar V, Reyon D, et al. Dimeric CRISPR RNA-guided FokI nucleases for highly specific genome editing. Nat Biotechnol. 2014;32:569–76.
133. Anguela XM, Sharma R, Doyon Y, Miller JC, Li H, Haurigot V, et al. Robust ZFN-mediated genome editing in adult hemophilic mice. Blood. 2013;122(19):3283–7.
134. Li L, Krymskaya L, Wang J, Henley J, Rao A, Cao LF, et al. Genomic editing of the HIV-1 coreceptor CCR5 in adult hematopoietic stem and progenitor cells using zinc finger nucleases. Mol Ther. 2013;21(6):1259–69.
135. Yusa K, Rashid ST, Strick-Marchand H, Varela I, Liu PQ, Paschon DE, et al. Targeted gene correction of alpha1-antitrypsin deficiency in induced pluripotent stem cells. Nature. 2011;478(7369):391–4.
136. Maier DA, Brennan AL, Jiang S, Binder-Scholl GK, Lee G, Plesa G, et al. Efficient clinical scale gene modification via zinc finger nuclease-targeted disruption of the HIV co-receptor CCR5. Hum Gene Ther. 2013;24(3):245–58.
137. Tebas P, Stein D, Tang WW, Frank I, Wang SQ, Lee G, et al. Gene editing of CCR5 in autologous CD4 T cells of persons infected with HIV. N Engl J Med. 2014;370:901–10.

The Use and Development of TAL Effector Nucleases

Alexandre Juillerat, Philippe Duchateau, Toni Cathomen, and Claudio Mussolino

Abstract In 2009 plant geneticists described a novel DNA binding domain derived from transcription activator-like effectors (TALEs) of the plant pathogen genus *Xanthomonas*. The DNA recognition domain was distinguished by a modular structure in which each building block binds to a single DNA nucleotide. The breakthrough was the identification of the key residues within each block that define its DNA binding properties and to show that specific alterations of these residues allow for the assembly of tailored DNA binding domains able to target any given sequence. This discovery set the stage for the generation of various designer proteins by fusing tailored TALE-based DNA binding domains, with either endonucleases, transcriptional modulators or chromatin remodeling domains, with the final purpose to modify the genome, the transcriptome or the epigenome. In the last few years, the exploitation of designer enzymes has expanded impressively with applications spanning from basic research to systems biology and human gene therapy.

Keywords Gene editing • Gene knockout • Genome engineering • Transcription activator-like effector nuclease • Zinc-finger nuclease • TALE engineering • TAL Effectors • TALE cloning • Golden Gate

Transcription Activator-Like Effectors

Transcription activator-like effectors (TALEs) are proteins originally identified in *Xanthomonas*, a genus of proteobacteria that includes a huge number of bacterial plant pathogens. During the infection process, a mixture of bacterial proteins,

A. Juillerat, Ph.D. • P. Duchateau, Ph.D.
Research Department, Cellectis SA, Paris, France

T. Cathomen, Ph.D.
Institute for Cell and Gene Therapy, Medical Center - University of Freiburg, Hugstetter Str. 55, 79106 Freiburg, Germany

C. Mussolino, Ph.D. (✉)
Center for Chronic Immunodeficiency, Medical Center - University of Freiburg, Hugstetter Str. 55, 79106 Freiburg, Germany
e-mail: claudio.mussolino@uniklinik-freiburg.de

T. Cathomen et al. (eds.), *Genome Editing*, Advances in Experimental Medicine and Biology 895, DOI 10.1007/978-1-4939-3509-3_3

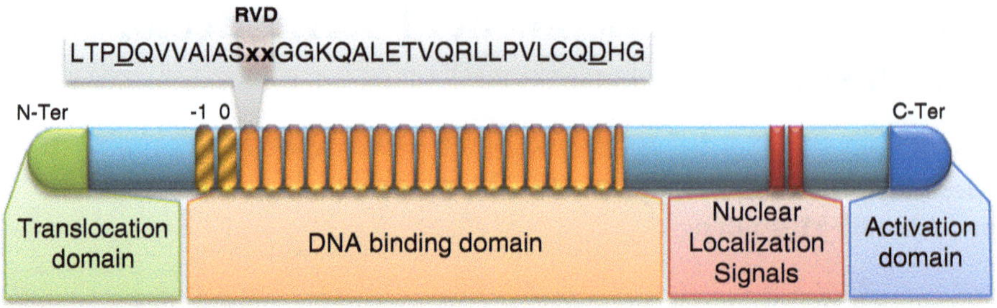

Fig. 1 Schematic of a TAL effector. A TAL effector (TALE) contains an N-terminal translocation domain and eukaryotic-like nuclear localization signals and activation domain within the C-terminal portion of the protein. The central part of the protein contains the DNA binding domain, which consists of a variable number of modules with DNA binding capacity. With the exception of the repeat-like structures 0 and -1, the protein sequence of the modules is highly conserved. The repeat variable di-residues (RVDs) within each module dictate the DNA binding specificity

including TALEs, are translocated into the cytoplasm of the plant host cells via a type III secretion system. After translocation into the nucleus, the TALEs mimic the function of eukaryotic transcription factors and bind to *cis*-regulatory elements of the host genome to control and manipulate cellular pathways, with the final goal of promoting bacterial replication [1].

TALEs are composed of N-terminal secretion and translocation signals, a central domain with DNA binding capability and a C-terminal acidic activator domain coupled to nuclear localization signals that enable translocation into nucleus [2] (Fig. 1).

Protein engineering strategies have been especially focused on the identification of N- or C-terminal truncations aimed at creating artificial TALE-based DNA binding domains that combine minimal size with efficient DNA binding activity. In the next paragraphs we will describe and summarize the development of different TALE scaffolds that have been engineered as versatile carriers for various effector domains.

DNA Binding Domain

The central DNA binding domain consists of a variable number of tandem repeats, generally between 15.5 and 19.5, in which the last repeat is shorter and usually referred to as "half-repeat". Each module constituting the DNA recognition domain is composed of 33–35 highly conserved amino acids, with the exception of those in positions 12 and 13 that are hyper-variable and referred to as repeat variable di-residues (RVDs) [2]. These two amino acids hold a key role in defining the nucleotide specificity that is a simple 'one-to-one code', in which a single RVD contacts a single nucleotide. Cracking this interaction code allowed researchers to infer unknown target sites of natural TALEs based on their protein sequence and, conversely, set the stage for targeting engineered TALEs to chosen genes by assembling the TALE tandem repeat modules in the appropriate sequence [3, 4]. The crystal structure of a natural TALE protein complexed to its target DNA was resolved in early 2012 and highlighted some interesting aspects on how TALEs recognize their

Table 1 RVDs specificities

RVD	Target nucleotide	Reference
NI	A	[3, 4]
HD	C	[3, 4]
NN	G/A	[3, 4]
NK	G	[3, 4, 7–9]
NH	G	[8, 9]
NG	T	[3, 4]
NG	5mC	[12]
N*	5mC	[13]
H*	5mC	[13]

cognate DNA [5, 6]. Each module is arranged in two α-helices connected by a loop that contains the RVD. The modules within an array are connected to form a right-handed superhelix structure with the RVD residues pointing inwards. This protein structure coils around the DNA double helix with the RVD residues directly contacting the major groove, remarkably without altering the structure of the DNA double helix. Moreover, the two amino acids of the RVD seem to have different roles: while the residue in position 12 stabilizes the loop the 13th residue makes the specific contact to the nucleotide of the target DNA.

Although the RVD–nucleotide association code was originally described for 15 different naturally occurring RVDs [3, 4], researchers have early on focused on the four most prominent RVDs present in *Xanthomonas* TALEs: HD, NN, NI and NG that specify the binding to a C, G or A, A and T, respectively; Table 1). This straightforward code (i.e. four different RVDs to target the four nucleotides) has allowed the creation of a multitude of molecular tools that were successfully used in various organisms, making TALEs one of the most promising platforms for DNA targeting. Nevertheless, a key issue when designing DNA targeting tools is specificity. Researchers have tried to improve the targeting specificity of engineered TALE arrays by using less frequent G-specific RVDs, such as NH or NK. However, despite the expected gain in specificity towards guanine, the implementation of the 'NK' RVD (instead of NN) in the context of TALE-based designer nucleases or transcription activators often compromised the activity of the final protein [7, 8]. This effect was particularly evident when the 'NK' RVDs were located in the N-terminal end of the array [7, 8]. Molecular modeling simulations using the available high resolution structure of PthXo1 bound to its cognate target DNA [6] revealed a higher affinity of NN-containing compared to NK-based arrays [7]. In addition potential context-dependence of these less frequent RVDs cannot be excluded. By screening the specificity of 23 natural RVDs, Zhang and colleagues reported that the 'NH' RVD can be a valuable alternative to NN as it showed an improved specificity for the guanine base while retaining similar levels of biological activity [9]. This interesting feature was simultaneously described in an independent study by Boch and colleagues [8]. However, because of the low numbers of NH-containing arrays that have been tested thus far, it is too early to advise scientists to switch from 'NN' to 'NH' RVDs when aiming to target a guanine. Hence, despite the dual preference for guanine and adenine, the 'NN' RVD is still used extensively to target guanine because of its high overall affinity to purines [10].

Recently, the use of alternative RVDs has been reported. We have highlighted that targeting specificity can be improved through the educated implementation of non-conventional RVDs, based on their exclusion capacities [11]. Miller and colleagues have explored the potential of novel RVDs in order to improve activity and specificity of previously characterized TALENs [12]. This study revealed a remarkable effect of position and sequence context on TALE-DNA recognition and provided novel non-canonical but binding-competent RVDs that can be employed to generate highly active and specific TALENs.

Particularly interesting in this context was evidence that in neural stem cells, TALE-based transcriptional activators have been reported to be unable to activate the silenced Oct4 promoter [13]. In the same study, a TALE-based transcriptional activator failed to activate an in vitro methylated reporter construct transfected in HEK293T cells. On the other hand, chemical inhibition of DNA methyltransferases using 5-aza-20-deoxycytidine enabled the recovery of its activity. These results highlighted the impact of cytosine methylation on TALE-based molecular tools. Structural studies of the T-specific RVD 'NG' have demonstrated that NG can accommodate interactions with 5-methylcytosine (5-mC), which suggested that TALEs could potentially be designed to recognize methylated CpG dinucleotides [14]. Valton and colleagues [15] have reported the implementation of 'N*' and 'H*' RVDs (the asterisks indicate the lack of the 13th residue) into TALE arrays to efficiently target 5-mC (Table 1). While this report confirmed the incompatibility of the 'HD' RVD with 5-methylcytosine, it emphasized the superiority of 'N*' and 'H*' over 'NG' to target this methylated base. As a conclusion, avoiding CpG dinucleotides or pre-screening the target site for the presence of methylated cytosines should improve the success rate in generating functional TALE-based molecular tools.

Exploration of the genome of different plant pathogens allowed for the identification of additional TALE-like proteins in *Ralstonia solanacearum* and *Burkholderia rhizoxinica*. The characterized *Ralstonia* proteins share many similarities with *Xanthomonas* TALEs, including their nuclear localization and the presence of an acidic activator domain at their C-terminus [16]. The DNA binding domain of *Ralstonia* TALE-like effectors (RTL proteins) is modular with each repeat unit composed of 35 moderately conserved residues that contain RVDs not previously observed in *Xanthomonas'* TALEs. RTLs hence offer a novel set of DNA binding modules with different nucleotide affinities and specificities. Moreover, in contrast to *Xanthomonas* TALEs, RTLs preferentially recognize a guanosine in position 0 of the target site. On the other hand, the generation of TALE arrays that target alternative nucleotides to the 'invariant' 5′-thymidine has been recently reported [17]. TALE-like proteins from *Burkholderia* (Bat proteins) bind to the DNA with the same code as *Xanthomonas* TALEs [18]. However, Bat proteins are usually shorter and are formed almost exclusively by the repeat-based DNA binding domain. Interestingly, these repeats are highly polymorphic, sharing less than 40 % sequence identity, and their overall affinity seems to be lower as compared to their *Xanthomonas* counterparts. We have reported the use of such TALE-like scaffolds from *Burkholderia* to create designer nucleases [19], and the access to the high resolution structure of such TALE-like proteins also highlighted new interesting DNA targeting features [20].

In summary, since the discovery of the 'one-to-one' code of TALEs in 2009, several technical advances have allowed scientist to considerably expand the targeting range and the biological application portfolio of molecular tools based on the TALE platform.

N-Terminal Domain

Early work on engineering TALE-effector proteins have demonstrated that the first 152 amino acids could be deleted from the N-terminus without affecting the protein activity, likely because this region is mainly responsible for the translocation into a plant cell [21, 22]. However, attempts to generate artificial TALEs with even shorter N-terminal portions failed to bind to DNA [23, 24] and the NΔ152 truncation version rapidly established as the reference scaffold. The crystal structure has revealed that portion encompassing residue 152 to the first regular repeat module is arranged in two repeat-like structures (usually named as repeats 0 and −1) in which a tryptophan in repeat −1 directly contacts the DNA target site at a 5′-thymidine, which is invariably found in nearly all the natural *Xanthomonas* TALE target sites [2]. The structure suggests that the first steps during binding of a TALE to the DNA target are mediated by this portion of the protein, likely contributing to the high binding energy that enables subsequent target recognition.

Various domains can be fused to the N-terminal portion of TALE proteins without affecting the binding capability of the final molecule; indeed, detection or purification tags (e.g. FLAG, HA, S) or localization signals (nuclear, mitochondrial) have been successfully fused to native TALEs and variants with truncated N-terminal domains. Interestingly, Yang and colleagues have provided evidence that also functional domains, such as the *Fok*I catalytic domain, can be fused to the N-terminus of native TALEs to generate moderately active nuclease pairs [25]. Based on the NΔ152 variant, Beurdeley and colleagues fused the catalytic domain of the I-*Tev*I homing endonuclease to the N-terminus of an engineered TALE. This "compact" TALE nuclease (cTALEN) couples the advantage of a partially selective catalytic domain with the programmable DNA targeting specificity of the TALE protein [26]. The impact of alternative N-terminal variants has been further investigated by Barbas and colleagues, using an incremental truncation-based library screening strategy, to demonstrate that NΔ120 or NΔ128 TALE variants are advantageous to create different designer enzymes, such as chimeric TALE recombinases [27].

C-Terminal Domain

Most of the studies that focused on introducing alterations in the C-terminal portion of TALE-based designer proteins showed that this domain is less critical for the DNA binding function of the protein. Indeed, swapping the natural transcriptional activation domain with heterologous activator domains to C-terminal TALE truncations resulted in functional protein, regardless of the extent of the truncations [23, 24, 28, 29]. However, in the context of designer nucleases, the length of the C-terminal 'linker' that connects the DNA binding domain with the FokI endonuclease domain has an impact on both nuclease activity and spacer length tolerability between the two target half sites. A TALEN scaffold retaining 10–70 amino acids of the C-terminal domain showed higher nuclease activity when compared to TALENs harboring the entire native C-terminal domain [7, 23, 29]. On the other hand, while variants with longer C-terminal 'linkers' (>40 residues) showed cleavage activity on a broad range of spacers (12–30 bp), TALENs harboring shorter 'linkers'—or completely lacking the C-terminal domain—exhibited activity over a more restricted range (13–16 bp) of DNA spacers [7, 23, 29]. Thus, depending on the spacer length of the DNA target site, a fitting C-terminal scaffold should be chosen, taking into consideration that minimizing spacer length tolerability with shorter C-terminal variants may help increase the specificity by limiting off-target cleavage. In conclusion, the versatility of the TALE-based DNA binding scaffold allows for the fusion of various effector domains to its C-terminus, such as transcriptional repressor [30] and activation domains [31], chromatin modifiers [32] and nucleases that have thus far been successfully used to modify the transcriptome, the epigenome and the genome of mammalian and plant cells (Fig. 2).

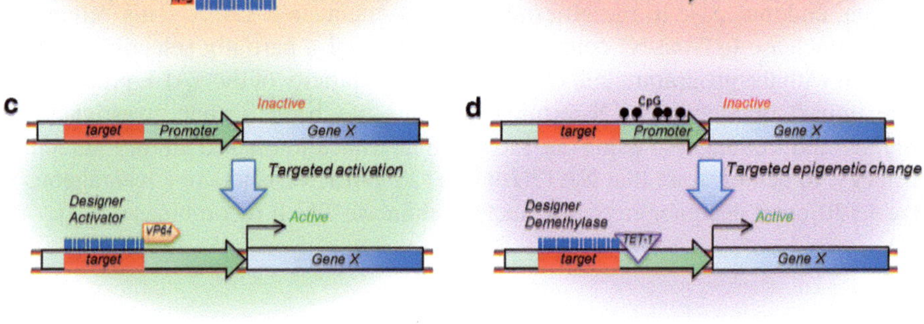

Fig. 2 TALE-based effectors. The fusion of various effector domains to a TALE DNA binding array allows for the generation of tailored effectors. (**a**) Designer nucleases are used to modify the genome by introducing targeted DNA double strand breaks. (**b**, **c**) Targeted regulation of the transcriptome is achieved through employment of artificial transcription factors that enable transcriptional activation or repression. (**d**) Targeted epigenetic changes are induced by using chromatin modifiers

Assembly of TALE Arrays

While the modular structure of DNA recognition by TALE permits the easy design of specific DNA targeting arrays, the physical assembly of nearly identical modules of ~100 bp turned out to be challenging using traditional cloning strategies. Since the advent of TALE-based molecular tools, several platforms enabling the rapid assembly of such targeting modules have been reported [33–42]. All these platforms vary in diverse key parameters, such as the number and preparation of starting building blocks, the flexibility of the final array length that can be assembled and finally their throughput. In the following sections we summarize different TALE array assembly strategies that have been developed in recent years and that can be divided into four categories based on their assembly methods (summarized in Fig. 3 and Table 2): (1) standard cloning, (2) Golden Gate based cloning, (3) solid phase assembly, and (4) ligation independent cloning. The vast majority of the protocols developed rely on the use of type IIS restriction enzymes that cleave DNA at a defined distance from their recognition sites, leaving a 4-bp overhang. Hence, if type IIS recognition sites are placed at the 5′ and 3′-ends of each DNA fragment in inverse orientation, various different four-nucleotide overhangs can be created using a single restriction enzyme. Upon restriction and ligation, the final construct is devoid of the original recognition sites ("seamless cloning").

Alternatively, TALE arrays can either be synthesized *de novo* or validated constructs can be purchased through commercial companies.

Standard Cloning Assembly

This strategy is conceptually the easiest way to assemble TALE arrays and relies on collections of plasmids encoding single or multiple building blocks, standard restriction/ligation enzymatic steps and amplification in *E. coli* to create intermediate arrays in a parallel hierarchical manner (Fig. 3a, left). The design of the starting constructs involves incorporation of either isocaudomers (e.g.: unit assembly) or type IIS restriction enzymes (e.g.: REAL, REAL-Fast). Depending on the method, the number of starting plasmids can strongly vary from less than ten [43], to a few dozen [35, 37, 44] or even several hundred [30, 34, 38, 45]. The size of the collection is related to the number of repeats incorporated in each starting building blocks (from one repeat up to four for the largest collections). The preparation of the starting blocks either involves PCR amplification or direct digestion from the plasmid collection. At each round of the assembly cycle, the intermediate products can be characterized by colony PCR or restriction digestion to validate a successful process. Depending on the design of the starting material, 2–8 [25] individual building blocks can be coupled in an ordered fashion in a single cloning reaction. Nevertheless, only some of these assembly methods [37, 38, 43, 45] offer large flexibility in the size of the final array length with more than one or two possibilities. The numerous and fastidious

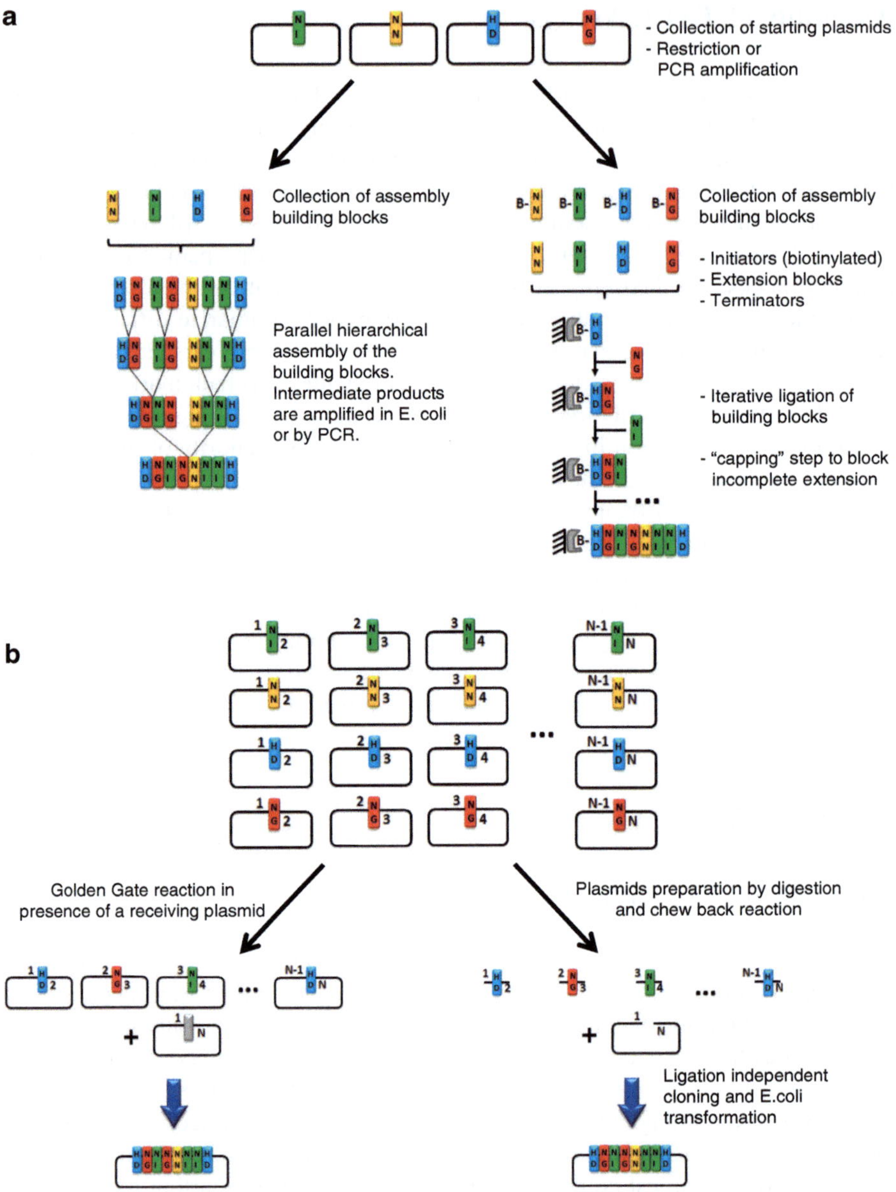

Fig. 3 Methods available to assemble TALE arrays. Sequences of single or multi-repeat units are encoded in a collection of starting plasmids. (**a**) Illustration of the standard cloning assembly using parallel hierarchical reactions (*left*) and of the solid phase assembly that is based on iterative enzymatic elongation (*right*). (**b**) Illustration of the Golden Gate "one pot–one step reaction" process (*left*) and of the ligation independent cloning (LIC) strategy (*right*)

Table 2 TALE array assembly methods

	Building blocks[a]	Building block preparation	Intermediate amplification step	Possible array length[b]	Estimated time lines[c]	Terminal Half repeat	Throughput	Ref.
Standard cloning assembly	~32	Digestion	Several	various	1–2 weeks	In final plasmid	Low	[31]
	~376	Digestion	Several	various	1–2 weeks	In final plasmid	Low	[29]
	~20	Digestion	Several	various	1–2 weeks	In final plasmid	Low	[28]
	~100	Digestion	No	12.5	2–3 days	In building blocks	Low/medium	[33]
	~48	PCR	1 (*E. coli*)	16.5 and 24.5	~1 week	In final plasmid	Low	[30]
	~832	Digestion	No	14.5	2–3 days	In building blocks	Low/medium	[32]
Golden Gate based assembly	~67	Plasmid	1 (*E. coli*)	17.5	~1 week	In final plasmid	Low/medium	[35]
	~80	Plasmid	1 (*E. coli*)	17.5 and 20.5	~1 week	In final plasmid	Low/medium	[39]
	~41	Plasmid	1 (*E. coli*)	0.5–23.5	~1 week	In building blocks	Low/medium	[38]
	~68–84	Plasmid	1 (*E. coli*)	11.5–30.5	~1 week	In building blocks	Low/medium	[37]
	~78–94	Plasmid	1 (*E. coli*)	8.5–31.5	~1 week	In building blocks	Low/medium	[40]
	~48	PCR/digestion	1 (PCR)	12.5	2–3 days	In final plasmid	Low/medium	[19]
	~72	PCR/digestion	1 (PCR)	18.5–24.5	4 days	In final plasmid	Low/medium	[43]
	~64	PCR/USER digestion	no	Described for 14.5 and 17.5	2–3 days	In building blocks	Medium	[41]

(continued)

Table 2 (continued)

	Building blocks[a]	Building block preparation	Intermediate amplification step	Possible array length[b]	Estimated time lines[c]	Terminal Half repeat	Throughput	Ref.
Solid phase assembly	~380	PCR or digestion	no	up to 19.5	3 days	In final plasmid	Medium/High	[46]
	unknown	unknown	no	16.5	2–3 days	In final plasmid?	Medium/High	[45]
	~30	PCR	no	up to 18.5	3 days	In building blocks	Medium/High	[44]
Ligation Independent cloning assembly	~64	Digestion and chew back reaction	1 (*E. coli*)	18.5 (9.5–18.5)[d]	3 days	In final plasmid	Medium/High	[47]
	~3072	Digestion and chew back reaction	1 (*E. coli*)	15.5 (9.5–18.5)[d]	3 days	In final plasmid	High	[47]

[a]Building blocks collections include intermediate cloning plasmids and final backbones. Building block number is also dependent on the number of different repeat type implemented in the assembly strategy

[b]Total array length includes the last terminal half repeat that could be incorporated in the final receiving TALE-backbone containing plasmid

[c]Estimated assembly time lines do not take in account preparation of the starting materials and sequencing of the final array in the final backbone

[d]Combination of both LIC-based collection and implementation of single LIC-compatible single repeat extend synthesis flexibility

molecular biology steps (restriction, ligation, plasmid DNA isolation and DNA fragment gel purification) of these methods clearly limit the production to low throughput. To assemble TALE arrays of a standard size for genome engineering tools (10–24 repeated units), up to 2 weeks are required (depending on the final array length). However, they present the advantage that the basic molecular biology techniques are already implemented in many laboratories.

'Golden Gate' Assembly

The Golden Gate cloning technology was primarily developed to allow enzymatic cloning of multiple DNA fragments in a defined linear order [41, 46]. The strength of this method relies on the fact that the whole cloning process (restriction and ligation) can be performed using multiple DNA fragments (e.g. PCR amplified or plasmids) in a single 'one step–one pot' reaction by cycling the experimental conditions (e.g. temperature) for both enzymatic steps (Fig. 3b, left). However, the cloning efficiency, i.e. the total number of positive clones obtained after *E. coli* transformation and plating, drastically decreases when more than nine fragments are assembled [47]. Thus, to generate a typical TALE array containing more than ten repeats, multiple parallel Golden Gate reactions are required to preassemble sub-arrays that have to be further amplified, either by PCR [24] or by plasmid amplification in *E. coli* transformation [36, 47–50], prior to their fusion with an additional Golden Gate reaction. While the original protocols are based on the use of type IIS restriction enzymes, Yang and colleagues [42] developed an alternative PCR-based method for the preparation of the building blocks. Their strategy relies on the use of uracil containing primers for amplification of the building blocks followed by the assembly of the array after a USER (Uracil-Specific Excision Reagent; [51]) digestion of the PCR products, in a single reaction. Another interesting variant of the Golden Gate strategy using PCR products was developed by Sanjana and colleagues [52] and relies on the circularization of the intermediate array followed by removal of unreacted and non-circular products by exonuclease treatment. Circular arrays are then amplified by PCR and further combined to give the final array. While the absence of an amplification step in *E. coli* and the possibility to eliminate side products represent valuable features in terms of throughput improvement, the necessity to purify intermediate PCR products might temper these advantages.

All Golden Gate based strategies used up to date to assemble TALE arrays rely on collections of a few dozen plasmids (24–78) or PCR amplified fragments that can be handled by most laboratories without particular instrumentation. They allow for the production of TALE arrays in a timeframe varying from a day to a week, depending on the array length (0.5–30.5) and the type of the intermediate step (PCR or plasmid amplification in *E. coli*). While the development of such 'one step–one pot' assembly methods clearly expands the possibilities beyond classical molecular biology techniques, the numerous intermediate steps, such as plating, colony picking, PCR screening and DNA isolation, clearly hamper their potential towards high-throughput automatization of the production.

Solid Phase Assembly

The solid phase assembly of TALE repeats was developed as a high-throughput alternative to Golden Gate cloning strategies (Fig. 3a, right). By analogy to chemical solid phase peptide or DNA/RNA synthesis, this method is based on the use of magnetic beads or coated wells as solid support. It allows for easy removal of excess material and change of reactant solutions. The arrays are thus bound to the support by an initial building block (or initiator) and then elongated enzymatically step-by-step in a reactant buffer solution. The building blocks are considered as protected when non-digested and as activated when a desired cohesive end is created upon enzymatic digestion. The coupling to the solid phase is brought about a biotin–streptavidin interaction and is easier to implement due to the commercial availability of the solid support and the ease to obtain biotinylated oligonucleotides.

As for the two previously described assembly methods, the solid support strategies rely on the use of either type IIS restriction enzymes [33, 38, 45] or isocaudomers [38, 40]. Collection of building blocks are composed of initiators that are biotinylated (5′-end coding strand), extension blocks and terminators. The Iterative Capped Assembly (ICA) method developed by Briggs and colleagues [33] introduced an important additional step that prevents yield decrease due to incomplete ligation efficiency by blocking unreacted products (blockers). Extension blocks are typically prepared by plasmid digestion or PCR amplification from a collection of single repeats [33] or pre-assembled multi-repeat modules [38, 40, 45]. Depending on the length of the array and the size of the starting blocks, the assembly of full size arrays is achievable within 3 days. Since the process can be easily automated using liquid handling workstations and thanks to the absence of intermediate cloning steps, these methods are amenable to medium and high-throughput production.

Ligation Independent Cloning (LIC)

The strategy released by Schmid-Burgk and colleagues [39] is the sole procedure not involving restriction and ligation steps for the coupling of the building blocks (Fig. 3b, right). Ligation independent cloning relies on the creation of long (up to 30 bp) non-palindromic overhangs that are generated taking advantage of the 3′–5′-exonuclease activity of the T4 DNA polymerase in the presence of only one of the four dNTP's. To increase the throughput of the assembly process, a collection of 3072 penta-repeats encoding plasmids was created and can replace or be combined with their original collection of di-repeats. Using these collections, a one-step LIC reaction enables the assembly of arrays of various sizes within 3 days. Additionally, a bacterial growth at limited dilution was implemented to rely only on liquid handling steps, so further improving the throughput of the LIC strategy. Interestingly, this improvement can also be implemented in most of the above-described assembly methods to improve their production throughput.

Designer TALEs and Their Use

The discovery of TALEs and their unique way to recognize DNA has had an extraordinary impact on life sciences. The modularity of the TALE DNA binding domain and the simple interaction code with DNA has boosted the development of customizable DNA binding domains for a variety of applications, spanning from basic research to applications in human gene therapy. Since 2009, when the TALE DNA recognition code was uncovered [3, 4], the number of published manuscripts that refer to TALE research, technical improvements of the technology and/or their applications has reached 231 in 2013, over 10 times more than 2009. As discussed above, different toolboxes are available to assemble an array of TALE modules to target a chosen DNA sequence. Naturally occurring TALEs contain from as little as 1.5 to 33.5 repeats in their DNA binding domain, with a median of 17.5 repeat modules [2]; however, a minimum of 6.5–10.5 repeats has been reported to be crucial to achieve a measurable biological activity [3]. Once assembled, the tailored DNA binding domain can be fused to different types of effector domains to create designer enzymes for a large variety of applications. The most successful class of artificial enzymes harboring a user-defined DNA binding domain is certainly represented by designer nucleases. These enzymes combine sequence specificity and cleavage activity, brought together by the fusion of a designer DNA binding domain with a nuclease domain, usually derived from the *Fok*I restriction enzyme. Correct dimerization at a given site allows for the introduction of a targeted double stranded break (DSB) in the genome of interest. Genome editing is the field that has benefitted the most from the introduction of designer nucleases as a mean to introduce targeted genomic modifications. Once the genomic DNA is naturally or artificially damaged, the cell relies on conserved repair mechanisms to promptly repair the insult and avoid apoptosis. In mammals, one of the two major DNA repair pathways is harnessed upon the introduction of a DSB to ensure DNA integrity: (1) non-homologous end joining (NHEJ) or (2) homology-directed repair which is based on homologous recombination (HR). NHEJ is active throughout the cell cycle and is the fastest way for the cell to repair a DSB; however, it is an error-prone mechanism that can lead to small insertions/deletions (indels) at the break sites with serious consequences, including the loss of gene function if the DSB occurs in a gene-encoding region. Conversely, the HR-based repair mechanism allows for precise correction of a DSB since it relies, in the natural situation, on the genetic information contained in the sister chromatid for DSB repair and it is thus restricted to the S/G_2 phases of the cell cycle. HR-based DNA repair is rare in mammalian cells with an event occurring in every 10^4–10^7 cells [53]. However, pioneering studies in Dr. Jasin's lab [54, 55] provided evidence that HR frequency can be increased by several orders of magnitude at a certain genomic position upon generation of a targeted DSB and the concomitant delivery of a donor DNA template homologous to the target site. Under these conditions, the genetic information is conveyed from the donor DNA to the target locus, allowing precise genomic modifications. Thus, by harnessing NHEJ or HR repair pathways at specific genomic locations, one can aim at diverse outcomes like gene disruption, gene correction or gene addition. With these tools in hands, scientists have boosted their knowledge of gene functions by expanding reverse genetics to a huge variety of organisms [53]. Besides basic research, genome editing has found broad applicability in other fields like biotechnology, systems biology or human gene therapy, where this technology has been employed to engineer crop species with novel traits, isogenic cell lines to model human diseases and human cells with a corrected genetic defect [56]. For more than 15 years, Cys_2-His_2 zinc finger-based DNA binding domains have been used to generate designer nucleases with remarkable success. Zinc finger nucleases (ZFNs) have represented a milestone in the genome engineering field, allowing genomic manipulations in new species for gene function studies [57–62] and in therapeutic contexts to correct genetic defects underlying human disorders [63–65]. The remarkable progress achieved using ZFNs is epitomized by a phase I clinical trial for the treatment of HIV [66] that demonstrated that gene disruption can be used to create resistance to HIV infection [67, 68]. However, the limited targeting capacity, the context-dependent effects on DNA binding specificity between the repeat units within a zinc finger-array that make the process of generating tailored DNA binding domains time consuming, and a certain degree of unspecific targeting (the so-called off-target cleveage events) have represented a major impediment for their widespread use [69–72].

The discovery of TALE-based DNA binding domains has provided a new customizable platform for the generation of designer nucleases. TALE-based nucleases (TALENs) combine high versatility and superior specificity as compared to the well-established ZFN pairs [23, 73]. TALEN with novel specificities can be designed in a reasonable time [48] to target any given DNA sequence with an average of 3 TALEN pairs per base pair of DNA [74]. The targeting range of ZFNs is much lower with an average of one available pair every 50–500 bp [75, 76]. It is hence easy to understand why TALENs have propelled a remarkable expansion of genome editing strategies in the last years with many academic labs employing this novel technology.

The first report of TALE-based designer nucleases dates back to 2010 [28] and subsequent improvement of the TALEN scaffold and their efficacy [23, 29, 77] has led to their use in a variety of cell lines and organisms, including zebrafish [78], mouse [79], rat [73], non-human primates [80] and human induced pluripotent stem cells (iPSCs) [81]. TALEN have also been successfully applied in human gene therapy models, including the targeted genetic correction of the sickle cell disease mutation in human cells [82], restore gp91phox expression in granulocytes derived from iPSC of chronic granulomatous disease patients [83], to restore Dystrophin expression in Duchenne muscular dystrophy patient-derived cells [84], and for the treatment of recessive dystrophic epidermolysis bullosa [85]. Interestingly, designer nucleases can cleave not only genomic DNA but, when using suitable localization signals, they can be targeted to destroy mitochondrial DNA [86, 87]. Although preliminary, this approach opens new opportunities for the treatment of mitochondrial disorders.

In addition to nucleases, other effector domains can be fused to a tailored DNA binding domain to extend the application portfolio from genome editing to the targeted modification of the transcriptome and epigenome. The concept of modulating gene expression at the transcriptional level using designer transcription factor was successfully addressed using zinc finger-based DNA binding domains. Expression levels of endogenous genes were effectively modulated in murine models of human disorders, highlighting the feasibility of using tailored transcription factors as new therapeutics [88, 89]. With the introduction of TALEs, the availability of an easy customizable DNA binding domain has boosted the use of tailored transcriptional activators and repressors to modulate endogenous gene expression [24, 30]. The use of high-throughput methods to generate huge numbers of TALE-based transcription factors [90] may further expand applications in systems biology for the transcriptional control of entire pathways and to model novel gene networks. To further broaden the use of tailored enzymes, customized DNA binding domains can be fused to histone deacetylases (HDACs) and methyltransferases (HMTs) to achieve targeted epigenetic modifications [91, 92].

Specificity of Tailored TALE-Based Effectors

The future of researchers planning to use designer effectors for permanent modifications of the genome, epigenome or transcriptome seems thriving. Yet, one caveat associated with the use of TALE-based designer enzymes is their genome-wide specificity. Lack of specificity may lead to unwanted side-effects through binding of the effectors to off-target sites that share a certain degree of nucleotide identity with the intended target site. This issue has represented a major obstacle when using first generation dimeric ZFNs [93], which was subsequently overcome by redesigning the *Fok*I dimerizing interface to avoid homodimerization [94, 95]. Most of the efforts to assess the specificity of TALE-based enzymes have focused on microarray analysis after delivering TALE-based repressors [30] or on high-throughput approaches developed to dissect the specificity profiles of ZFNs. Screening of in vitro cleaved libraries or the ability of integrase defective lentiviral vectors (IDLVs) to be trapped in DNA double strand breaks have allowed e.g. to profile the specificity of *CCR5*-specific ZFNs [70, 72]. Both studies exposed a non-trivial degree of off-target cleavage activity of these ZFNs [71]. The invention of alternative platforms for the generation of designer nucleases, such as RNA-guided nucleases (RGNs) and TALENs, has provided novel substance to the genome engineering field. However, while RGNs can show a high degree of off-target activity [96–98], the use of second generation TALEN scaffolds [99] and variant CRISPR/Cas9 designs [100, 101] turned out to be less cytotoxic when compared to second and third generation ZFNs [23, 102]. Importantly, we recently demonstrated that higher specificity is directly linked to lower cytotoxicity [103]. In particular for approaches aimed at clinical translation, these results clearly underline the importance of working with a highly specific nuclease platform, such as TALENs.

Evidently, the intrinsic ability of some TALE repeats to recognize more than a single nucleotide poses concerns regarding their specificity. While NG, NI and

HD modules show a prominent preference for a single nucleotide [9], the most commonly used G-specific 'NN' module can also bind to adenine. As discussed above, systematic studies identified novel and potentially more stringent TALE modules, which may help to further improve the high specificity of TALENs [104]. An open question is whether the risk of genotoxicity can be reduced by using more specific cleavage domains. Based on this notion, fusion of a TALE-based DNA binding domain to the cleavage domain of the I-*Tev*I homing endonuclease to form a monomeric compact TALEN (cTALEN) have been explored [26]. In this scenario, a second level of safety is intrinsically provided by the I-*Tev*I cleavage domain that is active only when a degenerate CNNNG sequence is present in the target site [105]. While the partial DNA sequence preference of the I-*Tev*I domain reduces the occurrence of potential target sites within a genome, the cTALENs certainly simplify the generation of catalytically active TALENs by overcoming the need to generate two monomers per target site. Additionally, as recently shown by Lin and colleagues [106], TALE-based DNA binding domains can be linked to re-engineered meganucleases to specifically target the human genome. With this approach, the engineered TALE-I-*Sce*I fusion protein targeted to the β-globin gene induced comparable HDR at the target locus as a conventional TALEN but showed a significantly lower toxicity. Similarly to what has been accomplished for ZFNs, adapting the obligate heterodimeric *Fok*I cleavage domain may provide additional benefit in terms of specificity [107], and implementing the use of improved or hyperactive *Fok*I domains could help to generate highly specific and highly efficient designer TALENs [108, 109]. Additionally, a rational target site choice to avoid target sequences that share a high degree of identity with other sites in the genome is probably the most simple way to minimize unwanted off-target cleavage [104], and a number of web-based tools assist researchers with this task [110].

Conclusions

The use of designer nucleases to induce permanent genomic modification is increasing exponentially. Since the first reports of chimeric ZFNs that were envisioned to work as customizable restriction enzymes [111], remarkable progress has boosted their use in human gene therapy. With the introduction of TALENs, the widespread use of these enzymes has increased steadily because of a combination of favorable features, like their ease of design, their efficacy and their specificity as a genome editing tool. The advent of RGNs, which are highly efficient in inducing targeted DBSs and even easier to engineer as TALENs, has further accelerated this trend [112]. Although impressive gene editing efficiencies have been reported in primary T cells [113], one limiting step for the researchers interested in modifying the genome of primary cells is the designer nuclease size. TALENs are rather large proteins and their delivery still represents a challenge, in particular when using viral vectors. Their size as well as their repetitive nature can be limiting parameter for viral vector systems, such as adeno-associated viral vectors and retroviral vectors.

However, Holkers and colleagues have recently reported that adenoviral vectors are able to transfer intact TALEN DNA into human cells [114] while Yang and colleagues have packaged TALEN genes into lentiviral vectors by diversifying the nucleotide sequence of TALE repeat modules [115]. While the efficiency of the recoded TALENs was lower as compared to the canonical counterpart, alternative methods have been explored in the meantime. Non-viral delivery methods, such as the transfection of mRNA molecules that contain the complete TALEN coding sequence, have proven exceptionally efficient in inducing gene knockout in primary T cells [99, 113], thereby setting the stage for the translation of TALEN-mediated genome editing in various clinically relevant cell types in the near future.

References

1. Buttner D, Bonas U. Regulation and secretion of Xanthomonas virulence factors. FEMS Microbiol Rev. 2010;34(2):107–33.
2. Boch J, Bonas U. Xanthomonas AvrBs3 family-type III effectors: discovery and function. Annu Rev Phytopathol. 2010;48:419–36.
3. Boch J, Scholze H, Schornack S, Landgraf A, Hahn S, Kay S, et al. Breaking the code of DNA binding specificity of TAL-type III effectors. Science. 2009;326(5959):1509–12.
4. Moscou MJ, Bogdanove AJ. A simple cipher governs DNA recognition by TAL effectors. Science. 2009;326(5959):1501.
5. Deng D, Yan C, Pan X, Mahfouz M, Wang J, Zhu JK, et al. Structural basis for sequence-specific recognition of DNA by TAL effectors. Science. 2012;335(6069):720–3.
6. Mak AN, Bradley P, Cernadas RA, Bogdanove AJ, Stoddard BL. The crystal structure of TAL effector PthXo1 bound to its DNA target. Science. 2012;335(6069):716–9.
7. Christian ML, Demorest ZL, Starker CG, Osborn MJ, Nyquist MD, Zhang Y, et al. Targeting G with TAL effectors: a comparison of activities of TALENs constructed with NN and NK repeat variable di-residues. PLoS One. 2012;7(9), e45383.
8. Streubel J, Blucher C, Landgraf A, Boch J. TAL effector RVD specificities and efficiencies. Nat Biotechnol. 2012;30(7):593–5.
9. Cong L, Zhou R, Kuo YC, Cunniff M, Zhang F. Comprehensive interrogation of natural TALE DNA-binding modules and transcriptional repressor domains. Nat Commun. 2012;3:968.
10. Meckler JF, Bhakta MS, Kim MS, Ovadia R, Habrian CH, Zykovich A, et al. Quantitative analysis of TALE-DNA interactions suggests polarity effects. Nucleic Acids Res. 2013; 41(7):4118–28.
11. Juillerat A, Pessereau C, Dubois G, Guyot V, Marechal A, Valton J, et al. Optimized tuning of TALEN specificity using non-conventional RVDs. Sci Rep. 2015;5:8150.
12. Miller JC, Zhang L, Xia DF, Campo JJ, Ankoudinova IV, Guschin DY, et al. Improved specificity of TALE-based genome editing using an expanded RVD repertoire. Nat Methods. 2015;12(5):465–71.
13. Bultmann S, Morbitzer R, Schmidt CS, Thanisch K, Spada F, Elsaesser J, et al. Targeted transcriptional activation of silent oct4 pluripotency gene by combining designer TALEs and inhibition of epigenetic modifiers. Nucleic Acids Res. 2012;40(12):5368–77.
14. Deng D, Yin P, Yan C, Pan X, Gong X, Qi S, et al. Recognition of methylated DNA by TAL effectors. Cell Res. 2012;22(10):1502–4.
15. Valton J, Dupuy A, Daboussi F, Thomas S, Marechal A, Macmaster R, et al. Overcoming transcription activator-like effector (TALE) DNA binding domain sensitivity to cytosine methylation. J Biol Chem. 2012;287(46):38427–32.

16. de Lange O, Schreiber T, Schandry N, Radeck J, Braun KH, Koszinowski J, et al. Breaking the DNA-binding code of Ralstonia solanacearum TAL effectors provides new possibilities to generate plant resistance genes against bacterial wilt disease. New Phytol. 2013;199(3):773–86.
17. Lamb BM, Mercer AC, Barbas 3rd CF. Directed evolution of the TALE N-terminal domain for recognition of all 5c bases. Nucleic Acids Res. 2013;41(21):9779–85.
18. de Lange O, Wolf C, Dietze J, Elsaesser J, Morbitzer R, Lahaye T. Programmable DNA-binding proteins from Burkholderia provide a fresh perspective on the TALE-like repeat domain. Nucleic Acids Res. 2014;42(11):7436–49.
19. Juillerat A, Bertonati C, Dubois G, Guyot V, Thomas S, Valton J, et al. BurrH: a new modular DNA binding protein for genome engineering. Sci Rep. 2014;4:3831.
20. Stella S, Molina R, Lopez-Mendez B, Juillerat A, Bertonati C, Daboussi F, et al. BuD, a helix-loop-helix DNA-binding domain for genome modification. Acta Crystallogr D Biol Crystallogr. 2014;70(Pt 7):2042–52.
21. Gurlebeck D, Szurek B, Bonas U. Dimerization of the bacterial effector protein AvrBs3 in the plant cell cytoplasm prior to nuclear import. Plant J. 2005;42(2):175–87.
22. Szurek B, Rossier O, Hause G, Bonas U. Type III-dependent translocation of the Xanthomonas AvrBs3 protein into the plant cell. Mol Microbiol. 2002;46(1):13–23.
23. Mussolino C, Morbitzer R, Lutge F, Dannemann N, Lahaye T, Cathomen T. A novel TALE nuclease scaffold enables high genome editing activity in combination with low toxicity. Nucleic Acids Res. 2011;39(21):9283–93.
24. Zhang F, Cong L, Lodato S, Kosuri S, Church GM, Arlotta P. Efficient construction of sequence-specific TAL effectors for modulating mammalian transcription. Nat Biotechnol. 2011;29(2):149–53.
25. Li T, Huang S, Jiang WZ, Wright D, Spalding MH, Weeks DP, et al. TAL nucleases (TALNs): hybrid proteins composed of TAL effectors and FokI DNA-cleavage domain. Nucleic Acids Res. 2011;39(1):359–72.
26. Beurdeley M, Bietz F, Li J, Thomas S, Stoddard T, Juillerat A, et al. Compact designer TALENs for efficient genome engineering. Nat Commun. 2013;4:1762.
27. Mercer AC, Gaj T, Fuller RP, Barbas 3rd CF. Chimeric TALE, recombinases with programmable DNA sequence specificity. Nucleic Acids Res. 2012;40(21):11163–72.
28. Christian M, Cermak T, Doyle EL, Schmidt C, Zhang F, Hummel A, et al. Targeting DNA double-strand breaks with TAL effector nucleases. Genetics. 2010;186(2):757–61.
29. Miller JC, Tan S, Qiao G, Barlow KA, Wang J, Xia DF, et al. A TALE nuclease architecture for efficient genome editing. Nat Biotechnol. 2011;29(2):143–8.
30. Mahfouz MM, Li L, Piatek M, Fang X, Mansour H, Bangarusamy DK, et al. Targeted transcriptional repression using a chimeric TALE-SRDX repressor protein. Plant Mol Biol. 2012;78(3):311–21.
31. Gao X, Yang J, Tsang JC, Ooi J, Wu D, Liu P. Reprogramming to pluripotency using designer TALE transcription factors targeting enhancers. Stem Cell Reports. 2013;1(2):183–97.
32. Maeder ML, Angstman JF, Richardson ME, Linder SJ, Cascio VM, Tsai SQ, et al. Targeted DNA demethylation and activation of endogenous genes using programmable TALE-TET1 fusion proteins. Nat Biotechnol. 2013;31(12):1137–42.
33. Briggs AW, Rios X, Chari R, Yang L, Zhang F, Mali P, et al. Iterative capped assembly: rapid and scalable synthesis of repeat-module DNA such as TAL effectors from individual monomers. Nucleic Acids Res. 2012;40(15), e117.
34. Li L, Piatek MJ, Atef A, Piatek A, Wibowo A, Fang X, et al. Rapid and highly efficient construction of TALE-based transcriptional regulators and nucleases for genome modification. Plant Mol Biol. 2012;78(4-5):407–16.
35. Li T, Huang S, Zhao X, Wright DA, Carpenter S, Spalding MH, et al. Modularly assembled designer TAL effector nucleases for targeted gene knockout and gene replacement in eukaryotes. Nucleic Acids Res. 2011;39(14):6315–25.
36. Morbitzer R, Elsaesser J, Hausner J, Lahaye T. Assembly of custom TALE-type DNA binding domains by modular cloning. Nucleic Acids Res. 2011;39(13):5790–9.

37. Reyon D, Khayter C, Regan MR, Joung JK, Sander JD. Engineering designer transcription activator-like effector nucleases (TALENs) by REAL or REAL-Fast assembly. Curr Protoc Mol Biol. 2012;Chapter 12:Unit 12.15.
38. Reyon D, Tsai SQ, Khayter C, Foden JA, Sander JD, Joung JK. FLASH assembly of TALENs for high-throughput genome editing. Nat Biotechnol. 2012;30(5):460–5.
39. Schmid-Burgk JL, Schmidt T, Kaiser V, Honing K, Hornung V. A ligation-independent cloning technique for high-throughput assembly of transcription activator-like effector genes. Nat Biotechnol. 2013;31(1):76–81.
40. Wang Z, Li J, Huang H, Wang G, Jiang M, Yin S, et al. An integrated chip for the high-throughput synthesis of transcription activator-like effectors. Angew Chem Int Ed Engl. 2012;51(34):8505–8.
41. Weber E, Gruetzner R, Werner S, Engler C, Marillonnet S. Assembly of designer TAL effectors by Golden Gate cloning. PLoS One. 2011;6(5), e19722.
42. Yang J, Yuan P, Wen D, Sheng Y, Zhu S, Yu Y, et al. ULtiMATE system for rapid assembly of customized TAL effectors. PLoS One. 2013;8(9), e75649.
43. Huang P, Xiao A, Zhou M, Zhu Z, Lin S, Zhang B. Heritable gene targeting in zebrafish using customized TALENs. Nat Biotechnol. 2011;29(8):699–700.
44. Sander JD, Cade L, Khayter C, Reyon D, Peterson RT, Joung JK, et al. Targeted gene disruption in somatic zebrafish cells using engineered TALENs. Nat Biotechnol. 2011;29(8):697–8.
45. Ding Q, Lee YK, Schaefer EA, Peters DT, Veres A, Kim K, et al. A TALEN genome-editing system for generating human stem cell-based disease models. Cell Stem Cell. 2013; 12(2):238–51.
46. Engler C, Gruetzner R, Kandzia R, Marillonnet S. Golden gate shuffling: a one-pot DNA shuffling method based on type IIs restriction enzymes. PLoS One. 2009;4(5), e5553.
47. Thieme F, Engler C, Kandzia R, Marillonnet S. Quick and clean cloning: a ligation-independent cloning strategy for selective cloning of specific PCR products from non-specific mixes. PLoS One. 2011;6(6), e20556.
48. Cermak T, Doyle EL, Christian M, Wang L, Zhang Y, Schmidt C, et al. Efficient design and assembly of custom TALEN and other TAL effector-based constructs for DNA targeting. Nucleic Acids Res. 2011;39(12), e82.
49. Geissler R, Scholze H, Hahn S, Streubel J, Bonas U, Behrens SE, et al. Transcriptional activators of human genes with programmable DNA-specificity. PLoS One. 2011;6(5), e19509.
50. Sakuma T, Hosoi S, Woltjen K, Suzuki K, Kashiwagi K, Wada H, et al. Efficient TALEN construction and evaluation methods for human cell and animal applications. Genes Cells. 2013;18(4):315–26.
51. Geu-Flores F, Nour-Eldin HH, Nielsen MT, Halkier BA. USER fusion: a rapid and efficient method for simultaneous fusion and cloning of multiple PCR products. Nucleic Acids Res. 2007;35(7), e55.
52. Sanjana NE, Cong L, Zhou Y, Cunniff MM, Feng G, Zhang F. A transcription activator-like effector toolbox for genome engineering. Nat Protoc. 2012;7(1):171–92.
53. Carroll D. Genome engineering with zinc-finger nucleases. Genetics. 2011;188(4):773–82.
54. Rouet P, Smih F, Jasin M. Expression of a site-specific endonuclease stimulates homologous recombination in mammalian cells. Proc Natl Acad Sci U S A. 1994;91(13):6064–8.
55. Smih F, Rouet P, Romanienko PJ, Jasin M. Double-strand breaks at the target locus stimulate gene targeting in embryonic stem cells. Nucleic Acids Res. 1995;23(24):5012–9.
56. Perez-Pinera P, Ousterout DG, Gersbach CA. Advances in targeted genome editing. Curr Opin Chem Biol. 2012;16(3-4):268–77.
57. Doyon Y, McCammon JM, Miller JC, Faraji F, Ngo C, Katibah GE, et al. Heritable targeted gene disruption in zebrafish using designed zinc-finger nucleases. Nat Biotechnol. 2008;26(6):702–8.
58. Geurts AM, Cost GJ, Freyvert Y, Zeitler B, Miller JC, Choi VM, et al. Knockout rats via embryo microinjection of zinc-finger nucleases. Science. 2009;325(5939):433.
59. Merlin C, Beaver LE, Taylor OR, Wolfe SA, Reppert SM. Efficient targeted mutagenesis in the monarch butterfly using zinc-finger nucleases. Genome Res. 2013;23(1):159–68.

60. Young JJ, Cherone JM, Doyon Y, Ankoudinova I, Faraji FM, Lee AH, et al. Efficient targeted gene disruption in the soma and germ line of the frog Xenopus tropicalis using engineered zinc-finger nucleases. Proc Natl Acad Sci U S A. 2011;108(17):7052–7.
61. Shukla VK, Doyon Y, Miller JC, DeKelver RC, Moehle EA, Worden SE, et al. Precise genome modification in the crop species Zea mays using zinc-finger nucleases. Nature. 2009;459(7245):437–41.
62. Zhang F, Maeder ML, Unger-Wallace E, Hoshaw JP, Reyon D, Christian M, et al. High frequency targeted mutagenesis in Arabidopsis thaliana using zinc finger nucleases. Proc Natl Acad Sci U S A. 2010;107(26):12028–33.
63. Li H, Haurigot V, Doyon Y, Li T, Wong SY, Bhagwat AS, et al. In vivo genome editing restores haemostasis in a mouse model of haemophilia. Nature. 2011;475(7355):217–21.
64. Soldner F, Laganiere J, Cheng AW, Hockemeyer D, Gao Q, Alagappan R, et al. Generation of isogenic pluripotent stem cells differing exclusively at two early onset Parkinson point mutations. Cell. 2011;146(2):318–31.
65. Zou J, Mali P, Huang X, Dowey SN, Cheng L. Site-specific gene correction of a point mutation in human iPS cells derived from an adult patient with sickle cell disease. Blood. 2011;118(17):4599–608.
66. Tebas P, Stein D, Tang WW, Frank I, Wang SQ, Lee G, et al. Gene editing of CCR5 in autologous CD4 T cells of persons infected with HIV. N Engl J Med. 2014;370(10):901–10.
67. Holt N, Wang J, Kim K, Friedman G, Wang X, Taupin V, et al. Human hematopoietic stem/progenitor cells modified by zinc-finger nucleases targeted to CCR5 control HIV-1 in vivo. Nat Biotechnol. 2010;28(8):839–47.
68. Perez EE, Wang J, Miller JC, Jouvenot Y, Kim KA, Liu O, et al. Establishment of HIV-1 resistance in CD4+ T cells by genome editing using zinc-finger nucleases. Nat Biotechnol. 2008;26(7):808–16.
69. Cathomen T, Joung JK. Zinc-finger nucleases: the next generation emerges. Mol Ther. 2008;16(7):1200–7.
70. Gabriel R, Lombardo A, Arens A, Miller JC, Genovese P, Kaeppel C, et al. An unbiased genome-wide analysis of zinc-finger nuclease specificity. Nat Biotechnol. 2011;29(9):816–23.
71. Mussolino C, Cathomen T. On target? Tracing zinc-finger-nuclease specificity. Nat Methods. 2011;8(9):725–6.
72. Pattanayak V, Ramirez CL, Joung JK, Liu DR. Revealing off-target cleavage specificities of zinc-finger nucleases by in vitro selection. Nat Methods. 2011;8(9):765–70.
73. Tesson L, Usal C, Menoret S, Leung E, Niles BJ, Remy S, et al. Knockout rats generated by embryo microinjection of TALENs. Nat Biotechnol. 2011;29(8):695–6.
74. Joung JK, Sander JD. TALENs: a widely applicable technology for targeted genome editing. Nat Rev Mol Cell Biol. 2013;14(1):49–55.
75. Bhakta MS, Henry IM, Ousterout DG, Das KT, Lockwood SH, Meckler JF, et al. Highly active zinc-finger nucleases by extended modular assembly. Genome Res. 2013;23(3):530–8.
76. Sander JD, Dahlborg EJ, Goodwin MJ, Cade L, Zhang F, Cifuentes D, et al. Selection-free zinc-finger-nuclease engineering by context-dependent assembly (CoDA). Nat Methods. 2011;8(1):67–9.
77. Bedell VM, Wang Y, Campbell JM, Poshusta TL, Starker CG, Krug 2nd RG, et al. In vivo genome editing using a high-efficiency TALEN system. Nature. 2012;491(7422):114–8.
78. Zu Y, Tong X, Wang Z, Liu D, Pan R, Li Z, et al. TALEN-mediated precise genome modification by homologous recombination in zebrafish. Nat Methods. 2013;10(4):329–31.
79. Sung YH, Baek IJ, Kim DH, Jeon J, Lee J, Lee K, et al. Knockout mice created by TALEN-mediated gene targeting. Nat Biotechnol. 2013;31(1):23–4.
80. Liu H, Chen Y, Niu Y, Zhang K, Kang Y, Ge W, et al. TALEN-mediated gene mutagenesis in rhesus and cynomolgus monkeys. Cell Stem Cell. 2014;14(3):323–8.
81. Menon T, Firth AL, Scripture-Adams DD, Galic Z, Qualls SJ, Gilmore WB, et al. Lymphoid regeneration from gene-corrected SCID-X1 subject-derived iPSCs. Cell Stem Cell. 2015; 16(4):367–72.
82. Sun N, Liang J, Abil Z, Zhao H. Optimized TAL effector nucleases (TALENs) for use in treatment of sickle cell disease. Mol Biosyst. 2012;8(4):1255–63.

83. Dreyer A-K, Hoffmann D, Lachmann N, Ackermann M, Steinemann D, Timm B, Siler U, et al. TALEN-mediated functional correction of X-linked chronic granulomatous disease in patient-derived induced pluripotent stem cells. Biomaterials 2015;69:191–200.
84. Ousterout DG, Perez-Pinera P, Thakore PI, Kabadi AM, Brown MT, Qin X, et al. Reading frame correction by targeted genome editing restores dystrophin expression in cells from Duchenne muscular dystrophy patients. Mol Ther. 2013;21(9):1718–26.
85. Osborn MJ, Starker CG, McElroy AN, Webber BR, Riddle MJ, Xia L, et al. TALEN-based gene correction for epidermolysis bullosa. Mol Ther. 2013;21(6):1151–9.
86. Bacman SR, Williams SL, Pinto M, Peralta S, Moraes CT. Specific elimination of mutant mitochondrial genomes in patient-derived cells by mitoTALENs. Nat Med. 2013;19(9):1111–3.
87. Minczuk M, Papworth MA, Miller JC, Murphy MP, Klug A. Development of a single-chain, quasi-dimeric zinc-finger nuclease for the selective degradation of mutated human mitochondrial DNA. Nucleic Acids Res. 2008;36(12):3926–38.
88. Mussolino C, Sanges D, Marrocco E, Bonetti C, Di Vicino U, Marigo V, et al. Zinc-finger-based transcriptional repression of rhodopsin in a model of dominant retinitis pigmentosa. EMBO Mol Med. 2011;3(3):118–28.
89. Rebar EJ, Huang Y, Hickey R, Nath AK, Meoli D, Nath S, et al. Induction of angiogenesis in a mouse model using engineered transcription factors. Nat Med. 2002;8(12):1427–32.
90. Maeder ML, Linder SJ, Reyon D, Angstman JF, Fu Y, Sander JD, et al. Robust, synergistic regulation of human gene expression using TALE activators. Nat Methods. 2013;10(3):243–5.
91. Konermann S, Brigham MD, Trevino AE, Hsu PD, Heidenreich M, Cong L, et al. Optical control of mammalian endogenous transcription and epigenetic states. Nature. 2013;500(7463):472–6.
92. Bernstein DL, Le Lay JE, Ruano EG, Kaestner KH. TALE-mediated epigenetic suppression of CDKN2A increases replication in human fibroblasts. J Clin Invest. 2015;125(5):1998–2006.
93. Alwin S, Gere MB, Guhl E, Effertz K, Barbas 3rd CF, Segal DJ, et al. Custom zinc-finger nucleases for use in human cells. Mol Ther. 2005;12(4):610–7.
94. Miller JC, Holmes MC, Wang J, Guschin DY, Lee YL, Rupniewski I, et al. An improved zinc-finger nuclease architecture for highly specific genome editing. Nat Biotechnol. 2007; 25(7):778–85.
95. Szczepek M, Brondani V, Buchel J, Serrano L, Segal DJ, Cathomen T. Structure-based redesign of the dimerization interface reduces the toxicity of zinc-finger nucleases. Nat Biotechnol. 2007;25(7):786–93.
96. Cradick TJ, Fine EJ, Antico CJ, Bao G. CRISPR/Cas9 systems targeting beta-globin and CCR5 genes have substantial off-target activity. Nucleic Acids Res. 2013;41(20):9584–92.
97. Fu Y, Foden JA, Khayter C, Maeder ML, Reyon D, Joung JK, et al. High-frequency off-target mutagenesis induced by CRISPR-Cas nucleases in human cells. Nat Biotechnol. 2013; 31(9):822–6.
98. Pattanayak V, Lin S, Guilinger JP, Ma E, Doudna JA, Liu DR. High-throughput profiling of off-target DNA cleavage reveals RNA-programmed Cas9 nuclease specificity. Nat Biotechnol. 2013;31(9):839–43.
99. Mock U, Machowicz R, Hauber I, Horn S, Abramowski P, Berdien B, et al. mRNA transfection of a novel TAL effector nuclease (TALEN) facilitates efficient knockout of HIV co-receptor CCR5. Nucleic Acids Res. 2015;43(11):5560–71.
100. Fu Y, Sander JD, Reyon D, Cascio VM, Joung JK. Improving CRISPR-Cas nuclease specificity using truncated guide RNAs. Nat Biotechnol. 2014;32(3):279–84.
101. Tsai SQ, Wyvekens N, Khayter C, Foden JA, Thapar V, Reyon D, et al. Dimeric CRISPR RNA-guided FokI nucleases for highly specific genome editing. Nat Biotechnol. 2014; 32(6):569–76.
102. Bogdanove AJ, Schornack S, Lahaye T. TAL effectors: finding plant genes for disease and defense. Curr Opin Plant Biol. 2010;13(4):394–401.
103. Mussolino C, Alzubi J, Fine EJ, Morbitzer R, Cradick TJ, Lahaye T, et al. TALENs facilitate targeted genome editing in human cells with high specificity and low cytotoxicity. Nucleic Acids Res. 2014;42(10):6762–73.
104. Kim Y, Kweon J, Kim A, Chon JK, Yoo JY, Kim HJ, et al. A library of TAL effector nucleases spanning the human genome. Nat Biotechnol. 2013;31(3):251–8.

105. Dean AB, Stanger MJ, Dansereau JT, Van Roey P, Derbyshire V, Belfort M. Zinc finger as distance determinant in the flexible linker of intron endonuclease I-TevI. Proc Natl Acad Sci U S A. 2002;99(13):8554–61.
106. Lin J, Chen H, Luo L, Lai Y, Xie W, Kee K. Creating a monomeric endonuclease TALE-I-SceI with high specificity and low genotoxicity in human cells. Nucleic Acids Res. 2015;43(2):1112–22.
107. Cade L, Reyon D, Hwang WY, Tsai SQ, Patel S, Khayter C, et al. Highly efficient generation of heritable zebrafish gene mutations using homo- and heterodimeric TALENs. Nucleic Acids Res. 2012;40(16):8001–10.
108. Doyon Y, Vo TD, Mendel MC, Greenberg SG, Wang J, Xia DF, et al. Enhancing zinc-finger-nuclease activity with improved obligate heterodimeric architectures. Nat Methods. 2011;8(1):74–9.
109. Guo J, Gaj T, Barbas 3rd CF. Directed evolution of an enhanced and highly efficient FokI cleavage domain for zinc finger nucleases. J Mol Biol. 2010;400(1):96–107.
110. Booher NJ, Bogdanove AJ. Tools for TAL effector design and target prediction. Methods. 2014;69(2):121–7.
111. Kim YG, Cha J, Chandrasegaran S. Hybrid restriction enzymes: zinc finger fusions to Fok I cleavage domain. Proc Natl Acad Sci U S A. 1996;93(3):1156–60.
112. Doudna JA, Charpentier E. Genome editing. The new frontier of genome engineering with CRISPR-Cas9. Science. 2014;346(6213):1258096.
113. Valton J, Guyot V, Marechal A, Filhol JM, Juillerat A, Duclert A, et al. A multidrug resistant engineered CAR T cell for allogeneic combination immunotherapy. Mol Ther. 2015; 23(9):1507–18.
114. Holkers M, Maggio I, Liu J, Janssen JM, Miselli F, Mussolino C, et al. Differential integrity of TALE nuclease genes following adenoviral and lentiviral vector gene transfer into human cells. Nucleic Acids Res. 2013;41(5), e63.
115. Yang L, Guell M, Byrne S, Yang JL, De Los AA, Mali P, et al. Optimization of scarless human stem cell genome editing. Nucleic Acids Res. 2013;41(19):9049–61.

Genome Editing for Neuromuscular Diseases

David G. Ousterout and Charles A. Gersbach

Abstract Neuromuscular diseases are a diverse range of conditions that include myopathic and neuropathic disorders related to muscular dysfunction. Inherited neuromuscular diseases are the result of a broad spectrum of genetic mutations, including point mutations, insertions and deletions, chromosomal rearrangements, epigenetic aberrations, and repeat expansions or contractions. Targeted genome editing is a promising method to correct the inherited mutations underlying these disorders. Over the last decade there have been many significant advances in engineering targeted DNA-binding proteins to manipulate specific sequences of complex genomes. These genome editing tools are rapidly becoming viable therapeutics that will allow the targeted addition, exchange, or removal of almost any genetic sequence in the human genome. In this chapter, selected neuromuscular diseases representing inherited myopathies or neuropathies are discussed. The genome editing tools available to create targeted genetic modifications are reviewed. Promising cell- and gene-based therapies are introduced in the context of the treatment of neuromuscular disorders in combination with genome editing therapies. Finally, specific examples of how genome editing may be applied to correct the genetic basis of particular neuromuscular disorders are presented and discussed.

Keywords Genome editing • Zinc finger nucleases • TALENs • CRISPR/Cas9-muscular dystrophy • Motor neuron disorders

Abbreviations

AAV	Adeno-associated virus
BMD	BECKER muscular dystrophy
CRISPR	Clustered regularly interspaced short palindromic repeats
DM	Myotonic dystrophy
DMD	Duchenne muscular dystrophy

D.G. Ousterout, Ph.D. • C.A. Gersbach, Ph.D. (✉)
Department of Biomedical Engineering, Duke University,
Room 136 Hudson Hall, Box 90281, Durham, NC 27708-0281, USA
e-mail: charles.gersbach@duke.edu

T. Cathomen et al. (eds.), *Genome Editing*, Advances in Experimental Medicine and Biology 895, DOI 10.1007/978-1-4939-3509-3_4

DSB	Double-strand break
FSHD	Fascioscapulohumeral dystrophy
gRNA	Guide RNA
HDR	Homology-directed repair
HT	Huntington's disease
iPSC	Induced pluripotent stem cell
LGMD2B	Limb-girdle muscular dystrophy type 2B
MGN	Meganuclease
MM	Miyoshi myopathy
NHEJ	Non-homologous end joining
RVD	Repeat variable diresidue
SMA	Spinal muscular atrophy
ssODN	Single-strand oligodeoxynucleotide
TALEN	Transcription activator-like effector nuclease
ZFN	Zinc finger nuclease

Introduction

Neuromuscular diseases include myopathic and neuropathic disorders that result in muscular dysfunction. These disorders can cause a range of symptoms from significant motor impairment to total paralysis and death. A subset of these diseases is caused by inherited genetic mutations that damage the function of an essential gene. Importantly, these diseases can be caused by a broad spectrum of genetic mutations, including point mutations, insertions and deletions, chromosomal rearrangements, epigenetic aberrations, and repeat expansions or contractions. Within the past decade, advances in efficiency, ease of use, and availability of genome editing technologies has enabled new approaches to study and potentially treat this class of diseases. These enabling technologies allow researchers to add, change and replace any genetic sequence of interest at will. In this chapter, neuromuscular diseases under investigation by genome editing will be overviewed and the methods and applications for novel treatment modalities and basic science research using genome editing tools will be discussed.

Overview of Common Neuromuscular Disorders

Muscular dystrophies are a class of myopathies that result directly from progressive degeneration of muscle fibers. This class of diseases results from mutations to an extensive panel of genes, of which at least 30 are currently known [1]. Motor neuron disorders indirectly affect muscle function by degrading the ability of the central nervous system to control muscle movement. These diseases are traditionally classified by the phenotype produced, including disease onset, severity, muscle groups affected, inheritance patterns, and other non-muscle phenotype changes [2, 3]. However, it is apparent that mutations to independent genes can result in similar phenotypes for a number of these disorders and mutations to the same gene can lead to phenotypically distinct conditions. This is the result of complicated and interdependent protein complexes present in muscle, making it necessary to reclassify some of these disorders by the molecular basis or common phenotype created, such as dysferlinopathy and limb-girdle muscular dystrophy, respectively. This section will focus on five major types of muscular dystrophies and three motor neuron disorders, illustrating a variety of genetic disruptions that will each require a unique gene editing approach.

Duchenne and Becker Muscular Dystrophies

Duchenne muscular dystrophy (DMD) is the most common monogenetic hereditary disorder, occurring in approximately 1 in 3500 male births [4]. This recessive, X-linked disorder is caused by mutations to the dystrophin gene. The primary function of the dystrophin protein is to provide a mechanical link between the cytoskeleton and extracellular matrix that protects the sarcolemma membrane from shearing forces during muscle contraction. Mutations that cause truncation of the dystrophin protein, thereby breaking this mechanical link, result in progressive muscle wasting and death by the third decade of life. The current standard of care for DMD is palliative and has focused on managing respiratory and cardiac failure with steroid and ACE inhibitor therapy. Similar to DMD, Becker muscular dystrophy (BMD) is a monogenic degenerative musculoskeletal disease that is caused by a mutation to the dystrophin gene. However in this disease, mutations to the dystrophin gene cause internal deletions, resulting in expression of partially functional dystrophin protein. BMD is typically less severe than DMD due to this partial dystrophin functionality. The symptoms and progression of BMD are more difficult to predict than that of DMD, but typically follow a similar progressive muscular degeneration pattern, albeit at a much slower rate [4]. As a result, with proper care and disease management, most BMD patients can live independently and have a close to normal life expectancy.

Several promising methods have emerged to correct the dystrophin gene. It has been challenging to deliver a functional copy of the dystrophin gene because of the large size of the gene itself (>11 kb coding sequence). Miniaturized versions of the dystrophin gene that can be packaged into adeno-associated virus vectors have been engineered to deliver a truncated, but functional, dystrophin gene to muscle tissue [5–7]. These minidystrophin genes may be sufficient to ameliorate the symptoms of DMD [5], though careful selection of appropriate minimal dystrophin proteins is still under investigation [8]. In contrast to these methods, restoration of the native dystrophin gene product may lead to improved clinical outcomes by salvaging protein functionality [9]. This is a relatively new area of gene and molecular therapy, in which strategies, dubbed “exon skipping”, have been developed to selectively exclude portions of a damaged, out-of-frame gene by selectively removing exons from dystrophin mRNA to restore the reading frame [10]. Currently, the overall efficacy of exon skipping approaches is still under investigation in human clinical trials [11–13].

Despite these hurdles, dystrophin remains one of the principal targets for gene therapies. Genome editing efforts have shown promise across three distinct approaches to rescue dystrophin expression, including the correction of point mutations [14], the creation of targeted frameshifts and deletions to restore the reading frame of an internally deleted dystrophin gene [15–18], the targeted addition of missing exons to the gene to address patient-specific mutations [18, 19], and the introduction of a functional dystrophin gene expression cassette to a predefined genomic location [20].

Limb-Girdle Muscular Dystrophy Type 2B

Dysferlinopathies are autosomal recessive disorders caused by loss of functional dysferlin protein expression and are the molecular basis of Limb-girdle muscular dystrophy type 2B (LGMD2B) and Miyoshi myopathy (MM) [21]. Dysferlin is a calcium-sensitive protein that is involved in membrane repair following damage to the sarcolemma in muscle fibers [22]. While the molecular basis of LGMD2B and MM is the same, the presentation of these diseases is distinct for unknown reasons. LGMD2B pathogenesis begins with shoulder and hip weakness (limb-girdle) that progress to proximal limb weakness [23]. Miyoshi myopathy caused by dysferlinopathy generally presents as distal limb weakness in the calves, forearms, hands and feet [23]. For both disorders, disease onset typically occurs by 30 years of age and life span is generally normal though ambulation is significantly impaired within two decades of prognosis. In both of these dysferlinopathies, monogenic mutations to the dysferlin gene disrupt protein function. Restoration of functional dysferlin gene expression is a promising approach to treating these disorders [24]. Possible genome editing strategies for dysferlinopathies would likely be similar to those for DMD, including targeted frameshifts to restore the reading frame around deleted non-essential areas of the gene, genetic knock-in of exons absent in the patient-specific mutation, and introduction of functional dysferlin gene cassettes by gene targeting.

Myotonic Dystrophy

Myotonic dystrophy is a disease caused by different autosomal dominant disorders, termed myotonic dystrophy type 1 (DM1) and myotonic dystrophy type 2 (DM2). These disorders are characterized as progressive multisystemic diseases, including progressive muscle wasting, myotonia, cataracts, endocrine changes, and other defects [25]. The onset of these diseases is variable, ranging from childhood to adulthood.

Generally, DM1 is a more severe form of myotonic dystrophy than DM2. The genetic basis of DM1 was discovered to be a trinucleotide cytosine-thymine-guanine (CTG) repeat expansion in the 3′ untranslated region of the myotonic dystrophy protein kinase (DMPK) gene [26]. Normally, there are approximately between 5 and 37 CTG repeats, however expansion to 50 or more CTG repeats creates instability in the DMPK gene that results in the onset of disease [27]. Furthermore, the size of the repeat expansion directly correlates to disease onset and progression. Interestingly, the expansion of CTG repeats has been shown to lead to chromatin condensation and gene silencing in this gene locus [25], though the extent of pathogenic contribution resulting from silencing this gene is unknown. A second myotonic dystrophy, DM2, is caused by an expansion of tetranucleotide cytosine-cytosine-thymine-guanine (CCTG) repeats in an intronic region of the ZNF9 gene [25]. Unlike DM1, the size of the DM2 CCTG repeat expansion does not seem to correlate with disease severity.

The pathogenic mechanisms of both DM1 and DM2 repeat expansion remain uncertain, though it is thought to be related to loss of DMPK or ZNF9 gene function, as well as toxic effects of the altered RNAs [25, 28]. Presently, the known mechanisms for toxic gain of function by these repeat expansions is likely related to disruption of RNA-binding proteins that can cause mis-splicing of several genes related to myotonia, insulin-resistance, and cardiac function [1]. Ongoing studies suggest that repeat expansions may have extended effects on other symptoms of myotonic dystrophies, such as sleep dysregulation, intellectual disability and other central nervous system defects [1]. Genome editing may be useful in ameliorating this disease by targeted genetic deletion or truncation of repeat expansions to restore gene function, or complete gene knockout or excision to reduce toxicity.

Fascioscapulohumeral Dystrophy

Fascioscapulohumeral dystrophy (FSHD) is an autosomal dominant disease that is the third most common muscular dystrophy, with a prevalence of about 1 in 7000 births [1]. The genetic mechanism underlying this disorder is complex, resulting from the convergence of aberrant genetic and epigenetic changes at a region of chromosome 4, termed the D4Z4 array, that is associated with the pathogenesis of this disease [1]. The D4Z4 array is a region of chromosome 4 with an array of repeated 3.3 kb *D4Z4* sequences, with each repeat containing an open reading frame encoding the protein DUX4. Two major events are known to occur in relation to FSHD onset [29]. First, the D4Z4 array undergoes a genetic contraction. This contraction brings the endogenous promoter and polyadenylation signal of the D4Z4 array in close proximity to a DUX4 reading frame contained within the D4Z4 repeats and stabilizes the DUX4 mRNA transcript. Second, concomitant with the contraction of the D4Z4 array, the surrounding region undergoes epigenetic relaxation that allows expression of the DUX4 gene product from this locus. The DUX4 protein then aberrantly activates a variety of genes that normally should only be expressed during early development, leading to general cellular toxicity in mature muscle tissues [30]. However, DUX4 expression alone cannot completely explain this disease state, as DUX4 is not always expressed in FSHD patient muscle tissue, and it is known to be expressed in unaffected individuals. This suggests that there may be another mechanism that causes this disease besides DUX4 expression, and may be related to chromosome 4p hypomethylation. Genome editing tools could enable novel studies and therapeutic strategies by editing the length and presence of the DUX4 array and associated genetic elements or potentially by altering the epigenetic state of chromosome 4p.

Spinal Muscular Atrophy

Spinal muscular atrophy (SMA) is an autosomal recessive disorder affecting approximately 1 in 11,000 births in the United States, and is the leading genetic cause of infant death [31]. Few treatment options currently exist, and severe cases of SMA often result in death during childhood. Two homologous genes on chromosome 5, *SMN1* and *SMN2*, encode the survival of motor neuron (SMN) protein and play a role in the genetic basis of this disorder. In normal individuals, *SMN2* is identical to *SMN1* except for a single nucleotide change (C840T) [31] in *SMN2* that results in alternative splicing, leading to full-length SMN protein (SMN-fl) encoded by 10–20 % of total *SMN2* transcripts and the remainder coding an altered form of SMN (SMNΔ7) that is missing exon 7. Importantly, the absence of exon 7 results in expression of an altered and rapidly degraded SMN protein lacking an essential oligomerization domain. SMA is caused by the loss of the *SMN1* gene, resulting in insufficient expression levels of SMN-fl protein from the remaining *SMN2* gene. Interestingly, there are known variations in *SMN2* that cause a greater percentage of SMN-fl production and are correlated with reduced disease severity. Analysis of full-length transcript expression in SMA patients with different disease severities suggests that a 20–25 % increase in SMA-fl transcript expression would correct the SMA phenotype [31]. There are a number of new therapies under investigation that aim to re-introduce functional SMN expression by *SMN1* gene replacement therapy, increased *SMN2* transcription, or alteration of *SMN2* splicing to increase SMN-fl expression [32]. Gene editing is an attractive way to address the genetic basis of the disease by introducing a functional *SMN1* gene by knock-in of an entire *SMN* gene cassette, or by editing of *SMN2* to alter the exon 7 splice junction.

Huntington's Disease

Huntington's disease (HD) is an autosomal dominant disorder caused by triplet repeat expansions in the huntingtin (*HTT*) gene. This disease is entirely genetic and is caused by unstable CAG triplet expansion in exon 1 of the *HTT* gene. This unstable triplet expansion results in expression of a mutant huntingtin protein with a polyglutamine tract that causes aggregation and abnormal degradation of huntingtin protein. These degradation products are subsequently ubiquitinated and transported to the cytosol, where their build up results in apoptosis [33]. Generally, 36–39 repeats can potentially cause huntingtin protein aggregation [34], while 40 or more repeats almost always causes aggregation and severe disease by age 65 [35]. The onset of disease is typically delayed well into mid-life, with the majority of patients experiencing initial symptoms by their mid-30s to mid-40s. While the exact mechanism of disease onset is still not understood, reduction or elimination of the repeat expansion may improve disease phenotype. Therefore, genome editing may be an attractive method to correct the *HTT* gene by deleting or reducing the number of CAG repeats. In fact, genome editing tools have been developed to target these repeat sequences [36].

Genome Editing Technologies

Extensive research over the past decade has led to rapid advances in genome engineering using a variety of different platforms to achieve site-specific sequence changes to chromosomal DNA in human cells. Notable advances have been made in methods based on engineered viral vectors, customizable DNA-binding enzymes, or oligonucleotides. Each advance has increased the specificity, activity, and/or simplicity in achieving efficient and specific genome editing. These advances have created a variety of gene editing tools [37, 38] that are available to modify DNA sequences including zinc finger nucleases (ZFNs) [39], transcription activator-like effector nucleases (TALENs) [40, 41] and more recently, the RNA-guided CRISPR/Cas system [42–47]. In addition, there are several others gene editing systems available, including oligonucleotide-mediated exon skipping [10, 11], meganucleases (MGNs) [48], triplex-forming oligonucleotide (TFO) complexes [49, 50], integrases [51, 52] and programmable recombinases based on zinc finger [53–55] or TALE DNA-binding domains [56]. This section will introduce these various genome editing technologies.

Introduction to Genome Editing Strategies

Targeted genome editing can be achieved by several distinct strategies. Most commonly, genome editing occurs by stimulating endogenous DNA repair pathways through gene targeting or generating site-specific double-strand breaks (DSBs) at the target locus (Fig. 1). Two distinct DNA repair processes can be used to modify a target DNA sequence – non-homologous end-joining (NHEJ) and homology-directed repair (HDR)—see chapter "Gene Editing: Double-Strand Break Induced Gene Targeting and Mutagenesis 20 Years Later" for a more complete discussion on these processes. Briefly, NHEJ is a stochastic, error-prone repair process that can be

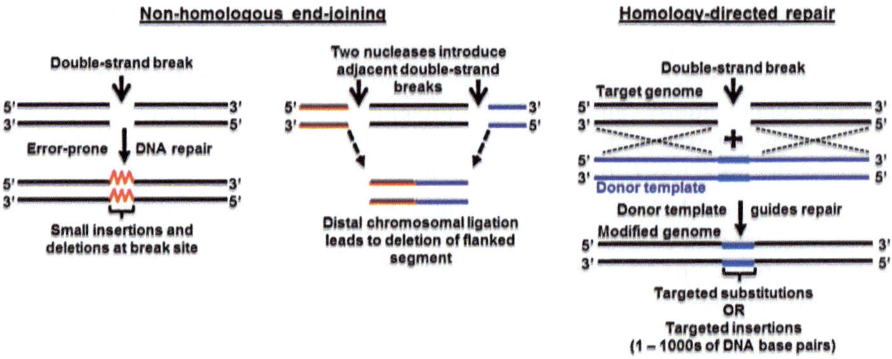

Fig. 1 Mechanisms of DNA repair following the creation of a double-strand break by an engineered nuclease

exploited to introduce random small insertions and deletions at the DNA breakpoint. NHEJ-based gene editing has been used in mammalian cells to disrupt genes [38, 57], delete chromosomal segments [58, 59], integrate gene cassettes by capture of linear DNA fragments at double-strand break sites [60], and restore aberrant reading frames [15, 61, 62]. HDR uses a donor DNA template to guide repair at a DSB and can be used to create specific sequence changes to the genome, including the targeted addition of whole genes. HDR has enabled integration of gene cassettes of up to 8 kb in the absence of selection at high frequency (~6 %) in human cells [63], although antibiotic selection is used in tandem with genome editing for gene correction in cell types with low levels of HDR repair [64–66].

Tools for Targeted Gene Modification

Advances in targeted gene editing have emerged from the development of novel designer nucleases, such as ZFNs and TALENs, as well as the re-engineering of naturally occurring DNA-binding enzymes, including MGNs and the more recently described CRISPR/Cas system. All of these engineered nucleases act by creating a targeted DNA double-strand break in the genome that stimulates cellular DNA repair through either HDR or NHEJ (Fig. 1, see section "Introduction to Genome Editing Strategies" above). Competing technologies to create genomic and mRNA-level gene correction based on engineered integrases, single-stranded oligonucleotides, and adeno-associated viral vectors are also discussed in this section.

Meganucleases

MGNs consist of overlapping DNA-binding and catalytic domains that simultaneously recognize and cleave a target DNA strand [67]. MGNs that are most commonly used for genome engineering, such as I-SceI and I-CreI, are from the

LAGLIDADG family that is named for a common peptide motif. MGNs operate either as homodimers or as long, single-chain proteins that recognize and cleave DNA. The interdependence of DNA-binding and catalytic activity results in high specificity, but also complicates the re-engineering of customized MGNs targeted to novel sequences. Despite this challenge, MGNs have been engineered to efficiently cleave new genomic target sites [68], including disease-related targets in human cells [19, 48, 69, 70].

Chimeric Nucleases Based on the FokI Domain

ZFNs and TALENs are chimeric nucleases that utilize an independent, programmable DNA-binding domain fused to the non-specific catalytic domain of the FokI endonuclease [71]. Site-specific double-strand breaks are created when two independent nucleases bind to adjacent predefined DNA sequences, thereby permitting dimerization of FokI and cleavage of the target DNA (Fig. 2a, b). The DNA-binding domains for ZFNs are based on the Cys2-His2 zinc finger domain, the most common DNA-binding motif in the human proteome. The DNA-binding specificity of synthetic zinc finger domains has been extensively engineered to recognize almost any DNA target through site-directed mutagenesis and rational design or the selection of large combinatorial libraries [38]. This work led to the establishment of ZFNs as one of the earliest and most widely used genome editing tools [72–76]. The DNA-binding domain for TALENs was adapted from a plant pathogen protein and consists of an array of repeat variable diresidue (RVD) modules, each of which specifically recognizes a single base pair of DNA [77, 78]. RVD modules can be arranged in any order to assemble an array that recognizes a defined sequence [77, 78]. The widespread adoption of TALEN technologies has dramatically advanced genome editing due to their high rate of successful and efficient genetic modification [40, 77–82].

CRISPR/Cas Systems

Recently, a new class of DNA-editing enzymes was adapted for use in mammalian cells from the CRISPR/Cas innate bacterial defense system [45, 47]. Unlike MGNs, ZFNs, or TALENs, CRISPR/Cas recognizes a target DNA sequence through the RNA-DNA interaction of a guide RNA sequence that tethers to the Cas9 nuclease and guides it to cleave a predefined DNA sequence (Fig. 2c). The CRISPR/Cas system has been successfully adapted to work in mammalian cells by co-expression of the Cas9 nuclease and a guide RNA (gRNA) targeted to a predefined genomic site via complementary base pairing between the 5′ end of the gRNA sequence and the chromosomal DNA [42–44, 46, 83]. This cognate target site must be adjacent to a unique protospacer-adjacent motif (PAM), a short sequence recognized by the Cas9 nuclease required for DNA cleavage. Several studies have studied the specificity of the *S. pyogenes* CRISPR/Cas system and observed positional dependence of mismatches in the protospacer on specificity [84–89]. Other CRISPR systems with

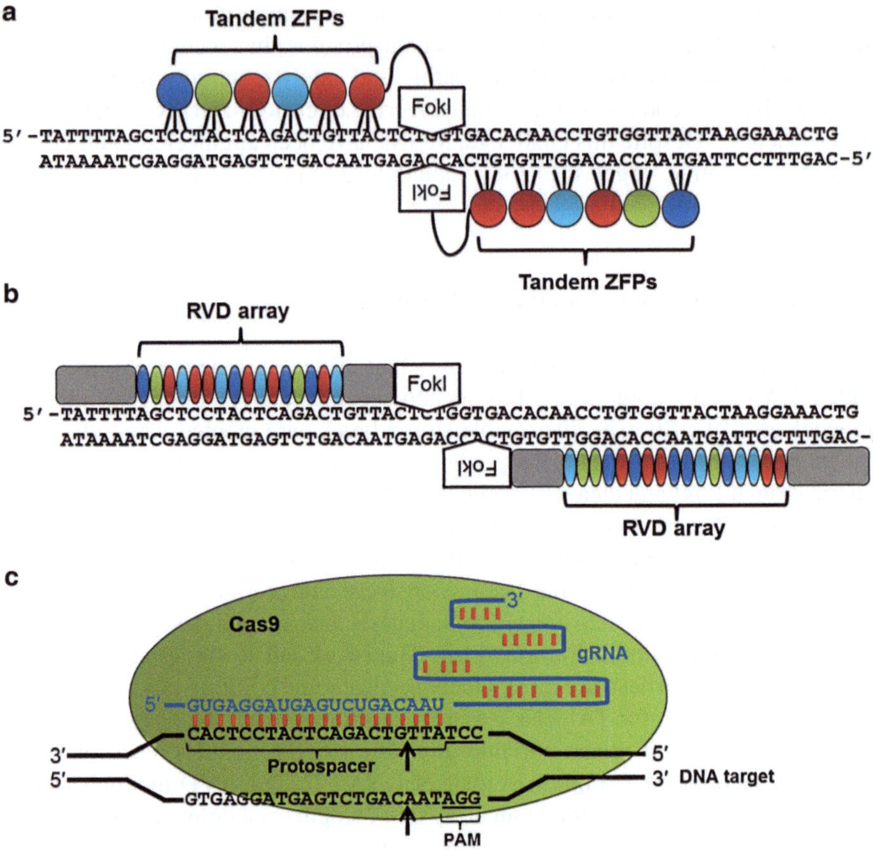

Fig. 2 Schematic of nuclease-based genome editing technologies, including (**a**) zinc finger nucleases, (**b**) TALENs, and (**c**) CRISPR/Cas9, with the DNA target in *black*, the gRNA in *blue*, and the Cas9 nuclease in *green*

smaller Cas genes have been described that are compatible size-restricted AAV vectors [90]. A unique capability of the CRISPR/Cas system is the ability to easily target multiple distinct genomic loci simultaneously, termed multiplexing, through expression of single Cas9 protein with two or more gRNAs [42, 91–93].

Integrases, Recombinases, and Transposases

Since nucleases act indirectly to cause DNA changes through endogenous DNA repair mechanisms, there has been significant interest in engineering targeted enzymes that can directly catalyze changes at a target DNA sequence. Integrases act on DNA through site-specific recombination and can efficiently insert large gene cassettes to predefined locations of the genome. The ΦC31 integrase has been demonstrated to have a limited number of integration sites in human and mouse genomes, allowing therapeutic genes to be integrated into known genomic loci [51, 52]. Programmable recombinases have also been engineered to work cooperatively with a distinct DNA-binding domain, such as zinc fingers [53, 55, 94, 95] or TALEs [56], to direct site-specific recombination at a desired genomic locus. Transposases have also been fused to ZFPs [96] and TALEs [97] as a novel to mediate site-specific integration of gene expression cassettes. Finally, retroviral integrases have been linked to ZFPs direct viral integration to a predefined genomic target [98, 99]. The continued development of these technologies will be valuable in achieving robust, reproducible and site-specific integration of therapeutic genes to treat neuromuscular disorders.

Adeno-Associated Virus-Mediated Gene Targeting

Adeno-associated virus (AAV) is a non-pathogenic virus with the capacity to deliver transgenes up to 4.7 kb in size [100] (see section "In Vivo Correction by Direct Delivery of Genome Editing Tools" below for more information). AAV-based delivery has been shown to increase the frequency of gene targeting through homologous recombination by several orders of magnitude compared to conventional, transfection-based techniques [100–108]. These gene targeting rates vary from 0.1 to 1 % of treated cells, and can generate small insertions or deletions, substitutions, and integration of entire gene cassettes [100–108]. AAV-based gene targeting has several advantages, including ease of designing gene targeting vectors, efficient integration as compared to transfection-based techniques, and apparent lack of cytotoxicity. Although the overall gene targeting efficiency of these vectors presents a challenge in strategies requiring greater gene correction rates, it can be further enhanced by the presence of nuclease-induced double-strand breaks [100, 103–108]. AAV vectors also may introduce non-specific genomic sequence changes by off-target, random chromosomal integration of the vector [107, 109–111].

Genetic Modification with Oligonucleotide Complexes

Permanent genetic modifications can be created by several types of oligonucleotides, including RNA–DNA complexes (chimeraplasts) [49] and chemically modified single-stranded oligonucleotides (ssODNs) [50, 112]. These oligonucleotides directly interact with the target DNA sequence and stimulate DNA repair responses. Generally, ssODNs are highly sensitive to the targeted strand, with a preference for the coding strand, as well as the introduced sequence change [50], potentially limiting the therapeutic potential of oligonucleotides. While the exact mechanism of oligonucleotide-mediated gene editing is still under investigation, more recent advances have focused on chemically modified ssODNs that have been shown to increase the efficiency of this approach [50, 112]. ssODNs have been shown to introduce targeted genetic corrections in transfected cells to address disease-related mutations for SMA [113] and DMD [114].

Introduction of Genetic Corrections In Vivo

Creating genetic changes that can correct disease phenotypes in vivo is the ultimate goal of novel gene therapies for neuromuscular disorders. There are numerous challenges in delivering the gene editing therapeutics specifically and efficiently to the desired target tissue. This section will discuss several key methods for introducing genetic corrections in vivo, including cell-based therapies and direct introduction of gene editing enzymes via viral and non-viral gene transfer.

Cell-Based Therapies to Introduce Genetically Corrected Cells In Vivo

Cell-based therapies aim to introduce functional expression of a therapeutic gene through engraftment of genetically corrected cells in patient tissue. These engrafted cells can then act as a depot for expressing a therapeutic gene or function by directly replacing diseased tissue. A myriad of cell types are currently under investigation for cell-based therapies for muscular dystrophies and may be applicable to other neuromuscular disorders, including induced pluripotent stem cells (iPSCs) [118, 119], bone marrow-derived progenitors [120], skeletal muscle progenitors [121], mesoangioblasts [122], CD133+ cells [123], and dermal fibroblasts [61, 124]. Additionally, advances in immortalization of human myogenic cells may greatly simplify clonal derivation of genetically corrected myogenic cells [125, 126]. These cell types can be modified by genome editing tools in vitro by transfection or electroporation of plasmid or mRNA, transduction by integrase-deficient lentivirus [127], or treatment with cell-permeable nucleases [128, 129].

Duchenne muscular dystrophy is an example of a prototypical disease that is addressable by gene therapy combined with cell therapy because muscle tissue readily incorporates donor progenitor cells during muscle regeneration. Initially, cell-based therapies for DMD focused on transplantation of allogeneic healthy donor muscle progenitors directly into DMD patient muscle by injection and was combined with immunosuppression [130]. Some of the challenges associated with combined genome editing and cell-based therapies include immune response to the donor cells or to the restored therapeutic gene product, insufficient donor cell engraftment, and poor migration from the injection site. One interesting approach to address these concerns is to create iPSCs from autologous patient cells, correct the dystrophin gene ex vivo, differentiate the iPSCs towards the myogenic lineage, and inject the cells to repopulate dystrophic tissue [15, 20, 118, 119, 122, 131]. This approach may reduce the potential for immune rejection of donor cells and allow for extensive characterization of corrected cells.

In Vivo Correction by Direct Delivery of Genome Editing Tools

Traditionally, gene therapy aims to replace mutated genes by gene transfer to a target tissue. Gene transfer can be efficiently accomplished through viral and non-viral based methods to directly deliver genes to target tissues following local or systemic injection. An important consideration for genome editing therapies is that the editing tools need only be present as long as necessary to create a permanent genetic change. Thus, transient exposure to the editing tools, either by direct treatment or transient expression of genes encoding the editing enzymes, is a desirable property, as reduced exposure time to gene editing nucleases may limit potential off-target effects [128, 132].

Viral gene transfer takes advantage of the innate tissue-tropic properties of viruses. One of the leading viral gene transfer vectors for in vivo gene therapy is the adeno-associated virus, which is a small, non-pathogenic dependovirus that typically encodes a single-strand DNA genome with a maximum packaging capacity of ~4.7 kb [133]. Importantly, AAV vectors can efficiently transduce non-dividing cells types, such as muscle fibers and neurons. The non-integrative nature of AAV reduces the chance for unintended integration into the genome that may result in oncogenesis or disruption of essential genes. For gene editing purposes, this may also be advantageous because expression of the editing enzyme is not permanent, potentially reducing long-term toxicity. Gene targeting may also be enhanced by using AAV-based delivery of donor templates and gene editing enzymes [100, 103–108, 134]. Moreover, AAV delivery of ZFNs with a donor template has been shown to mediate efficient gene targeting in vivo [60, 138]. AAV vectors based on AAV2 pseudotyped with alternative muscle-tropic AAV capsids, such as AAV2/1, AAV2/6, AAV2/7, AAV2/8, AAV2/9, and the more recent AAV2.5, AAV/SASTG, and AAV2i8G9 vectors have been shown to efficiently transduce a variety of musculoskeletal tissues following systemic and local delivery [7, 139–149] with sustained gene transfer for at least several months in post-mitotic tissues. A number of these AAV vectors have also been shown to efficiently transduce neurons, glia, and astrocytes [150–152]. More recently, the first AAV-based gene therapy was approved as a product in Europe and is also administered by intramuscular delivery [153].

Non-viral methods are also an attractive method for efficient, tissue-specific gene transfer. A major advantage of this method is the transient delivery of a therapeutic payload in a highly controlled manner [154]. Nanoparticles can also be programmed to deliver a payload to specific tissues by conjugating targeting peptides to the nanoparticle complex, enabling versatile control over delivery of gene editing therapeutics. Moreover, these complexes can be used to deliver therapeutic peptides or RNA in addition to conventional gene transfer via DNA. Notably, it has been demonstrated that nanoparticles can mediate in vivo gene editing by simultaneously delivering both the gene editing agent and a suitable donor template [155]. Thus, non-viral methods may be a promising method to deliver short bursts of gene editing drugs.

Genome Editing Methods Applied to Neuromuscular Diseases

Genome editing tools can be utilized to generate an array of manipulations of the chromosomal DNA to study and treat neuromuscular diseases. This section will focus on three major strategies that are of interest for genome editing in the context of these disorders: restoration of the native gene, targeted addition of a replacement gene, and genetic disruption of a disease-related target.

Correction of a Native Gene In Situ

A powerful application of gene editing is to directly restore a patient's own gene in situ, effectively correcting the underlying genetic cause of a disease. This methodology presents several potential advantages, including permanent gene restoration and appropriate physiologic and tissue-specific expression under the control of the natural genomic regulatory elements. This section will emphasize using genome editing to generate targeted large genomic deletions, small insertions and deletions to create intra-exonic frameshifts, or insertions of missing exons using a DNA donor template (see examples in Tables 1 and 2).

Reading Frame Restoration by Targeted Genomic Modifications

Engineered nucleases can efficiently introduce targeted insertions and deletions that restore the reading frame around a non-essential region of the dystrophin gene [15, 18]. Importantly, targeted frameshifts can be accomplished by nuclease activity alone, without a DNA repair template, through the error-prone NHEJ DNA repair pathway that generates random small insertions and deletions. However, the resulting restored proteins will have heterogeneous protein sequences at the targeted region, owing to the different size and composition of the insertions or deletions generated. This may be important in determining the overall functionality of restored proteins, and the variability of the epitopes created may cause immunogenicity and regulatory concerns.

An alternative approach would utilize targeted substitutions [50], insertions, or deletions to disrupt sequences that are essential for inclusion or exclusion of entire exons during pre-mRNA splicing. A parallel and potentially more flexible approach is to co-express two nucleases that cleave at adjacent genomic loci and mediate genomic deletions with high efficiency [58, 157, 158], creating the possibility of selective excision of aberrant sequences from the genome [16–18]. Either of these methods would likely result in predictable and reproducible changes to the mRNA sequence and thereby to the resulting restored protein.

Table 1 Representative strategies to utilize genome editing as a novel therapeutic to address neuromuscular disorders

Gene correction strategy	Types of mutation(s) addressed	Applicable gene editing technology	Relevant diseases	Example
Alternative exon splicing	Aberrant exon splicing	Single-strand oligonucleotides, nucleases, AAV	SMA	Remove sequences that cause unstable or pathogenic exon splicing
Intraexonic gene editing	Premature stop codons, frameshift mutations	Nucleases, AAV, Recombinases/ Integrases/Transposases	DMD, BMD, LGMD2B	Introduce small insertions and deletions to restore frame of internally deleted gene
Exclusion of non-essential exon	Premature stop codons, frameshift mutations	Single-strand oligonucleotides, Nucleases, AAV, Recombinases/ Integrases/Transposases	DMD, BMD, LGMD2B	Disrupt splice sites or remove entire sequence of non-essential exons to restore frame of internally deleted gene
Targeted knock-in of absent exons in native gene	Frameshift mutations, deletions containing essential coding regions	Nucleases, AAV, Recombinases/ Integrases/Transposases	DMD, BMD, LGMD2B	Integrate cassette containing missing exons in appropriate intron
Targeted knock-in of replacement gene	All coding region mutations	Nucleases, AAV, Recombinases/ Integrases/Transposases	DMD, BMD, LGMD2B, SMA	Integrate replacement gene at native promoter or other transcriptionally active chromosomal locus
Targeted deletion	Repeat expansions	Nucleases, Recombinases	HD, FSHD, DM1/DM2	Targeted large chromosomal deletion of repeat expansions
Compensatory gene knockout	Modulation of disease-related phenotype	Nucleases, AAV, Recombinases/ Integrases/Transposases	DMD, BMD	Genetic disruption of coding region in a disease-related gene

Table 2 Select studies demonstrating the use of genome editing as a novel tool to correct the genetic basis of hereditary neuromuscular disorders

Disease	Gene target	Method of correction	Technology used	Refs
Duchenne muscular dystrophy	*DMD*	Small intraexonic insertions and deletions to correct aberrant reading frames	TALENs, CRISPR	[15, 17, 18]
	DMD	Deletions of complete exons to restore reading frames	ZFNs, TALENs, CRISPR	[16–18]
	DMD	Correction of point mutations by homology-directed repair	CRISPR	[14]
	DMD	Transient exclusion of non-essential exons to correct aberrant reading frames	Single-strand oligonucleotides	[10] [11–13]
	DMD	Targeted knock-in of absent exons	Meganucleases, TALENs, CRISPR	[18, 19]
	DMD	Targeted knock-in of replacement dystrophin gene cassette	ZFNs	[20]
	MSTN	Gene disruption of myostatin to generate a compensatory phenotype	TALENs	[171]
Limb-girdle muscular dystrophy	*DYSF*	Transient exclusion of non-essential exons to correct aberrant reading frames	Single-strand oligonucleotides	[117]
Spinal muscular atrophy	*SMN2*	Prevention of recognition of splice-altering SNP in intron 7 to restore proper exon 7 splicing	Single-strand oligonucleotides	[115] [116, 156]
	SMN2	Genetic conversion of *SMN2* to *SMN1* by alteration of SNP in intron 7 to wild-type	Single-strand oligonucleotides	[113]

Complete Gene Restoration by Knock-In of Essential Gene Sequences Absent in Mutant Genes

In contrast to targeted frameshifts, the complete wild-type gene product can be restored by the introduction and correct integration of a homologous donor template carrying the missing exons [18]. For example, meganucleases were used to stimulate homologous recombination-mediated knock-in of exons 45–52 in cells from a DMD patient where these exons were absent [19]. Alternatively, as discussed above, AAV vectors alone have been shown to mediate gene targeting much more efficiently than traditional plasmid-based methods, though in many cases this is significantly less efficient than nuclease-mediated stimulation of homology directed-repair. Importantly, the use of AAV vectors may significantly reduce the risk of off-target effects as compared to current nuclease technologies. Knock-in of essential sequences may be useful in addressing deletions in the dystrophin gene for DMD, dysferlin for LGMD2B, *SMN1* for SMA, and any disease where the patient has a genetic mutation that includes a loss of essential coding sequences.

Correction of Point Mutations

Premature stop codons, aberrant substitutions of single base pairs, and splice-altering SNPs are common genetic causes of disease. A number of the gene editing tools discussed in the section "Overview of Common Neuromuscular Disorders" may be useful in correcting these types of mutations. For example, peptide nucleic acids (PNAs) were shown to effectively introduce point substitutions in vivo and restore dystrophin expression in an animal model of DMD [50]. AAV-mediated gene targeting offers a similar "one-shot" approach to correct mutations in vivo [159], as demonstrated by correction of several mouse models for diseases including hereditary tyrosinemia [160]. Notably, the efficacy of these approaches is poor, except in cases where correction of a disease-related mutation confers a selective growth advantage to the corrected cells, as in hereditary tyrosinemia. Targeted genome editing with nuclease-mediated homology-directed repair is a high-efficiency strategy to correct mutations with high fidelity in a target gene of interest. In this method, both the nuclease and a homologous donor template carrying the desired substitution are introduced into cells to achieve genetic correction. This approach has been used to correct the dystrophin point mutation in the *mdx* mouse model of DMD [14]. In vivo delivery of ZFNs and donor templates by AAV gene transfer has been shown to be a highly efficient and robust method to stably integrate a therapeutic transgene that corrects the phenotype of hemophilic mice [60, 138]. Notably, these studies demonstrated the persistence of genomic modifications following induced liver regeneration. A primary challenge to the in vivo translation of this approach will be the efficiency of donor template integration, which may be dependent on cell type and cell cycle state. However, this approach has been successfully applied in vitro to correct point mutations in genes for a number of monogenic human diseases, including severe combined immunodeficiency [73], sickle cell anemia [64, 65], alpha-1 antitrypsin [66], and others.

Wholesale Replacement of Mutated Genes

Genome editing technologies can enable the introduction of complete gene expression cassettes to functionally replace damaged endogenous genes through high-efficiency gene targeting [127, 161]. In this method, a replacement gene is inserted into the genome to functionally replace the damaged endogenous gene. Exogenous gene cassettes typically contain the gene of interest driven by a synthetic promoter that can constitutively, inducibly, or tissue-specifically drive expression across a range of tunable levels. DNA donor templates containing a replacement gene coding sequence are synthesized with flanking stretches of DNA sequence that are homologous to a predefined region of the genome, thereby promoting site-specific recombination at the target locus. The efficiency of this process is enhanced several orders of magnitude when stimulated by the creation of targeted DSBs with nucleases or by direct integration with site-specific integrases.

Promoters to Drive Transgene Expression

Most engineered gene cassettes typically contain a strong synthetic promoter to drive expression of the desired transgene, and can utilize a number of available promoters to achieve tissue-specific or global expression. A commonly used constitutive promoter based on the cytomegalovirus (CMV) promoter can drive very high levels of expression, but is sometimes subject to silencing in certain cell types [162]. Other common constitutive promoters with varying levels of transgene expression include simian virus 40 (SV40) early promoter, human Ubiquitin C (hUbC) promoter, human elongation factor 1-α (EF1-α) promoter, and phosphoglycerate kinase 1 (PGK-1) promoter. Generally, constitutive promoters can drive robust levels of transgene expression, though expression levels can vary across specific cell types.

In contrast to global expression, it may be desirable to limit the expression of a therapeutic transgene to the intended tissue. Inducible promoter systems, such as the tet-on or tet-off systems, can control gene expression by addition of a small molecule, such as doxycycline, that can tune expression in a dose-dependent manner. Although the tet-inducible system may not be directly translatable to human patients, similar chemically inducible gene expression systems are in clinical trials [163]. Tissue-specific expression may be particularly useful for genome editing applications in order to reduce potential off-target effects. Endogenous promoters specific to muscle have been adapted to drive transgene expression at physiologic or high levels only in muscle tissue, including the desmin promoter, muscle creatine kinase promoter, and the myoglobin promoter [164, 165]. Similarly, neuron-specific transgene expression can be driven by synthetic promoters such as synapsin I (SYN1) and neuron-specific enolase (NSE/RU5) [166, 167].

Alternatively, gene targeting can be designed to integrate a transgene immediately downstream of an endogenous promoter, placing it under control of that promoter. The choice of endogenous promoter will determine tissue-specific, physiologic expression of the gene construct similar to synthetic tissue-specific promoters. Additionally, because the donor repair template does not include a promoter, this approach reduces the size of the donor template and reduces the potential for oncogenesis by random integration of a construct containing a strong synthetic promoter.

"Safe Harbor" Sites for Integration

The choice of integration site is critical to ensure appropriate expression of a therapeutic transgene while ensuring minimal effects on adjacent genes. Several integration sites have been characterized as so-called "safe harbor" loci for inserting exogenous gene expression constructs. Ideally, safe harbor loci do not express essential genes, are distant from known oncogenic or disruptive genes, and enable reliable and efficient transgene expression in tissues of interest [168]. One such safe harbor site in the human genome is AAVS1 [169], which is transcriptionally active in virtually all tissues and appears to be non-essential for normal cellular physiology.

AAVS1 is therefore an attractive safe harbor locus to drive expression of a gene through its native promoter. Additionally, the CCR5 receptor gene has been suggested as an alternative safe harbor locus for expressing therapeutic transgenes [20, 127, 161]. It may also be of interest to directly integrate replacement genes at the native locus of the corresponding gene, thereby effectively replacing the mutant portion of the coding region [60, 138]. Notably, combining AAV gene transfer with genome editing may enable highly efficient integration of gene cassettes in dividing and non-dividing cells through both enhanced homology-directed repair and non-homologous end-joining capture of AAV genomes [60, 108, 134–137].

Disease Modulation by Modification of Compensatory Gene Targets

A potential alternative to correcting disease-causing genes is to modulate disease phenotypes by manipulating other independent genes broadly related to a disease, termed here as compensatory gene targets. The myostatin gene and signaling pathway is a commonly studied compensatory gene target because it regulates muscle hypertrophy. Several studies have shown that the knockdown or knockout of myostatin [170, 171], or blocking of the myostatin receptor ACVRIIB [172], leads to increased hypertrophy and may be able to improve phenotypes related to musculoskeletal disorders, particularly for DMD. Genome editing has been demonstrated to be a potentially useful tool to permanently knockout this signaling pathway by disrupting the coding sequence of myostatin [171]. Inflammatory responses are often common features of neuromuscular disorders and may also be attractive compensatory targets to enhance cell therapies for neuromuscular disorders or to alleviate symptoms related to inflammatory response. Similarly, disruption of disease-related growth factors, receptors or other gene targets may be of interest in creating versatile therapies for a broad range of neuromuscular disorders.

Discussion and Conclusion

Genome editing comprises a powerful set of tools to correct a variety of the underlying genetic causes for many neuromuscular disorders. The rapid pace of changes in genome editing technologies will continue to yield more efficient, diverse, and specific gene editing tools. In particular, the recently described CRISPR/Cas system is a promising method to easily generate targeted mutations in human cells. This system has continued to evolve through the identification or engineering of novel Cas9 proteins and gRNA architectures with improved specificity. Additionally, advancements in the design, targeting, and specificity of site-specific integrases may constitute a second generation of genome editing that favors designer enzymes with intrinsic capacity to directly manipulate target DNA. This may enable efficient genome engineering for large inserts and cell types that are refractory to homology-directed repair. Finally, recent studies have demonstrated the capacity of programmable DNA enzymes to create site-specific epigenetic modifications [173–180]. This may be particularly useful in understanding and/or treating complex genetic and epigenetic disorders, such as FSHD.

Introduction of corrected genes or gene editing enzymes to diseased tissue presents another major hurdle to translation of genome editing to bench-side therapies. Adeno-associated virus is a promising gene transfer vector with favorable safety and expression profiles to deliver transgenes to musculoskeletal tissues. Future advances in targeted non-viral and viral delivery methods promises to increase efficacy and safety of genome editing therapies by confining gene editing to the cells of interest. Additionally, genetic reprogramming presents an exciting avenue for autologous cell therapies for neuromuscular disorders. Ex vivo manipulation of autologous stem cells in a lab would allow isogenic generation and extensive characterization of genetically corrected cell populations that can be differentiated into cell lineages of interest to repopulate damaged tissues. Future studies will need to focus on how to optimally combine genome editing tools with delivery strategies for specific neuromuscular disorders.

Ultimately, the goal of genome editing therapies is to halt disease progression or prevent disease pathogenesis. There have been extraordinary efforts across the past 5 years to develop novel molecular tools that can control both genetic content and epigenetic states in complex human genomes. These tools have enabled rapid manipulation of genome sequences in human cells in vitro and in small animal models in vivo to study and potentially treat disease. These genome editing technologies are synergizing with advances in gene delivery and stem cell therapies to enable genetic correction as a feasible approach to treating neuromuscular disorders.

References

1. Leung DG, Wagner KR. Therapeutic advances in muscular dystrophy. Ann Neurol. 2013;74(3):404–11.
2. Peeters K, Chamova T, Jordanova A. Clinical and genetic diversity of SMN1-negative proximal spinal muscular atrophies. Brain. 2014;137(Pt 11):2879–96.
3. Farrar MA, Kiernan MC. The genetics of spinal muscular atrophy: progress and challenges. Neurotherapeutics. 2015;12(2):290–302.
4. Konieczny P, Swiderski K, Chamberlain JS. Gene and cell-mediated therapies for muscular dystrophy. Muscle Nerve. 2013;47(5):649–63.
5. Wang B, Li J, Xiao X. Adeno-associated virus vector carrying human minidystrophin genes effectively ameliorates muscular dystrophy in mdx mouse model. Proc Natl Acad Sci U S A. 2000;97(25):13714–9.
6. Pichavant C, Aartsma-Rus A, Clemens PR, Davies KE, Dickson G, Takeda S, et al. Current status of pharmaceutical and genetic therapeutic approaches to treat DMD. Mol Ther. 2011;19(5):830–40.
7. Qiao C, Koo T, Li J, Xiao X, Dickson JG. Gene therapy in skeletal muscle mediated by adeno-associated virus vectors. Methods Mol Biol. 2011;807:119–40.

8. Harper SQ, Hauser MA, DelloRusso C, Duan D, Crawford RW, Phelps SF, et al. Modular flexibility of dystrophin: implications for gene therapy of Duchenne muscular dystrophy. Nat Med. 2002;8(3):253–61.
9. Ruszczak C, Mirza A, Menhart N. Differential stabilities of alternative exon-skipped rod motifs of dystrophin. Biochim Biophys Acta. 2009;1794(6):921–8.
10. Lu QL, Yokota T, Takeda S, Garcia L, Muntoni F, Partridge T. The status of exon skipping as a therapeutic approach to Duchenne muscular dystrophy. Mol Ther. 2010;19(1):9–15.
11. Cirak S, Arechavala-Gomeza V, Guglieri M, Feng L, Torelli S, Anthony K, et al. Exon skipping and dystrophin restoration in patients with Duchenne muscular dystrophy after systemic phosphorodiamidate morpholino oligomer treatment: an open-label, phase 2, dose-escalation study. Lancet. 2011;378(9791):595–605.
12. Goemans NM, Tulinius M, van den Akker JT, Burm BE, Ekhart PF, Heuvelmans N, et al. Systemic administration of PRO051 in Duchenne's muscular dystrophy. N Engl J Med. 2011;364(16):1513–22.
13. van Deutekom JC, Janson AA, Ginjaar IB, Frankhuizen WS, Aartsma-Rus A, Bremmer-Bout M, et al. Local dystrophin restoration with antisense oligonucleotide PRO051. N Engl J Med. 2007;357(26):2677–86.
14. Long C, McAnally JR, Shelton JM, Mireault AA, Bassel-Duby R, Olson EN. Prevention of muscular dystrophy in mice by CRISPR/Cas9-mediated editing of germline DNA. Science. 2014;345(6201):1184–8.
15. Ousterout DG, Perez-Pinera P, Thakore PI, Kabadi AM, Brown MT, Qin X, et al. Reading frame correction by targeted genome editing restores dystrophin expression in cells from Duchenne muscular dystrophy patients. Mol Ther. 2013;21(9):1718–26.
16. Ousterout DG, Kabadi AM, Thakore PI, Perez-Pinera P, Brown MT, Majoros WH, et al. Correction of dystrophin expression in cells from duchenne muscular dystrophy patients through genomic excision of exon 51 by zinc finger nucleases. Mol Ther. 2015;23(3):523–32.
17. Ousterout DG, Kabadi AM, Thakore PI, Majoros WH, Reddy TE, Gersbach CA. Multiplex CRISPR/Cas9-based genome editing for correction of dystrophin mutations that cause Duchenne muscular dystrophy. Nat Commun. 2015;6:6244.
18. Li HL, Fujimoto N, Sasakawa N, Shirai S, Ohkame T, Sakuma T, et al. Precise correction of the dystrophin gene in duchenne muscular dystrophy patient induced pluripotent stem cells by TALEN and CRISPR-Cas9. Stem Cell Reports. 2015;4(1):143–54.
19. Popplewell L, Koo T, Leclerc X, Duclert A, Mamchaoui K, Gouble A, et al. Gene correction of a duchenne muscular dystrophy mutation by meganuclease-enhanced exon knock-in. Hum Gene Ther. 2013;24(7):692–701.
20. Benabdallah BF, Duval A, Rousseau J, Chapdelaine P, Holmes MC, Haddad E, et al. Targeted gene addition of microdystrophin in mice skeletal muscle via human myoblast transplantation. Mol Ther Nucleic Acids. 2013;2, e68.
21. Barthelemy F, Wein N, Krahn M, Levy N, Bartoli M. Translational research and therapeutic perspectives in dysferlinopathies. Mol Med. 2011;17(9-10):875–82.
22. Bansal D, Miyake K, Vogel SS, Groh S, Chen CC, Williamson R, et al. Defective membrane repair in dysferlin-deficient muscular dystrophy. Nature. 2003;423(6936):168–72.
23. Rahimov F, Kunkel LM. The cell biology of disease: cellular and molecular mechanisms underlying muscular dystrophy. J Cell Biol. 2013;201(4):499–510.
24. Aartsma-Rus A, Singh KH, Fokkema IF, Ginjaar IB, van Ommen GJ, den Dunnen JT, et al. Therapeutic exon skipping for dysferlinopathies? Eur J Hum Genet. 2010;18(8):889–94.
25. Nelson DL, Orr HT, Warren ST. The unstable repeats—three evolving faces of neurological disease. Neuron. 2013;77(5):825–43.
26. Mahadevan M, Tsilfidis C, Sabourin L, Shutler G, Amemiya C, Jansen G, et al. Myotonic dystrophy mutation: an unstable CTG repeat in the 3′ untranslated region of the gene. Science. 1992;255(5049):1253–5.
27. Turner C, Hilton-Jones D. The myotonic dystrophies: diagnosis and management. J Neurol Neurosurg Psychiatry. 2010;81(4):358–67.

28. Kumar A, Agarwal S, Agarwal D, Phadke SR. Myotonic dystrophy type 1 (DM1): a triplet repeat expansion disorder. Gene. 2013;522(2):226–30.
29. Lemmers RJ, van der Vliet PJ, Klooster R, Sacconi S, Camano P, Dauwerse JG, et al. A unifying genetic model for facioscapulohumeral muscular dystrophy. Science. 2010;329(5999):1650–3.
30. Snider L, Geng LN, Lemmers RJ, Kyba M, Ware CB, Nelson AM, et al. Facioscapulohumeral dystrophy: incomplete suppression of a retrotransposed gene. PLoS Genet. 2010;6(10), e1001181.
31. Arnold WD, Burghes AH. Spinal muscular atrophy: development and implementation of potential treatments. Ann Neurol. 2013;74(3):348–62.
32. Seo J, Howell MD, Singh NN, Singh RN. Spinal muscular atrophy: an update on therapeutic progress. Biochim Biophys Acta. 2013;1832(12):2180–90.
33. Jonson I, Ougland R, Larsen E. DNA repair mechanisms in Huntington's disease. Mol Neurobiol. 2013;47(3):1093–102.
34. Rubinsztein DC, Leggo J, Coles R, Almqvist E, Biancalana V, Cassiman JJ, et al. Phenotypic characterization of individuals with 30-40 CAG repeats in the Huntington disease (HD) gene reveals HD cases with 36 repeats and apparently normal elderly individuals with 36-39 repeats. Am J Hum Genet. 1996;59(1):16–22.
35. Ross CA, Tabrizi SJ. Huntington's disease: from molecular pathogenesis to clinical treatment. Lancet Neurol. 2011;10(1):83–98.
36. Mittelman D, Moye C, Morton J, Sykoudis K, Lin Y, Carroll D, et al. Zinc-finger directed double-strand breaks within CAG repeat tracts promote repeat instability in human cells. Proc Natl Acad Sci U S A. 2009;106(24):9607–12.
37. Gaj T, Gersbach CA, Barbas 3rd CF. ZFN, TALEN, and CRISPR/Cas-based methods for genome engineering. Trends Biotechnol. 2013;31(7):397–405.
38. Perez-Pinera P, Ousterout DG, Gersbach CA. Advances in targeted genome editing. Curr Opin Chem Biol. 2012;16(3-4):268–77.
39. Urnov FD, Rebar EJ, Holmes MC, Zhang HS, Gregory PD. Genome editing with engineered zinc finger nucleases. Nat Rev Genet. 2010;11(9):636–46.
40. Joung JK, Sander JD. TALENs: a widely applicable technology for targeted genome editing. Nat Rev Mol Cell Biol. 2013;14(1):49–55.
41. Mussolino C, Cathomen T. TALE nucleases: tailored genome engineering made easy. Curr Opin Biotechnol. 2012;23(5):644–50.
42. Cong L, Ran FA, Cox D, Lin S, Barretto R, Habib N, et al. Multiplex genome engineering using CRISPR/Cas systems. Science. 2013;339(6121):819–23.
43. Mali P, Yang L, Esvelt KM, Aach J, Guell M, DiCarlo JE, et al. RNA-guided human genome engineering via Cas9. Science. 2013;339(6121):823–6.
44. Jinek M, East A, Cheng A, Lin S, Ma E, Doudna J. RNA-programmed genome editing in human cells. Elife. 2013;2, e00471.
45. Doudna JA, Charpentier E. Genome editing. The new frontier of genome engineering with CRISPR-Cas9. Science. 2014;346(6213):1258096.
46. Cho SW, Kim S, Kim JM, Kim JS. Targeted genome engineering in human cells with the Cas9 RNA-guided endonuclease. Nat Biotechnol. 2013;31(3):230–2.
47. Cox DB, Platt RJ, Zhang F. Therapeutic genome editing: prospects and challenges. Nat Med. 2015;21(2):121–31.
48. Redondo P, Prieto J, Munoz IG, Alibes A, Stricher F, Serrano L, et al. Molecular basis of xeroderma pigmentosum group C DNA recognition by engineered meganucleases. Nature. 2008;456(7218):107–11.
49. Bertoni C, Rando TA. Dystrophin gene repair in mdx muscle precursor cells in vitro and in vivo mediated by RNA-DNA chimeric oligonucleotides. Hum Gene Ther. 2002;13(6):707–18.
50. Kayali R, Bury F, Ballard M, Bertoni C. Site-directed gene repair of the dystrophin gene mediated by PNA-ssODNs. Hum Mol Genet. 2010;19(16):3266–81.
51. Quenneville SP, Chapdelaine P, Rousseau J, Beaulieu J, Caron NJ, Skuk D, et al. Nucleofection of muscle-derived stem cells and myoblasts with phiC31 integrase: stable expression of a full-length-dystrophin fusion gene by human myoblasts. Mol Ther. 2004;10(4):679–87.

52. Bertoni C, Jarrahian S, Wheeler TM, Li Y, Olivares EC, Calos MP, et al. Enhancement of plasmid-mediated gene therapy for muscular dystrophy by directed plasmid integration. Proc Natl Acad Sci U S A. 2006;103(2):419–24.
53. Gaj T, Mercer AC, Gersbach CA, Gordley RM, Barbas 3rd CF. Structure-guided reprogramming of serine recombinase DNA sequence specificity. Proc Natl Acad Sci U S A. 2011;108(2):498–503.
54. Gersbach CA, Gaj T, Gordley RM, Barbas 3rd CF. Directed evolution of recombinase specificity by split gene reassembly. Nucleic Acids Res. 2010;38(12):4198–206.
55. Gersbach CA, Gaj T, Gordley RM, Mercer AC, Barbas 3rd CF. Targeted plasmid integration into the human genome by an engineered zinc-finger recombinase. Nucleic Acids Res. 2011;39(17):7868–78.
56. Mercer AC, Gaj T, Fuller RP, Barbas 3rd CF. Chimeric TALE, recombinases with programmable DNA sequence specificity. Nucleic Acids Res. 2012;40(21):11163–72.
57. Perez EE, Wang J, Miller JC, Jouvenot Y, Kim KA, Liu O, et al. Establishment of HIV-1 resistance in CD4+ T cells by genome editing using zinc-finger nucleases. Nat Biotechnol. 2008;26(7):808–16.
58. Sollu C, Pars K, Cornu TI, Thibodeau-Beganny S, Maeder ML, Joung JK, et al. Autonomous zinc-finger nuclease pairs for targeted chromosomal deletion. Nucleic Acids Res. 2010;38(22):8269–76.
59. Lee HJ, Kim E, Kim JS. Targeted chromosomal deletions in human cells using zinc finger nucleases. Genome Res. 2010;20(1):81–9.
60. Anguela XM, Sharma R, Doyon Y, Miller JC, Li H, Haurigot V, et al. Robust ZFN-mediated genome editing in adult hemophilic mice. Blood. 2013;122(19):3283–7.
61. Peault B, Rudnicki M, Torrente Y, Cossu G, Tremblay JP, Partridge T, et al. Stem and progenitor cells in skeletal muscle development, maintenance, and therapy. Mol Ther. 2007;15(5):867–77.
62. Chapdelaine P, Pichavant C, Rousseau J, Paques F, Tremblay JP. Meganucleases can restore the reading frame of a mutated dystrophin. Gene Ther. 2010;17(7):846–58.
63. Moehle EA, Rock JM, Lee Y-L, Jouvenot Y, DeKelver RC, Gregory PD, et al. Targeted gene addition into a specified location in the human genome using designed zinc finger nucleases. Proc Natl Acad Sci U S A. 2007;104(9):3055–60.
64. Sebastiano V, Maeder ML, Angstman JF, Haddad B, Khayter C, Yeo DT, et al. In situ genetic correction of the sickle cell anemia mutation in human induced pluripotent stem cells using engineered zinc finger nucleases. Stem Cells. 2011;29(11):1717–26.
65. Zou J, Mali P, Huang X, Dowey SN, Cheng L. Site-specific gene correction of a point mutation in human iPS cells derived from an adult patient with sickle cell disease. Blood. 2011;118(17):4599–608.
66. Yusa K, Rashid ST, Strick-Marchand H, Varela I, Liu PQ, Paschon DE, et al. Targeted gene correction of alpha1-antitrypsin deficiency in induced pluripotent stem cells. Nature. 2011;478(7369):391–4.
67. Arnould S, Delenda C, Grizot S, Desseaux C, Paques F, Silva GH, et al. The I-CreI meganuclease and its engineered derivatives: applications from cell modification to gene therapy. Protein Eng Des Sel. 2011;24(1-2):27–31.
68. Silva G, Poirot L, Galetto R, Smith J, Montoya G, Duchateau P, et al. Meganucleases and other tools for targeted genome engineering: perspectives and challenges for gene therapy. Curr Gene Ther. 2011;11(1):11–27.
69. Arnould S, Perez C, Cabaniols JP, Smith J, Gouble A, Grizot S, et al. Engineered I-CreI derivatives cleaving sequences from the human XPC gene can induce highly efficient gene correction in mammalian cells. J Mol Biol. 2007;371(1):49–65.
70. Smith J, Grizot S, Arnould S, Duclert A, Epinat JC, Chames P, et al. A combinatorial approach to create artificial homing endonucleases cleaving chosen sequences. Nucleic Acids Res. 2006;34(22), e149.
71. Miller JC, Tan S, Qiao G, Barlow KA, Wang J, Xia DF, et al. A TALE nuclease architecture for efficient genome editing. Nat Biotechnol. 2011;29(2):143–8.

72. Porteus MH, Baltimore D. Chimeric nucleases stimulate gene targeting in human cells. Science. 2003;300(5620):763.
73. Urnov F, Miller J, Lee Y, Beausejour C, Rock J, Augustus S, et al. Highly efficient endogenous human gene correction using designed zinc-finger nucleases. Nature. 2005;435:2.
74. Alwin S, Gere MB, Guhl E, Effertz K, Barbas 3rd CF, Segal DJ, et al. Custom zinc-finger nucleases for use in human cells. Mol Ther. 2005;12(4):610–7.
75. Bibikova M, Beumer K, Trautman JK, Carroll D. Enhancing gene targeting with designed zinc finger nucleases. Science. 2003;300(5620):764.
76. Maeder ML, Thibodeau-Beganny S, Osiak A, Wright DA, Anthony RM, Eichtinger M, et al. Rapid "open-source" engineering of customized zinc-finger nucleases for highly efficient gene modification. Mol Cell. 2008;31(2):294–301.
77. Moscou MJ, Bogdanove AJ. A simple cipher governs DNA recognition by TAL effectors. Science. 2009;326(5959):1501.
78. Boch J, Scholze H, Schornack S, Landgraf A, Hahn S, Kay S, et al. Breaking the code of DNA binding specificity of TAL-type III effectors. Science. 2009;326(5959):1509–12.
79. Hockemeyer D, Wang H, Kiani S, Lai CS, Gao Q, Cassady JP, et al. Genetic engineering of human pluripotent cells using TALE nucleases. Nat Biotechnol. 2011;29(8):731–4.
80. Mussolino C, Morbitzer R, Lutge F, Dannemann N, Lahaye T, Cathomen T. A novel TALE nuclease scaffold enables high genome editing activity in combination with low toxicity. Nucleic Acids Res. 2011;39(21):9283–93.
81. Reyon D, Tsai SQ, Khayter C, Foden JA, Sander JD, Joung JK. FLASH assembly of TALENs for high-throughput genome editing. Nat Biotechnol. 2012;30(5):460–5.
82. Cermak T, Doyle EL, Christian M, Wang L, Zhang Y, Schmidt C, et al. Efficient design and assembly of custom TALEN and other TAL effector-based constructs for DNA targeting. Nucleic Acids Res. 2011;39(12), e82.
83. Hwang WY, Fu Y, Reyon D, Maeder ML, Tsai SQ, Sander JD, et al. Efficient genome editing in zebrafish using a CRISPR-Cas system. Nat Biotechnol. 2013;31(3):227–9.
84. Cradick TJ, Fine EJ, Antico CJ, Bao G. CRISPR/Cas9 systems targeting beta-globin and CCR5 genes have substantial off-target activity. Nucleic Acids Res. 2013;41(20):9584–92.
85. Fu Y, Foden JA, Khayter C, Maeder ML, Reyon D, Joung JK, et al. High-frequency off-target mutagenesis induced by CRISPR-Cas nucleases in human cells. Nat Biotechnol. 2013; 31(9):822–6.
86. Hsu PD, Scott DA, Weinstein JA, Ran FA, Konermann S, Agarwala V, et al. DNA targeting specificity of RNA-guided Cas9 nucleases. Nat Biotechnol. 2013;31(9):827–32.
87. Mali P, Aach J, Stranges PB, Esvelt KM, Moosburner M, Kosuri S, et al. CAS9 transcriptional activators for target specificity screening and paired nickases for cooperative genome engineering. Nat Biotechnol. 2013;31(9):833–8.
88. Pattanayak V, Lin S, Guilinger JP, Ma E, Doudna JA, Liu DR. High-throughput profiling of off-target DNA cleavage reveals RNA-programmed Cas9 nuclease specificity. Nat Biotechnol. 2013;31(9):839–43.
89. Ran FA, Hsu PD, Lin CY, Gootenberg JS, Konermann S, Trevino AE, et al. Double nicking by RNA-guided CRISPR Cas9 for enhanced genome editing specificity. Cell. 2013; 154(6):1380–9.
90. Ran FA, Cong L, Yan WX, Scott DA, Gootenberg JS, Kriz AJ, et al. In vivo genome editing using Staphylococcus aureus Cas9. Nature. 2015;520(7546):186–91.
91. Maeder ML, Linder SJ, Cascio VM, Fu Y, Ho QH, Joung JK. CRISPR RNA-guided activation of endogenous human genes. Nat Methods. 2013;10(10):977–9.
92. Perez-Pinera P, Kocak DD, Vockley CM, Adler AF, Kabadi AM, Polstein LR, et al. RNA-guided gene activation by CRISPR-Cas9-based transcription factors. Nat Methods. 2013;10(10):973–6.
93. Yang L, Guell M, Byrne S, Yang JL, De Los AA, Mali P, et al. Optimization of scarless human stem cell genome editing. Nucleic Acids Res. 2013;41(19):9049–61.
94. Gordley RM, Gersbach CA, Barbas 3rd CF. Synthesis of programmable integrases. Proc Natl Acad Sci U S A. 2009;106(13):5053–8.

95. Gaj T, Mercer AC, Sirk SJ, Smith HL, Barbas 3rd CF. A comprehensive approach to zinc-finger recombinase customization enables genomic targeting in human cells. Nucleic Acids Res. 2013;41(6):3937–46.
96. Kettlun C, Galvan DL, George Jr AL, Kaja A, Wilson MH. Manipulating piggyBac transposon chromosomal integration site selection in human cells. Mol Ther. 2011;19(9):1636–44.
97. Owens JB, Mauro D, Stoytchev I, Bhakta MS, Kim MS, Segal DJ, et al. Transcription activator like effector (TALE)-directed piggyBac transposition in human cells. Nucleic Acids Res. 2013;41(19):9197–207.
98. Tan W, Dong Z, Wilkinson TA, Barbas 3rd CF, Chow SA. Human immunodeficiency virus type 1 incorporated with fusion proteins consisting of integrase and the designed polydactyl zinc finger protein E2C can bias integration of viral DNA into a predetermined chromosomal region in human cells. J Virol. 2006;80(4):1939–48.
99. Lim KI, Klimczak R, Yu JH, Schaffer DV. Specific insertions of zinc finger domains into Gag-Pol yield engineered retroviral vectors with selective integration properties. Proc Natl Acad Sci U S A. 2010;107(28):12475–80.
100. Deyle DR, Russell DW. Adeno-associated virus vector integration. Curr Opin Mol Ther. 2009;11(4):442–7.
101. Inoue N, Dong R, Hirata RK, Russell DW. Introduction of single base substitutions at homologous chromosomal sequences by adeno-associated virus vectors. Mol Ther. 2001;3(4):526–30.
102. Hirata R, Chamberlain J, Dong R, Russell DW. Targeted transgene insertion into human chromosomes by adeno-associated virus vectors. Nat Biotechnol. 2002;20(7):735–8.
103. Miller DG, Petek LM, Russell DW. Human gene targeting by adeno-associated virus vectors is enhanced by DNA double-strand breaks. Mol Cell Biol. 2003;23(10):3550–7.
104. Miller DG, Petek LM, Russell DW. Adeno-associated virus vectors integrate at chromosome breakage sites. Nat Genet. 2004;36(7):767–73.
105. Porteus MH, Cathomen T, Weitzman MD, Baltimore D. Efficient gene targeting mediated by adeno-associated virus and DNA double-strand breaks. Mol Cell Biol. 2003;23(10):3558–65.
106. Hirsch ML, Green L, Porteus MH, Samulski RJ. Self-complementary AAV mediates gene targeting and enhances endonuclease delivery for double-strand break repair. Gene Ther. 2010;17(9):1175–80.
107. Gellhaus K, Cornu TI, Heilbronn R, Cathomen T. Fate of recombinant adeno-associated viral vector genomes during DNA double-strand break-induced gene targeting in human cells. Hum Gene Ther. 2010;21(5):543–53.
108. Asuri P, Bartel MA, Vazin T, Jang JH, Wong TB, Schaffer DV. Directed evolution of adeno-associated virus for enhanced gene delivery and gene targeting in human pluripotent stem cells. Mol Ther. 2012;20(2):329–38.
109. Miller DG, Rutledge EA, Russell DW. Chromosomal effects of adeno-associated virus vector integration. Nat Genet. 2002;30(2):147–8.
110. Nakai H, Montini E, Fuess S, Storm TA, Grompe M, Kay MA. AAV serotype 2 vectors preferentially integrate into active genes in mice. Nat Genet. 2003;34(3):297–302.
111. Donsante A, Miller DG, Li Y, Vogler C, Brunt EM, Russell DW, et al. AAV vector integration sites in mouse hepatocellular carcinoma. Science. 2007;317(5837):477.
112. Bertoni C, Rustagi A, Rando TA. Enhanced gene repair mediated by methyl-CpG-modified single-stranded oligonucleotides. Nucleic Acids Res. 2009;37(22):7468–82.
113. DiMatteo D, Callahan S, Kmiec EB. Genetic conversion of an SMN2 gene to SMN1: a novel approach to the treatment of spinal muscular atrophy. Exp Cell Res. 2008;314(4):878–86.
114. Maguire K, Suzuki T, DiMatteo D, Parekh-Olmedo H, Kmiec E. Genetic correction of splice site mutation in purified and enriched myoblasts isolated from mdx5cv mice. BMC Mol Biol. 2009;10:15.
115. Douglas AG, Wood MJ. Splicing therapy for neuromuscular disease. Mol Cell Neurosci. 2013;56:169–85.
116. Passini MA, Bu J, Richards AM, Kinnecom C, Sardi SP, Stanek LM, et al. Antisense oligonucleotides delivered to the mouse CNS ameliorate symptoms of severe spinal muscular atrophy. Sci Transl Med. 2011;3(72):72ra18.

117. Wein N, Avril A, Bartoli M, Beley C, Chaouch S, Laforet P, et al. Efficient bypass of mutations in dysferlin deficient patient cells by antisense-induced exon skipping. Hum Mutat. 2010;31(2):136–42.
118. Darabi R, Arpke RW, Irion S, Dimos JT, Grskovic M, Kyba M, et al. Human ES-and iPS-derived myogenic progenitors restore dystrophin and improve contractility upon transplantation in dystrophic mice. Cell Stem Cell. 2012;10(5):610–9.
119. Tedesco FS, Gerli MF, Perani L, Benedetti S, Ungaro F, Cassano M, et al. Transplantation of genetically corrected human iPSC-derived progenitors in mice with limb-girdle muscular dystrophy. Sci Transl Med. 2012;4(140):140ra89.
120. Dezawa M, Ishikawa H, Itokazu Y, Yoshihara T, Hoshino M, Takeda S, et al. Bone marrow stromal cells generate muscle cells and repair muscle degeneration. Science. 2005; 309(5732):314–7.
121. Cerletti M, Jurga S, Witczak CA, Hirshman MF, Shadrach JL, Goodyear LJ, et al. Highly efficient, functional engraftment of skeletal muscle stem cells in dystrophic muscles. Cell. 2008;134(1):37–47.
122. Tedesco FS, Hoshiya H, D'Antona G, Gerli MF, Messina G, Antonini S, et al. Stem cell-mediated transfer of a human artificial chromosome ameliorates muscular dystrophy. Sci Transl Med. 2011;3(96):96ra78.
123. Negroni E, Riederer I, Chaouch S, Belicchi M, Razini P, Di Santo J, et al. In vivo myogenic potential of human CD133+ muscle-derived stem cells: a quantitative study. Mol Ther. 2009;17(10):1771–8.
124. Kimura E, Han JJ, Li S, Fall B, Ra J, Haraguchi M, et al. Cell-lineage regulated myogenesis for dystrophin replacement: a novel therapeutic approach for treatment of muscular dystrophy. Hum Mol Genet. 2008;17(16):2507–17.
125. Zhu CH, Mouly V, Cooper RN, Mamchaoui K, Bigot A, Shay JW, et al. Cellular senescence in human myoblasts is overcome by human telomerase reverse transcriptase and cyclin-dependent kinase 4: consequences in aging muscle and therapeutic strategies for muscular dystrophies. Aging Cell. 2007;6(4):515–23.
126. Berghella L, De Angelis L, Coletta M, Berarducci B, Sonnino C, Salvatori G, et al. Reversible immortalization of human myogenic cells by site-specific excision of a retrovirally transferred oncogene. Hum Gene Ther. 1999;10(10):1607–17.
127. Lombardo A, Genovese P, Beausejour CM, Colleoni S, Lee YL, Kim KA, et al. Gene editing in human stem cells using zinc finger nucleases and integrase-defective lentiviral vector delivery. Nat Biotechnol. 2007;25(11):1298–306.
128. Gaj T, Guo J, Kato Y, Sirk SJ, Barbas 3rd CF. Targeted gene knockout by direct delivery of zinc-finger nuclease proteins. Nat Methods. 2012;9(8):805–7.
129. Liu J, Gaj T, Patterson JT, Sirk SJ, Barbas Iii CF. Cell-penetrating peptide-mediated delivery of TALEN proteins via bioconjugation for genome engineering. PLoS One. 2014;9(1), e85755.
130. Palmieri B, Tremblay JP. Myoblast transplantation: a possible surgical treatment for a severe pediatric disease. Surg Today. 2010;40(10):902–8.
131. Filareto A, Parker S, Darabi R, Borges L, Iacovino M, Schaaf T, et al. An ex vivo gene therapy approach to treat muscular dystrophy using inducible pluripotent stem cells. Nat Commun. 2013;4:1549.
132. Pruett-Miller SM, Reading DW, Porter SN, Porteus MH. Attenuation of zinc finger nuclease toxicity by small-molecule regulation of protein levels. PLoS Genet. 2009;5(2), e1000376.
133. Seto JT, Ramos JN, Muir L, Chamberlain JS, Odom GL. Gene replacement therapies for duchenne muscular dystrophy using adeno-associated viral vectors. Curr Gene Ther. 2012;12(3):139–51.
134. Ellis BL, Hirsch ML, Porter SN, Samulski RJ, Porteus MH. Zinc-finger nuclease-mediated gene correction using single AAV vector transduction and enhancement by Food and Drug Administration-approved drugs. Gene Ther. 2013;20(1):35–42.
135. Hirsch ML, Green L, Porteus MH, Samulski RJ. Self-complementary AAV mediates gene targeting and enhances endonuclease delivery for double-strand break repair. Gene Ther. 2010;17(9):1175–80.

136. Handel EM, Gellhaus K, Khan K, Bednarski C, Cornu TI, Muller-Lerch F, et al. Versatile and efficient genome editing in human cells by combining zinc-finger nucleases with adeno-associated viral vectors. Hum Gene Ther. 2012;23(3):321–9.
137. Rahman SH, Bobis-Wozowicz S, Chatterjee D, Gellhaus K, Pars K, Heilbronn R, et al. The nontoxic cell cycle modulator indirubin augments transduction of adeno-associated viral vectors and zinc-finger nuclease-mediated gene targeting. Hum Gene Ther. 2013;24(1):67–77.
138. Li H, Haurigot V, Doyon Y, Li T, Wong SY, Bhagwat AS, et al. In vivo genome editing restores haemostasis in a mouse model of haemophilia. Nature. 2011;475(7355):217–21.
139. Asokan A, Conway JC, Phillips JL, Li C, Hegge J, Sinnott R, et al. Reengineering a receptor footprint of adeno-associated virus enables selective and systemic gene transfer to muscle. Nat Biotechnol. 2010;28(1):79–82.
140. Messina E, Nienaber J, Daneshmand M, Villamizar N, Samulski RJ, Milano CA, et al. Adeno-associated viral vectors based on serotype 3b use components of the fibroblast growth factor receptor signaling complex for efficient transduction. Hum Gene Ther. 2012;23(10):1031–42.
141. Bowles DE, McPhee SWJ, Li C, Gray SJ, Samulski JJ, Camp AS, et al. Phase 1 gene therapy for duchenne muscular dystrophy using a translational optimized AAV vector. Mol Ther. 2012;20(2):443–55.
142. Qiao C, Li J, Zheng H, Bogan J, Li J, Yuan Z, et al. Hydrodynamic limb vein injection of adeno-associated virus serotype 8 vector carrying canine myostatin propeptide gene into normal dogs enhances muscle growth. Hum Gene Ther. 2009;20(1):1–10.
143. Wang Z, Zhu T, Qiao C, Zhou L, Wang B, Zhang J, et al. Adeno-associated virus serotype 8 efficiently delivers genes to muscle and heart. Nat Biotechnol. 2005;23(3):321–8.
144. Watchko J, O'Day T, Wang B, Zhou L, Tang Y, Li J, et al. Adeno-associated virus vector-mediated minidystrophin gene therapy improves dystrophic muscle contractile function in mdx mice. Hum Gene Ther. 2002;13(12):1451–60.
145. Yang L, Jiang J, Drouin LM, Agbandje-McKenna M, Chen C, Qiao C, et al. A myocardium tropic adeno-associated virus (AAV) evolved by DNA shuffling and in vivo selection. Proc Natl Acad Sci U S A. 2009;106(10):3946–51.
146. Liu M, Yue Y, Harper SQ, Grange RW, Chamberlain JS, Duan D. Adeno-associated virus-mediated microdystrophin expression protects young mdx muscle from contraction-induced injury. Mol Ther. 2005;11(2):245–56.
147. Odom GL, Gregorevic P, Allen JM, Finn E, Chamberlain JS. Microutrophin delivery through rAAV6 increases lifespan and improves muscle function in dystrophic dystrophin/utrophin-deficient mice. Mol Ther. 2008;16(9):1539–45.
148. Wang Z, Kuhr CS, Allen JM, Blankinship M, Gregorevic P, Chamberlain JS, et al. Sustained AAV-mediated dystrophin expression in a canine model of Duchenne muscular dystrophy with a brief course of immunosuppression. Mol Ther. 2007;15(6):1160–6.
149. Shen S, Horowitz ED, Troupes AN, Brown SM, Pulicherla N, Samulski RJ, et al. Engraftment of a galactose receptor footprint onto adeno-associated viral capsids improves transduction efficiency. J Biol Chem. 2013;288(40):28814–23.
150. Gray SJ. Gene therapy and neurodevelopmental disorders. Neuropharmacology. 2013; 68:136–42.
151. Hester ME, Foust KD, Kaspar RW, Kaspar BK. AAV as a gene transfer vector for the treatment of neurological disorders: novel treatment thoughts for ALS. Curr Gene Ther. 2009;9(5):428–33.
152. Ojala DS, Amara DP, Schaffer DV. Adeno-associated virus vectors and neurological gene therapy. Neuroscientist. 2015;21(1):84–98.
153. Gruber K. Europe gives gene therapy the green light. Lancet. 2012;380(9855), e10.
154. Lorden ER, Levinson HM, Leong KW. Integration of drug, protein, and gene delivery systems with regenerative medicine. Drug Deliv Transl Res. 2015;5(2):168–86.
155. McNeer NA, Schleifman EB, Cuthbert A, Brehm M, Jackson A, Cheng C, et al. Systemic delivery of triplex-forming PNA and donor DNA by nanoparticles mediates site-specific genome editing of human hematopoietic cells in vivo. Gene Ther. 2013;20(6):658–69.

156. Porensky PN, Mitrpant C, McGovern VL, Bevan AK, Foust KD, Kaspar BK, et al. A single administration of morpholino antisense oligomer rescues spinal muscular atrophy in mouse. Hum Mol Genet. 2012;21(7):1625–38.
157. Lee HJ, Kweon J, Kim E, Kim S, Kim JS. Targeted chromosomal duplications and inversions in the human genome using zinc finger nucleases. Genome Res. 2012;22(3):539–48.
158. Xiao A, Wang Z, Hu Y, Wu Y, Luo Z, Yang Z, et al. Chromosomal deletions and inversions mediated by TALENs and CRISPR/Cas in zebrafish. Nucleic Acids Res. 2013;41(14), e141.
159. Miller DG, Wang PR, Petek LM, Hirata RK, Sands MS, Russell DW. Gene targeting in vivo by adeno-associated virus vectors. Nat Biotechnol. 2006;24(8):1022–6.
160. Paulk NK, Wursthorn K, Wang Z, Finegold MJ, Kay MA, Grompe M. Adeno-associated virus gene repair corrects a mouse model of hereditary tyrosinemia in vivo. Hepatology. 2010;51(4):1200–8.
161. Lombardo A, Cesana D, Genovese P, Di Stefano B, Provasi E, Colombo DF, et al. Site-specific integration and tailoring of cassette design for sustainable gene transfer. Nat Methods. 2011;8(10):861–9.
162. Brooks AR, Harkins RN, Wang P, Qian HS, Liu P, Rubanyi GM. Transcriptional silencing is associated with extensive methylation of the CMV promoter following adenoviral gene delivery to muscle. J Gene Med. 2004;6(4):395–404.
163. Huang C, Ramakrishnan R, Trkulja M, Ren X, Gabrilovich DI. Therapeutic effect of intratumoral administration of DCs with conditional expression of combination of different cytokines. Cancer Immunol Immunother. 2012;61(4):573–9.
164. Himeda CL, Tai PW, Hauschka SD. Analysis of muscle gene transcription in cultured skeletal muscle cells. Methods Mol Biol. 2012;798:425–43.
165. Salva MZ, Himeda CL, Tai PW, Nishiuchi E, Gregorevic P, Allen JM, et al. Design of tissue-specific regulatory cassettes for high-level rAAV-mediated expression in skeletal and cardiac muscle. Mol Ther. 2007;15(2):320–9.
166. Yaguchi M, Ohashi Y, Tsubota T, Sato A, Koyano KW, Wang N, et al. Characterization of the properties of seven promoters in the motor cortex of rats and monkeys after lentiviral vector-mediated gene transfer. Hum Gene Ther Methods. 2013;24(6):333–44.
167. Delzor A, Dufour N, Petit F, Guillermier M, Houitte D, Auregan G, et al. Restricted transgene expression in the brain with cell-type specific neuronal promoters. Hum Gene Ther Methods. 2012.
168. Papapetrou EP, Lee G, Malani N, Setty M, Riviere I, Tirunagari LM, et al. Genomic safe harbors permit high beta-globin transgene expression in thalassemia induced pluripotent stem cells. Nat Biotechnol. 2011;29(1):73–8.
169. Hockemeyer D, Soldner F, Beard C, Gao Q, Mitalipova M, DeKelver RC, et al. Efficient targeting of expressed and silent genes in human ESCs and iPSCs using zinc-finger nucleases. Nat Biotechnol. 2009;27(9):851–7.
170. McPherron AC, Lawler AM, Lee SJ. Regulation of skeletal muscle mass in mice by a new TGF-beta superfamily member. Nature. 1997;387(6628):83–90.
171. Xu L, Zhao P, Mariano A, Han R. Targeted myostatin gene editing in multiple mammalian species directed by a single pair of TALE nucleases. Mol Ther Nucleic acids. 2013;2, e112.
172. Lee SJ, Reed LA, Davies MV, Girgenrath S, Goad ME, Tomkinson KN, et al. Regulation of muscle growth by multiple ligands signaling through activin type II receptors. Proc Natl Acad Sci U S A. 2005;102(50):18117–22.
173. Maeder ML, Angstman JF, Richardson ME, Linder SJ, Cascio VM, Tsai SQ, et al. Targeted DNA demethylation and activation of endogenous genes using programmable TALE-TET1 fusion proteins. Nat Biotechnol. 2013;31(12):1137–42.
174. Konermann S, Brigham MD, Trevino AE, Hsu PD, Heidenreich M, Cong L, et al. Optical control of mammalian endogenous transcription and epigenetic states. Nature. 2013; 500(7463):472–6.
175. Mendenhall EM, Williamson KE, Reyon D, Zou JY, Ram O, Joung JK, et al. Locus-specific editing of histone modifications at endogenous enhancers. Nat Biotechnol. 2013;31(12):1133–6.

176. Siddique AN, Nunna S, Rajavelu A, Zhang Y, Jurkowska RZ, Reinhardt R, et al. Targeted methylation and gene silencing of VEGF-A in human cells by using a designed Dnmt3a-Dnmt3L single-chain fusion protein with increased DNA methylation activity. J Mol Biol. 2013;425(3):479–91.
177. de Groote ML, Verschure PJ, Rots MG. Epigenetic Editing: targeted rewriting of epigenetic marks to modulate expression of selected target genes. Nucleic Acids Res. 2012;40(21):10596–613.
178. Kearns NA, Pham H, Tabak B, Genga RM, Silverstein NJ, Garber M, et al. Functional annotation of native enhancers with a Cas9-histone demethylase fusion. Nat Methods. 2015;12(5):401–3.
179. Hilton IB, D'Ippolito AM, Vockley CM, Thakore PI, Crawford GE, Reddy TE, et al. Epigenome editing by a CRISPR-Cas9-based acetyltransferase activates genes from promoters and enhancers. Nat Biotechnol. 2015;33(5):510–7.
180. Polstein L, Perez-Pinera P, Kocak D, Vockley C, Bledsoe P, Song L, et al. Genome-wide specificity of DNA-binding, gene regulation, and chromatin remodeling by TALE- and CRISPR/Cas9-based transcriptional activators. Genome Res. 2015;25(8):1158–69.

[illegible]

Phage Integrases for Genome Editing

Michele P. Calos

Abstract Phage integrases are prokaryotic site-specific recombinases that perform precise cut-and-paste recombination between their short *attB* and *attP* recognition sequences. These enzymes work in cellular environments ranging from bacteria to mammalian cells and have become useful genome engineering tools. PhiC31 was the first phage integrase to be developed for use in mammalian cells. This integrase has the useful property of being able to recombine its own *attB* and *attP* sites. In addition, phiC31 integrase performs recombination at related native sequences called pseudo *att* sites present in large genomes, which has allowed integration into unmodified genomes. PhiC31 integrase can also be used in conjunction with another phage integrase, Bxb1, which has different recognition sequences and does not recombine at pseudo *att* sites. The properties of these phage integrases have led to a range of applications, summarized here, from creation of transgenic organisms and in vivo gene therapy, to cellular reprogramming and precise genome editing by cassette exchange. The latest system, dual integrase cassette exchange (DICE), uses target phiC31 and Bxb1 *attP* sequences precisely placed in genomes by homologous recombination and is especially useful for iterative genome engineering in pluripotent stem cells.

Keywords *attB* site • *attP* site • Bxb1 integrase • Cassette exchange • Embryonic stem cell • Homologous recombination • Induced pluripotent stem cell • phiC31 integrase • Reprogramming • TALEN

Utility of Phage Integrases

It was reported in 1998 that phage phiC31integrase could recombine its *attB* and *attP* recognition sites in *E. coli* and in buffer, independent of its native *Streptomyces* cellular environment [1]. These data suggested that the enzyme was self-contained

M.P. Calos, Ph.D. (✉)
Department of Genetics, Stanford University School of Medicine,
300 Pasteur Drive, Alway Bldg., Room M334, Stanford, CA 94305-5120, USA
e-mail: calos@stanford.edu

T. Cathomen et al. (eds.), *Genome Editing*, Advances in Experimental Medicine and Biology 895, DOI 10.1007/978-1-4939-3509-3_5

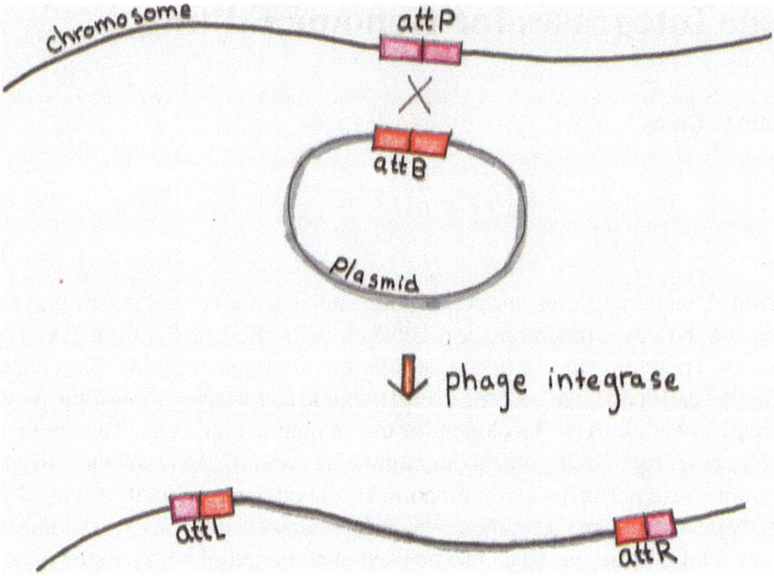

Fig. 1 Basic phage integrase reaction. Phage integrases in the serine family possess *attB* and *attP* sites originating in the host bacterial and the phage genomes, respectively, that are approximately 34-bp long partially palindromic sequences. Dimers of integrase bind to each *att* site, and upon synapsis of the two sites (upper part of drawing), recombination takes place by a concerted cut-and-paste reaction at the center of the *att* sites, resulting in covalent strand exchange. By this means, a plasmid carrying an *attB* site can be integrated into a chromosome carrying the cognate *attP* site (lower). Hybrid *att* sites consisting of half of an *attB* and half of an *attP* site flank the integration. These sites are not substrates for integrase, so the reaction is unidirectional

and did not require host-specific cofactors for activity. On this basis, we hypothesized that phiC31 integrase might work in foreign cellular environments, such as mammalian cells, where site-specific integration systems were needed. Furthermore, the recombinational behavior reported for various combinations of *att* sites indicated that the recombination reaction was unidirectional [1], which was a desirable feature for high efficiency integration into mammalian genomes.

These predictions were realized in our 2000 study demonstrating activity of phiC31 integrase at its wild-type *attB* and *attP* sites in mammalian cells [2]. The length of the *att* sites was also defined in that study for the first time, at approximately 34-bp. As illustrated in Fig. 1, integrase mediates recombination across the center of the *attP* and *attB* sites, producing recombinant *attL* and *attR* sites that are no longer substrates for integrase, rendering a unidirectional reaction. It is now understood that each *att* site is bound by a dimer of integrase molecules, which mediate a concerted cut-and-paste recombination reaction [3].

Integration into Pseudo Sites Versus into attP Sites

The *att* site length of ~30-bp suggested the possibility that rare native sequences with partial sequence identity might be adequate to catalyze recombination by phiC31 integrase. While a perfect match to a 30-bp sequence would not be statistically expected, even in large genomes, matches of 16-bp would be expected, and we predicted that this level of identity might be adequate for reaction. This prediction was fulfilled in studies in human and mouse cell lines, which revealed phiC31 integrase-mediated recombination at native sequences, named pseudo *att* sites [4] (Fig. 2).

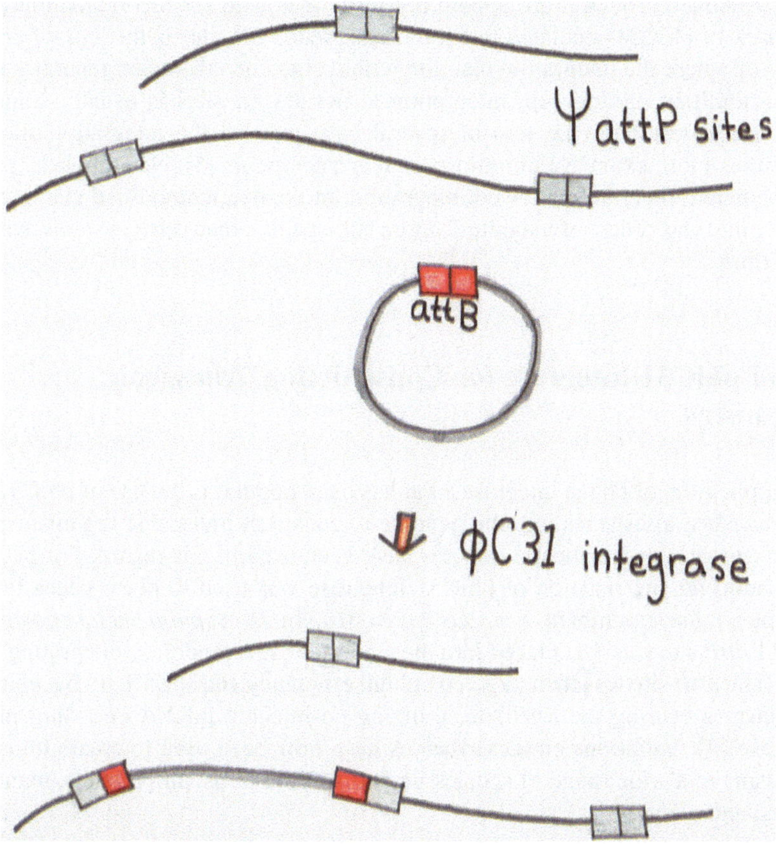

Fig. 2 The pseudo site integration reaction. A subset of serine phage integrases, notably phiC31 integrase, can interact with native chromosomal sequences having only partial identity with their true *attP* site. These sequences are known as pseudo *att* sites. Typically, a number of potential pseudo sites exist in the genome, defined by both DNA sequence and genomic context (three pseudo *att* sites are schematically illustrated in the upper part of the drawing). In the presence of phiC31 integrase and a plasmid bearing a phiC31 *attB* site, integration of the plasmid can occur at a pseudo *att* site, usually in single copy (lower). Integration at pseudo *att* sites is less precise than at genuine *attP* sites, and small deletions in the vicinity of the integration site often occur

The pseudo *att* site recombination reaction mediated by phiC31 integrase was the first instance of this type of "semi-specific" integration behavior into a mammalian genome. The level of specificity attained compared favorably with the integration specificity of the systems available at the time, including random integration of DNA and retrovirus- or transposon-mediated integration, also largely random. PhiC31 integrase-mediated integration at pseudo *att*-sites had an immediate appeal for situations involving integration into unmodified genomes, including in vivo gene therapy and construction of transgenic organisms.

Genomes typically harbor numerous potential pseudo *att* sites, and these sites appear to be utilized in a manner that depends on both the extent of DNA sequence identity and the genomic context; those pseudo *att* sites present in open, transcriptionally active chromatin are apparently more available for recombination [5]. A drawback of phiC31-mediated integration at pseudo *att* sites is the lack of predictability of where the integration reaction will occur, since there are generally multiple possibilities. In addition, integration at pseudo *att* sites is usually somewhat imprecise, often involving loss of several base pairs at the integration sites and sometimes more extensive chromosome rearrangements [5]. Nevertheless, pseudo site-mediated integration allowed integration into native, unmodified genomes in a manner that was orders of magnitude more site-specific than other systems available at the time.

Use of phiC31 Integrase for Constructing Transgenic Organisms

One application of phage integrases that has been popular is the use of phiC31 integrase to place transgenes into the genome to construct transgenic organisms. Both pseudo *att* sites and authentic *att* sites have been used in this regard. For example, the pseudo *att* site reaction of phiC31 integrase was used to place genes into the genome of the amphibian *Xenopus laevis* [6]. In *Drosophila melanogaster*, the phiC31 *attP* site was first placed into the genome with a randomly integrating P element. The *attP* site was then targeted at high efficiency and specificity by an incoming plasmid bearing the *attB* site, utilizing co-injected mRNA encoding phiC31 integrase [7]. Variations on these themes have now been used to create transgenic organisms in a wide range of species, including fish, birds, amphibians, mammals, insects, and plants [8].

Gene Therapy Studies Utilizing Integration into Pseudo Sites

The ability of phiC31 integrase to integrate DNA into unmodified mammalian genomes at a relatively low number of positions opened up new possibilities for in vivo gene therapy. The first studies utilizing phiC31 integrase for gene therapy took advantage of a relatively simple and effective in vivo delivery method in mice, hydrodynamic injection, to place plasmids encoding integrase and a human factor IX gene into the liver [9]. Therapeutic levels of factor IX were produced after just one injection of small amounts of plasmid DNA. Site-specific integration of the therapeutic plasmid in hepatocytes was verified, and the integration specificity observed was impressive, since most integration occurred at one hotspot in the genome. Furthermore, the integration was stable, and factor IX production persisted long-term [10]. These features suggested that use of phiC31 integrase for correction of hemophilia might be clinically translatable. Studies in disease model mice for hemophilia A and B were carried out, with long-term expression of human factor VIII and IX observed [11, 12]. Unfortunately, efforts to translate the hydrodynamic DNA delivery method to the livers of larger animals have not been sufficiently effective to date to achieve therapeutic factor levels and enable clinical translation.

Many other gene therapy studies in animals have been carried out utilizing phiC31 integrase for genomic integration of plasmids carrying therapeutic genes in a variety of tissues and species (reviewed in [8, 13, 14]). For example, DNA was delivered by electroporation to rat retina, and long-term expression of a marker gene was observed [15]. Electroporation was also utilized to deliver plasmid DNA carrying the *DYSTROPHIN* gene to mouse muscle in a model of Duchenne muscular dystrophy [16]. While proof of principle for successful long-term delivery of plasmid DNA and site-specific integration into the chromosomes was demonstrated in these rodent studies, their clinical translation awaits effective delivery methods for these plasmid-based strategies that can be translated to large animals.

Utilizing phiC31 Integrase for Reprogramming Mammalian Cells

One approach to circumvent the difficulty of effective delivery of plasmid DNA to the body is to deliver DNA first to cells in vitro, since effective transfection methods exist for most cell types, then deliver the transfected cells carrying integrated transgenes to the body. If the cells are immortal, then single cells with defined integration sites can be cloned, permitting the integration site to be defined, an important safety feature.

Cells that can be cloned include pluripotent stem cells, such as embryonic stem cells (ESC). In 2006, induced pluripotent stem cells (iPSC) were described, in which ESC-like cells could be derived from somatic cells by addition of four transcription factor genes [17, 18]. These iPSC hold extensive potential for regenerative medicine, because they are immunologically matched to the patient, can be grown to large numbers, are susceptible to genetic engineering methods, and are free of political or ethical issues. For genetic diseases, the iPSC from a patient can be corrected in vitro, for example by integration of the relevant therapeutic gene, then used in a therapeutic strategy involving in vitro differentiation, followed by transplantation to the appropriate tissue or organ.

Reprogramming therefore opened up a vast number of potential therapeutic strategies. However, the initial methods to generate iPSC utilized retroviruses to integrate the four transcription factors into the genome. This methodology generally resulted in numerous, random integration events scattered about the genome, increasing the risk of insertional mutagenesis and resulting tumorigenesis [17, 18]. To overcome this problem, we devised a strategy utilizing phiC31 integrase to introduce one copy of a reprogramming cassette carrying all four transcription factors at a single, safe site in the mouse genome [19].

In this method, a reprogramming plasmid (p4FLR, 11.9-kb) was constructed that carried all four transcription factors identified by Yamanaka for reprogramming, along with strategically located recombinase recognition sites. The cDNA sequences for the murine *cMyc*, *Klf4*, *Oct4*, and *Sox2* genes were linked by 2A sequences, facilitating their polycistronic transcription from the CAG promoter. The plasmid also carried a phiC31 integrase *attB* site to mediate integration into the genome at pseudo *att* sites (Fig. 2). Two *loxP* sites flanked the reprogramming cassette, so that it could be deleted after reprogramming by transient transfection with a plasmid expressing Cre resolvase [19]. This step was important, to render the iPSC less tumorigenic and more amenable to differentiation.

The phiC31-mediated method was successful for reprogramming mouse embryonic fibroblasts and adult mesenchymal stem cells at efficiencies comparable to retroviral methods, without the handling hazards and random integration risks associated with viruses. Individual iPSC clones were analyzed by ligation-mediated PCR to determine the integration site, with the result that approximately one-third of the iPSC carried a single integration of p4FLR. Many different mouse pseudo *attP* sites were utilized in the collection of IPSC analyzed. Six of 14 sites were intergenic, and of those, two were considered to be safe sites, in terms of being distant from known cancer genes and other hazards. A plasmid expressing Cre resolvase was transfected into two representative iPSC clones, and precise deletion of the reprogramming cassette was demonstrated. Pluripotency of the iPSC before and after Cre excision was demonstrated, including ability of the iPSC to generate teratomas, as well as chimeric mice [19].

Site-Specific Integration of a Therapeutic Gene at a Pre-integrated attP Site

We built further on the concepts in the initial reprogramming study in order to create a stronger reprogramming cassette that would supply a higher percentage of single-copy iPSC clones. We also included sequences on the reprogramming plasmid to permit us to integrate a therapeutic gene site-specifically into the iPSC [20]. The new reprogramming plasmid, pCOBLW, carried a more favorable order of the reprogramming genes, placing the Oct gene first and the Myc gene last (OSKM; Fig. 3). We also added the WPRE element to enhance transcription of the reprogramming cassette. pCOBLW was able to reprogram cells more effectively with

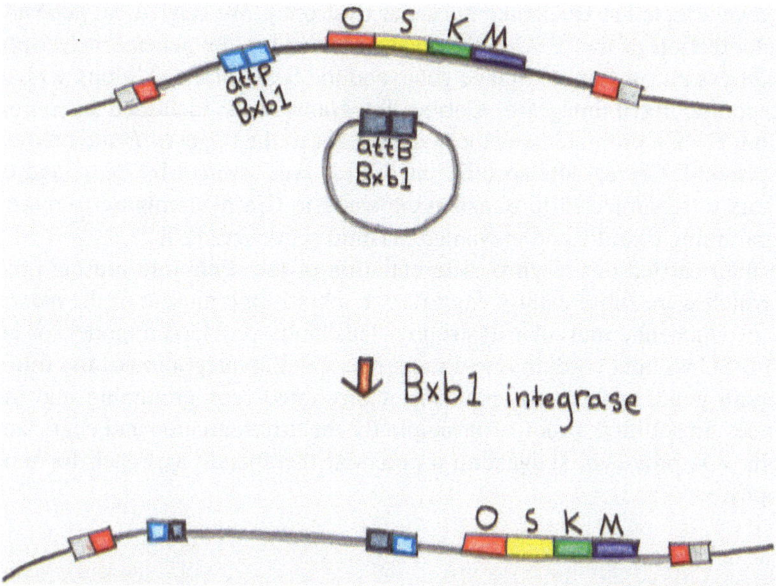

Fig. 3 Integration into a Bxb1 *attP* site placed by phiC31 integrase. We developed a reprogramming plasmid carrying the Oct-Sox-Klf-Myc (OSKM) reprogramming cassette and also a Bxb1 *attP* site. PhiC31 integrase was used to integrate the reprogramming plasmid at a pseudo *attP* site, producing iPSC (*upper part* of diagram). After determining that the integration location was safe by DNA sequencing and bioinformatics analysis, a therapeutic gene borne on a plasmid carrying a Bxb1 *attB* site was integrated precisely at the Bxb1 *attP* site resident in the integrated plasmid in the iPSC, resulting in site-specific integration of the therapeutic gene at a safe location. Cre resolvase was then applied to remove unwanted sequences, including the reprogramming genes and plasmid backbone sequences, by recombining between strategically located *loxP* sites in the plasmids

only a single copy, reflected by 93 % of iPSC generated with this plasmid being single-copy [20].

Along with the *loxP* sites we previously included for use in excision of reprogramming genes and plasmid sequences, we included the *attP* site of Bxb1 integrase, as a target for addition of a therapeutic gene to the integrated reprogramming plasmid (Fig. 3). Bxb1 is a serine integrase related to phiC31, but its *att* sites are completely distinct and do not cross-react with those of phiC31. Bxb1 is active on its own *att* sites in mammalian cells, but does not recognize native pseudo *att* sites at a measurable frequency [21]. Insulator sequences were included, such that they would flank the therapeutic gene after integration, to reduce position effects on the therapeutic gene and on neighboring genomic sequences.

To carry out this reprogramming strategy, we nucleofected pCOBLW into fibroblasts derived from the *mdx* mouse, along with a plasmid encoding phiC31 integrase. Integration of the vector through its phiC31 *attB* site occurred at pseudo *att* sites. An iPSC line with integration at a safe site was chosen for addition of a therapeutic gene. In this case, we used the full-length cDNA for mouse dystrophin, which is the gene affected in Duchenne muscular dystrophy. We carried out gene addition by nucleofection of the iPSC with a plasmid carrying the genes for dystrophin, a promoterless puromycin resistance gene, and the Bxb1 *attB* site, along with a plasmid encoding Bxb1 integrase. Correct integrants were identified by puromycin selection, since a promoter was located adjacent to the target *attP* site on the pCO-BLW plasmid. Correct site-specific integration was verified by PCR, and correct integrants were subjected to transient exposure to Cre recombinase to remove the reprogramming cassette and unwanted plasmid sequences [20].

We then carried out in vitro differentiation of the iPSC into muscle precursor cells, which were subsequently engrafted in a hind limb muscle of the *mdx* mouse model of Duchenne muscular dystrophy. This study provided a model for generation of iPSC without random integration, site-specific integration of the full-length dystrophin gene, and precise excision of unwanted reprogramming and plasmid sequences. In addition, proof-of-principle for the differentiation and engraftment of the cells was provided, suggesting a potential therapeutic approach for muscular dystrophy.

The DICE System: Combining Homologous Recombination with phiC31 and Bxb1 Integrases for Cassette Exchange

While phiC31-mediated integration into pseudo *att* sites provides a convenient method for genomic integration into unmodified genomes, it is laborious to analyze a set of clones to find an integration site that is safe and desirable. In pluripotent stem cells, such as ESC and iPSC, as well as in immortalized cell lines, another strategy is available that allows the user to control the site of integration precisely, via homologous recombination. If homologous recombination is used to position *attP* sites for integrases, these sites can then be targeted precisely for integration and /or cassette exchange of incoming genes. We developed a strategy called Dual Integrase Cassette Exchange, or DICE that utilizes precise placement of *attP* sites for cassette exchange [22].

In the DICE method, a "landing pad" carrying *attP* sites for phiC31 integrase and Bxb1 integrase, is positioned in the genome by homologous recombination. In our study, we used an intergenic, safe, transcriptionally active site called H11 on human chromosome 22 as the destination for the landing pad. Neomycin resistance and GFP genes placed between the *attP* sites served as markers for selection and screening of clones carrying the landing pad (Fig. 4). We utilized TALENs targeted to H11 to stimulate the frequency of homologous recombination. This strategy was particularly valuable in iPSC having significant disease pathology that depressed the rate of homologous recombination.

Once a line has been constructed that carries a landing pad, it can be used to position incoming genes at the desired site by an efficient and site-specific cassette exchange reaction mediated by phiC31 and Bxb1 integrases. To obtain cassette exchange, the landing pad line is nucleofected with a donor plasmid carrying the

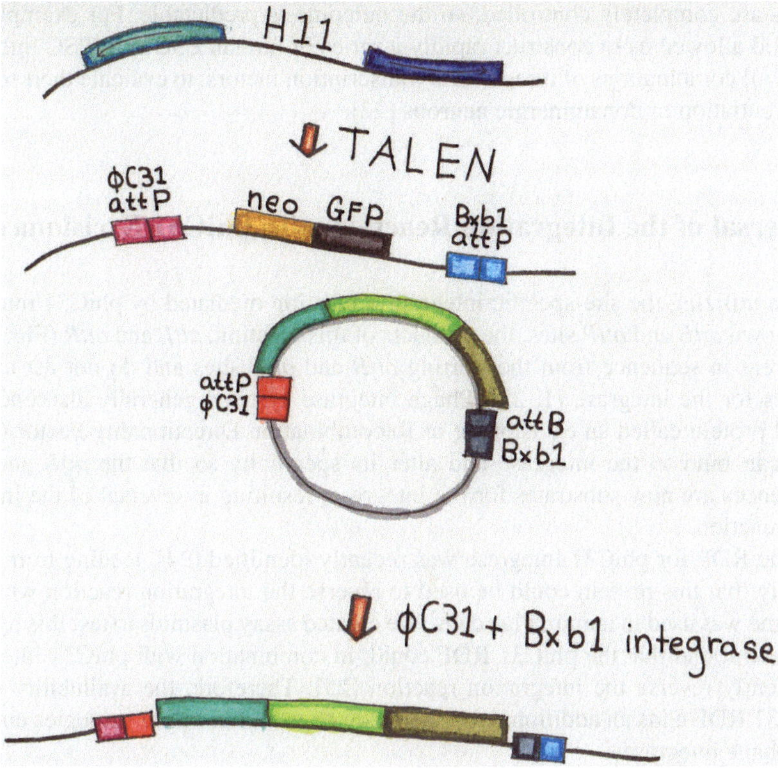

Fig. 4 DICE, dual integrase cassette exchange. To carry out integration by DICE, a landing pad bearing phiC31 and Bxb1 integrase *attP* sites is placed into the genome at a desired location by homologous recombination. We inserted a landing pad in human ESC and iPSC at the H11 locus on chromosome 22, a safe site where transcription is ubiquitous. The frequency of homologous recombination can be stimulated by the use of TALENS to make double-strand breaks at the target site. Selectable and screenable markers such as neomycin resistance and green fluorescent protein can be used to facilitate identification of correct integrants (*upper part* of drawing). Once the landing pad is inserted in the genome, it can be targeted readily by introducing plasmids carrying the genes one wants to integrate, such as a therapeutic gene, marker genes, or transcription factors, flanked by phiC31 and Bxb1 *attB* sites, along with plasmids encoding the phiC31 and Bxb1 integrases. Site-specific cassette exchange occurs, with the genetic information between the *attB* sites now present in the chromosome (*lower*)

genes of interest flanked by *attB* sites for phiC31 and Bxb1 integrases, along with plasmids encoding the two integrases. During cassette exchange, the neomycin and GFP genes will be removed from the landing pad and the donor genes will be inserted in their place (Fig. 4). Donor genes can include therapeutic genes and/or genes for selection, screening, tracking engraftment, and so on.

The DICE system is particularly valuable when there is a need to construct a series of parallel cell lines with different genes inserted into the same precise location. During cassette exchange, the content, direction, and position of the incoming genes are completely controlled, so the outcome is predictable. For example, the method allowed us to construct rapidly a series of human ESC and iPSC lines carrying all combinations of three neural transcription factors, to evaluate their roles in differentiation of dopaminergic neurons [22].

Reversal of the Integration Reaction with phiC31 Excisionase

When utilizing the site-specific integration reaction mediated by phiC31 integrase at its own *attB* and *attP* sites, the products of this reaction, *attL* and *attR* (Fig. 1) are different in sequence from the starting *attB* and *attP* sites and do not act as substrates for the integrase [1, 23]. Phage integrase systems generally also encode a small protein called an excisionase or Recombination Directionality Factor (RDF) that can bind to the integrase and alter its specificity so that the *attL* and *attR* sequences are now substrates for the integrase, resulting in reversal of the integration reaction.

The RDF for phiC31 integrase was recently identified [24], leading to the possibility that this protein could be used to reverse the integration reaction when the enzyme was used in mammalian cells. We created assay plasmids to test this hypothesis and found that the phiC31 RDF could, in combination with phiC31 integrase, efficiently reverse the integration reaction [25]. Therefore, the availability of the phiC31 RDF adds an additional tool that may be useful in future strategies employing phage integrases.

Acknowledgements M.P.C. thanks Victoria Ellis for creating the figures and the California Institute for Regenerative Medicine for financial support.

References

1. Thorpe HM, Smith MCM. In vitro site-specific integration of bacteriophage DNA catalyzed by a recombinase of the resolvase/invertase family. Proc Natl Acad Sci U S A. 1998;95:5505–10.
2. Groth AC, Olivares EC, Thyagarajan B, Calos MP. A phage integrase directs efficient site-specific integration in human cells. Proc Natl Acad Sci U S A. 2000;97:5995–6000.
3. Rutherford K, Van Duyne G. The ins and outs of serine integrase site-specific recombination. Curr Opin Struct Biol. 2014;24:125–31.
4. Thyagarajan B, Olivares EC, Hollis RP, Ginsburg DS, Calos MP. Site-specific genomic integration in mammalian cells mediated by phage phiC31 integrase. Mol Cell Biol. 2001;21: 3926–34.
5. Chalberg TC, et al. Integration specificity of phage phiC31 integrase in the human genome. J Mol Biol. 2006;357:28–48.
6. Allen BG, Weeks DL. Transgenic Xenopus laevis embryos can be generated using phiC31 integrase. Nat Methods. 2005;2:975–9.
7. Groth AC, Fish M, Nusse R, Calos MP. Creation of transgenic *Drosophila* by using the site-specific integrase from phage phiC31. Genetics. 2004;166:1775–82.

8. Geisinger J, Calos MP. Site-specific recombination using phiC31 integrase. In: Renault S, Duchateau P, editors. Site-directed insertion of transgenes. Dordrecht: Springer Science; 2013. p. 211–39.
9. Olivares EC, et al. Site-specific genomic integration produces therapeutic factor IX levels in mice. Nat Biotechnol. 2002;20:1124–8.
10. Chavez C, et al. Kinetics and longevity of phiC31 integrase in mouse liver and cultured cells. Hum Gene Ther. 2010;21:1287–97.
11. Keravala A, et al. Long-term phenotypic correction in factor IX knockout mice by using phiC31 integrase-mediated gene therapy. Gene Ther. 2011;18:842–8.
12. Chavez C, et al. Long-term expression of human coagulation factor VIII in a tolerant mouse model using the phiC31 integrase system. Hum Gene Ther. 2012;23:390–8.
13. Chavez C, Calos M. Therapeutic applications of the phiC31 integrase system. Curr Gene Ther. 2011;11:375–81.
14. Karow M, Calos M. The therapeutic potential of phiC31 integrase as a gene therapy system. Expert Opin Biol Ther. 2011;11:1287–96.
15. Chalberg TC, Genise HL, Vollrath D, Calos MP. PhiC31 integrase confers genomic integration and long-term transgene expression in rat retina. Invest Ophthalmol Vis Sci. 2005;46:2140–6.
16. Bertoni C, et al. Enhancement of plasmid-mediated gene therapy for muscular dystrophy by directed plasmid integration. Proc Natl Acad Sci U S A. 2006;103:419–24.
17. Takahashi K, et al. Induction of pluripotent stem cells from adult human fibroblasts by defined factors. Cell. 2007;131:861–72.
18. Takahashi K, Yamanaka S. Induction of pluripotent stem cells from mouse embryonic and adult fibroblast cultures by defined factors. Cell. 2006;126:663–76.
19. Karow M, et al. Site-specific recombinase strategy to create induced pluripotent stem cells efficiently with plasmid DNA. Stem Cells. 2011;29:1696–704.
20. Zhao C, et al. Recombinase-mediated reprogramming and dystrophin gene addition in mdx mouse induced pluripotent stem cells. PLoS One. 2014;9(4), e96279.
21. Keravala A, et al. A diversity of serine phage integrases mediate site-specific recombination in mammalian cells. Mol Genet Genomics. 2006;276:135–46.
22. Zhu F, et al. DICE, an efficient system for iterative genomic editing in human pluripotent stem cells. Nucleic Acids Res. 2014;42(5), e34.
23. Thorpe HM, Wilson SE, Smith MCM. Control of directionality in the site-specific recombination system of the *Streptomyces* phage phiC31. Mol Microbiol. 2000;38:232–41.
24. Khaleel T, Younger E, McEwan A, Varghese A, Smith M. A phage protein that binds phiC31 integrase to switch its directionality. Mol Microbiol. 2011;80:1450–63.
25. Farruggio A, Chavez C, Mikell C, Calos M. Efficient reversal of phiC31 integrase recombination in mammalian cells. Biotechnol J. 2012;7(11):1332–6.

Precise Genome Modification Using Triplex Forming Oligonucleotides and Peptide Nucleic Acids

Raman Bahal, Anisha Gupta, and Peter M. Glazer

Abstract Many genetic disorders are caused by single base pair mutations which lead to defective protein synthesis. In addition to gene replacement therapy, modification of genomic DNA sequences at specific sites has been employed to manipulate the function and expression of various genes, which are implicated in various genetic disorders. On this front, triplex technology has been used to alter the expression of different genes by correcting mutations site specifically via homologous recombination (HR) or targeted mutagenesis based mechanisms. In this chapter we will discuss the advances made in triplex technology involving triplex forming oligonucleotides (TFOs) and peptide nucleic acids (PNAs) for site specific genome editing.

Keywords TFOs • PNA • Recombination • Repair • Mutagenesis • Genome modification

Introduction

Many genetic disorders are caused by single base pair mutations which lead to defective protein synthesis. In addition to gene replacement therapy, modification of genomic DNA sequences at specific sites has been employed to manipulate the function and expression of various genes, which are implicated in various genetic disorders. On this front, triplex technology has been used to alter the expression of

R. Bahal, Ph.D. • A. Gupta, Ph.D.
Department of Therapeutic Radiology, Yale University School of Medicine, New Haven, CT, USA

P.M. Glazer, M.D., Ph.D. (✉)
Department of Therapeutic Radiology, Yale University, New Haven, CT, USA

Department of Genetics, Yale University, New Haven, CT, USA
e-mail: peter.glazer@yale.edu

T. Cathomen et al. (eds.), *Genome Editing*, Advances in Experimental Medicine and Biology 895, DOI 10.1007/978-1-4939-3509-3_6

different genes by correcting mutations site specifically via homologous recombination (HR) or targeted mutagenesis based mechanisms. In this chapter we will discuss the advances made in triplex technology involving triplex forming oligonucleotides (TFOs) and peptide nucleic acids (PNAs) for site specific genome editing.

Triplex Forming Oligonucleotides

In addition to Watson–Crick base pairing, knowledge of alternative binding interactions between the nucleobases known as Hoogsteen base pairing led to the design of another scaffold for the molecular recognition of DNA known as TFOs. The process of triple helix formation was first suggested by Pauling and Corey in 1953 whereas the formation of triplex structures was first reported in 1957 by Felsenfield et al. [1]. Through RNA diffraction studies it was observed that stretches of poly(U) polyuridylate and poly(A) (polyadenylate) sequences hybridize in a 2:1 binding ratio in presence of divalent magnesium ions [1]. The high density of target homopurine sequences in the genome coupled with the sequence specificity of TFOs makes them attractive molecules to target individual genes and modulate gene function [2].

Binding Code

In the case of polypyrimidine TFOs binding to polypurine DNA targets, 'pyrimidine motif' T binds AT, and C binds GC (forming T–A–T and C^{+}–G–C base triplets) via Hoogsteen hydrogen bonding (Fig. 1) resulting in parallel orientation with respect to the polypurine site of the target duplex [3,4]. The orientation of the 'purine motif' was also demonstrated by Peter Dervan and his co-workers, wherein G binds GC and A binds AT (forming A–A–T and G–G–C base triplets) through reverse hydrogen bonding resulting in antiparallel orientation with respect to the polypurine site in target B-DNA duplex [5–7]. Through X-ray diffraction [8], nuclear magnetic resonance [9], and chemical probing studies [10], it was revealed that TFO binding to the genomic DNA leads to helical distortions [11], which evoke the repair system of cells leading to manifold applications by inducing mutagenesis and enhancing homologous recombination.

Chemical Modifications to Improve TFO Binding Affinity and Stability

The binding of TFOs to duplex DNA is influenced by a number of different factors including ionic conditions, sequence composition and accessibility to the target site [12]. In the case of polypyrimidine TFOs that bind in a parallel orientation, the N3 of the cytosine must be protonated in order to form optimum Hoogsteen hydrogen

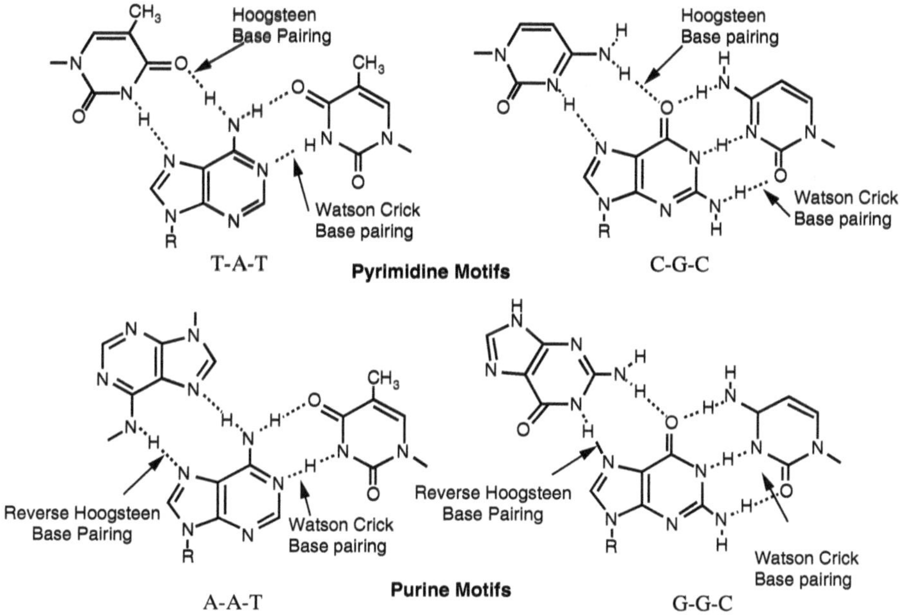

Fig. 1 Motifs for triple helix formation. (*Top*) In the pyrimidine motif the third strand binds parallel to the purine strand of DNA via Hoogsteen bonds. (*Bottom*) In the purine motif, the third strand binds antiparallel to the purine strand via reverse Hoogsteen hydrogen bonds. The canonical base triplets are shown for each motif

bonds with N7 of the guanine [13]. Hence, the pH dependence of pyrimidine TFOs limits their utility in intracellular targeting. On the other hand, polypurine TFOs bind to the target DNA in an antiparallel orientation without any dependence on pH. However G-rich purine TFOs are limited in application because the guanine rich sequences tend to form G-quadruplexes at physiological conditions in presence of high K^+ concentrations [14–17]. The intracellular K^+ promotes the aggregation of TFOs due to the formation of stable secondary structures known as G-tetrads. In order to form stable triplex structures, continuous polypurine runs are required [18]. The interruptions by pyrimidines or single base pair mismatches can significantly destabilize the binding of TFO. Because TFOs are limited to homopyrimidine or homopurine sequences, triplex binding could not be extended to the human genome sequences containing mixed sequence ds-B-DNA. Furthermore the charge repulsions due the three polyanionic strands and the accessibility to the binding site in the cellular environment pose significant challenges in success of the TFO design [19].

Hence, chemical modification of TFOs is essential to increase triplex stability and to protect TFOs from enzymatic degradation. Many efforts have been made to optimize the TFO design. In case of pyrimidine rich TFOs, the replacement of cytosine with 5-methyl cytosine or pseudoisocytosine have been done to overcome the dependence on pH [20,21]. Different cytosine substitutions including 8-oxoadenine [22], 7,8-dihyro-8-oxoadenine [23], 6-oxocytidine [24], 8-oxo-2′deoxyadenosine

[25,26] and 8-aminoguanine [27] have improved the binding affinity. Likewise the modification of thymidine to 2′deoxyuridine (dU), 5-propargylamino- and 2′-aminoethoxy,5-propargylamino-dU bis-substituted derivatives have shown to improve triplex binding stability [28–32]. Modified sequences comprising of 2′-deoxy-6-thioguanosine or 7-deaza-2′-deoxyxanthosine in the G-rich TFOs have been shown to inhibit the G-Quartets formation [15–17]. Many different modifications in the sugar involving 2′-O-ribonucleotides [33,34], and 2'-O-aminoethyl ribose [35–37] have been shown to confer nuclease resistance and promote triplex stability. Different chemical modifications at the backbone or bases have been also explored to improve the binding affinity and biological applications of the TFOs. It has been demonstrated that by replacing the sugar phosphate backbone with morpholino oligonucleotides or by cationionc phosphoramidate linkages such as N,N-diethethyleneamine (DEED) or N,N-dimethyl-aminopropyl amine, binding affinity of the TFO can be increased in vitro [38–41]. Further, by replacing the backbone to uncharged, achiral peptide units in case of PNAs, significant progress has been made in the area of triplex research [42–44]. The different studies involving PNA as a triplex forming agent are discussed in detail in another section of this chapter.

Triplex Mediated Genome Modification

Targeted Mutagenesis via Triplex Formation

By inducing mutations at specific sites in the genome, triplex technology can be employed to induce heritable changes in gene function and expression. The TFOs have the potential to invoke the DNA repair by directly binding with a segment of a gene or by delivering a mutagen at the site of repair. The conjugation of psoralen (pso), a photoreactive mutagenic agent to a TFO has been shown to induce damage to the DNA [45–48]. Upon irradiation with UVA light, psoralen intercalates into the DNA and covalently crosslink thymines on both strands [49]. The characterized mutations in mammalian cells involved T:A to A:T transversions [46,50–52].

Plasmid Based Assays for Detecting Mutagenesis in Mammalian Cells

To analyze TFO induced mutagenesis, a reporter system based on supF gene was constructed. The supF reporter gene cloned into an SV40 vector encodes an amber suppressor tyrosine tRNA and contains a TFO binding site. Upon treatment with pso-TFO and UV A irradiation, the vector was transected into monkey COS-7 cells. The mutation frequencies were quantified by isolating the plasmids and co-transform in bacterial cells. The results suggested that 6 % of the plasmids underwent targeted mutagenesis out of which 55 % were T:A to A:T transversions [45]. Furthermore, it was shown that a TFO alone without conjugating to a mutagen can also induces site directed mutagenesis [53].

TFO Induced Mutagenesis at Chromosomal Sites in Mammalian Cells

TFOs have been shown to induce mutagenesis at chromosomal target sites in cell culture. Pso-TFOs induced mutations at the endogenous hypoxanthine phosphoribosyl transferase (*hprt*) gene in CHO cells have been reported by Majumdar et al. [54]. They showed that 80 % of the mutations occurred at the triplex target region. Vasquez et al. also constructed a transgenic mouse containing multiple copies of lambda vector containing a 30-bp triplex target site within the supF mutation reporter gene (Fig. 2) [55]. Upon treatment of the cells with pso-TFOs, a tenfold induction of site specific mutations above the background were observed. The mutations were observed both in presence and absence of UV A irradiations suggesting that the TFO alone has the potential to induce mutations.

TFOs in Homologous Recombination

The modification of human genome by HR is widely employed in several different applications. However the use of HR is limited due to low frequency in mammalian cells and random integration at non-targeted sites (reviewed in ref [56]). In order to improve the frequency of HR, several different approaches have been employed which include DNA damage near the target site by UV irradiation, alkylation and psoralen induced cross linking or double-strand DNA breaks by endonucleases [57–59]. TFO directed DNA damage either by formation of a triplex alone or by a mutagenic agent has been shown to improve recombination and gene modification frequencies. The studies involving different TFOs in genome modification via HR are summarized in Table 1.

Intramolecular Recombination

To detect the enhancement in homologous recombination, different reporter constructs have been used. Our lab has used Pso-TFOs to induce recombination between tandem mutant copies of a reporter gene in plasmids and in chromosomal sites [60]. In order to study targeted recombination via TFO directed intrastrand crosslinks, Faruqi et al. used SV40 shuttle vectors with two tandem supF genes containing different point mutations and a TFO binding site between the mutated genes [61]. Later, Luo et al. studied a similar construct in chromosomal sites in mice [62]. The TFO-induced recombination was detected through beta galactosidase screening assay using bacterial colonies. The studies involved the use of unconjugated TFOs as well as TFOs conjugated to psoralen. It was observed that the recombination events stimulated by intrastrand crosslinks via TFO conjugated psoralen occurred at a higher frequency in comparison to unconjugated TFOs. Further, the mechanisms involved in these intrachromosomal recombination events were studied. It was observed the increase in triplex induced recombination was dependent on

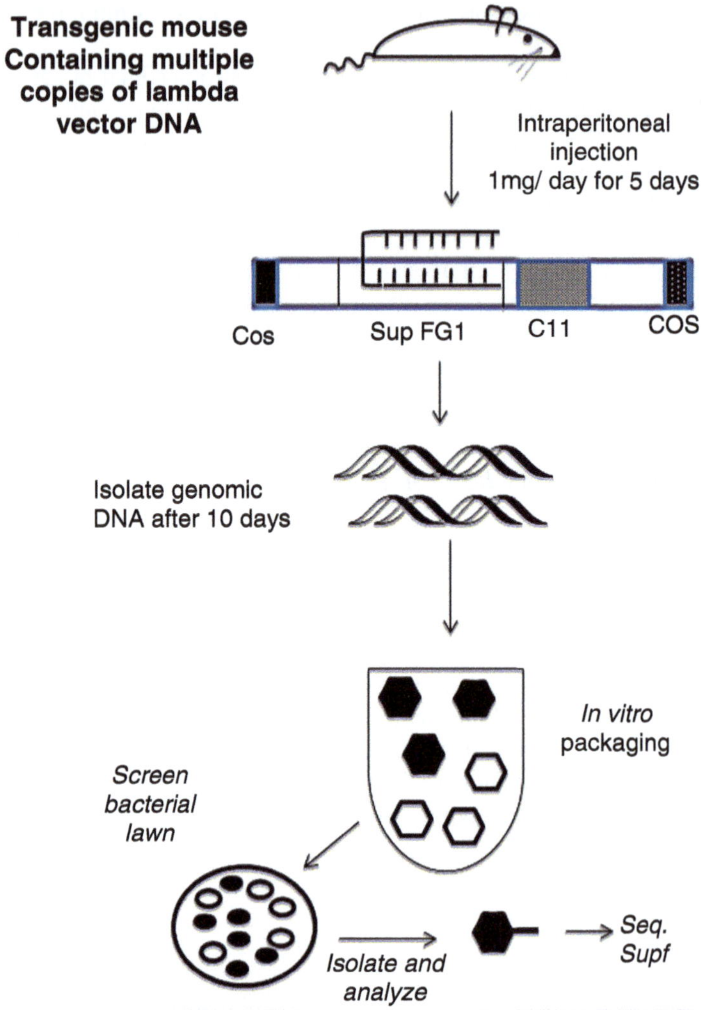

Fig. 2 Experimental protocol for chromosomal mutations detection after TFO administration in mice. Transgenic mice containing many copies of chromosomally integrated A supFG 1 vector, which contains the 30 bp triplex binding site was used. TFOs were injected intraperitoneally. After treatment, tissues were harvested, and genomic DNA was isolated and analyzed. In vitro packaging of the phage vector led to mutagenesis detection via plating on a bacterial lawn. If no mutations occur, the plaques will appear *blue* in the presence of IPTG and X-Gal. If a mutation does occur, the resulting plaques will be *white*

nucleotide excision repair (NER) pathway [61]. No significant recombination was observed in cell lines deficient in Xeroderma Pigmentosum Group A (XPA) repair factor [63]. The psoralen conjugated TFO directed recombination was partially dependent on XPA demonstrating the involvement of multiple repair pathways.

Table 1 HR induced by TFOs in mammalian cells

TFO	Type of recombination	Reporter gene	Recombination frequency (%)	Cell line
AG-30	Intramolecular	supF (bacterial suppressor tRNA)	O.37	COS-7
Pso-AG30	Intramolecular	supF	0.58	COS-7
AG30	Intrachromosomal	TK (thymidine kinase)	1.2	LTK-
AG30+51 mer donor DNA	Intermolecular	Fluc (Firefly luciferase gene)	0.05	CHO derived
22 mer purine rich TFO	Intermolecular	supF	~0.05	Jurkat T lymphoblastoid
Psoralen conjugated 19mer purine TFOs	Intra molecular	supF	0.002	HeLa

Intermolecular Recombination

Chan et al. used an approach based on TFOs etethered to donor DNA fragments homologous to the target site (except at the base pair to be corrected) and were able to demonstrate intermolecular recombination events [64]. In this strategy, the formation of the triplex induces repair and sensitizes the target site for recombination [65]. The TFO domain also positions the donor DNA for recombination. Subsequent studies have shown that the conjugation of the donor DNA to the TFOs not necessary [66–69].

Peptide Nucleic Acids (PNAs)

One of the major issues related with DNA based TFOs is enzymatic degradation. In the past, in order to improve the enzymatic stability, numerous synthetic nucleic acid analogues have been developed. One such promising class of nucleic acid mimic is known as PNA. Structurally, PNA is a homomorphous unit of DNA where phosphodiester backbone is replaced with an uncharged aminoethylglycine backbone [42]. Due to its achiral, neutral polyamide backbone, PNAs are resistant to enzymatic degradation (Fig. 3). PNA hybridizes with complementary DNA/RNA targets in a sequence specific manner to form highly stable duplexes (PNA/DNA or PNA/RNA). PNAs also form triplex complexes where one strand, invades the DNA duplex through Watson-Crick base pairing and another strand is capable of forming Hoogsteen bonds with the PNA–DNA duplex [70]. Two PNA strands linked by a flexible linker forms a clamp and form highly stable complexes with the target dsDNA.

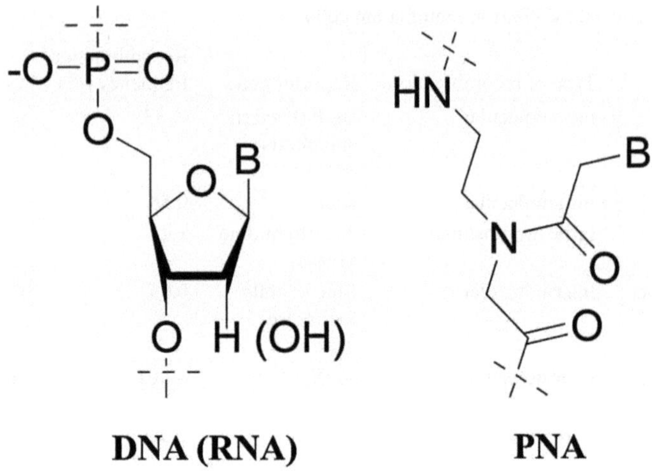

Fig. 3 Chemical structure of DNA (RNA) and PNA

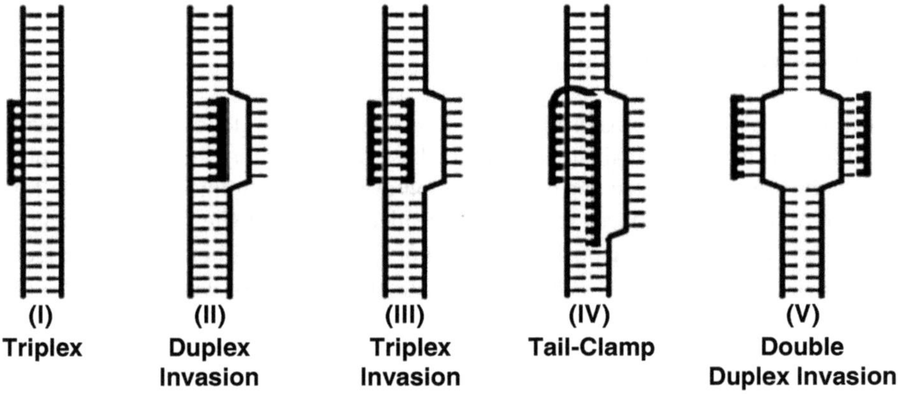

Fig. 4 Different binding modes of PNA with dsDNA

Recently it was found that PNA has potential to invade certain regions of genomic dsDNA through strand invasion based mechanisms. The mechanism by which PNA binds to dsDNA relies on the sequence composition of the target region. At least five different binding modes have been established (Fig. 4). For cytosine-rich PNA, binding takes place in the major groove through Hoogsteen base-pairing in a 1:1 ratio. Homopurine PNA also binds dsDNA in a 1:1 ratio, but through Watson-Crick base-pairing resulting in local displacement of the homologous DNA strand, which creates D-loop formation [71]. Homopyrimidine PNA, on the other hand, binds dsDNA in a 2:1 (PNA:DNA) stoichiometry in which one strand forms Watson-Crick base pairs whereas the second strand forms Hoogsteen base-pairing, resulting in a PNA_2–DNA triplex and a locally displaced D-loop. A combined homopyrimidine/homopurine binding mode has also been exploited in the development of "tail-clamp" hairpins for recognition of partially mixed-sequence containing dsDNA [44]. Peter Nielsen and coworkers have demonstrated a fifth binding mode that relies on the use of pseudo-complementary (pcPNAs) PNAs to bind simultaneously to both strands of the DNA double helix. This strategy is also known as double duplex invasion strategy [70]. This approach possess greater flexibility in sequence selection, but it also complicates the targeting strategy by use of two separate strands of pcPNAs for binding because of their propensity to interact with each other, which greatly limits its utility.

PNAs have demonstrated enormous potential by their activity as transcriptional regulators, inhibitors of protein binding and DNA polymerization blockers. However cellular delivery of PNAs has been the major issue for its therapeutic application.

Use of PNA in Repair and Recombination

Due to high binding affinity with their complementary base pairs, PNAs have potential to induce mutagenesis and HR based repair mechanisms. In the past, Faruqi et al. employed a single dimeric PNA to target a site in the supFG1 mutation reporter gene within a chromosomally integrated recoverable λ phage shuttle vector in mouse fibroblasts [72]. The designed PNA forms a clamp via both double- and triple-helix formation within 8- or a 10-bp site in the supFG1 gene and induce mutations at frequencies in the range of 0.1 %, tenfold above the background [72].

Rogers et al. have done comprehensive studies to demonstrate that dimeric bis-PNAs have potential to promote site directed recombination [73]. Generally, bis-PNAs undergo both strand invasion as well as triplex formation, which lead to formation of clamp structures on target dsDNA. In this study, they have shown that a bis-PNA conjugated to a 40-nucleotide donor DNA is able to induce site directed recombination. PNA-donor DNA conjugates were prepared by employing maleimide based chemistry. The PNA–DNA conjugate mediated sequence specific changes within the supFG1 reporter gene in vitro in human cell free extracts, resulting in correction of a mutation at a frequency at least 60-fold above background [73]. Similarly induced site-specific recombination also achieved by using bis-PNA and the donor DNA together without any covalent linkage. The bis-PNA and the bis-PNA-donor DNA conjugate were also found to induce DNA repair with high specificity in the target plasmid. It was shown that both PNA-induced recombination as well as repair was found to be dependent on the nucleotide excision repair factor, XPA (Xeroderma pigmentosum complementation group A protein) [73]. Due to the formation of clamp structures with duplex DNA, PNA creates a helical distortion that strongly provokes DNA repair and thereby sensitizes the target DNA sites to recombination.

Wang et al. designed a series of dimeric PNAs to form clamps at a 10 bp homopurine/homopyrimidine site of the E. coli supFG1 gene [74]. To study the recombination frequency based mechanisms, mouse fibroblasts with 15 chromosomally integrated copies of the lambda phage shuttle vector containing the supFG1 gene were employed. SupFG1 gene mutagenesis determined by counting colonies after infection of E. coli ClacZ125 (am) bacteria. It was demonstrated that, dimeric PNAs induced mutations in the chromosomal supFG1 gene in the mouse cells at a frequency of 0.1 % (tenfold above background) [74]. Sequence analysis also substantiates that the majority of mutations were located within the PNA binding site. Further exploring the utility of PNAs for site-specific gene modification various groups have shown that PNAs conjugated with DNA modifying agents, such as benzophenone, anthraquinone, or psoralen, can be used for DNA modification in vitro.

In another studies, dimeric bis-PNAs-psoralen conjugates were employed. The designed PNA-psoralen conjugates were able to bind the target site on the supFLSG3 reporter gene by triplex invasion- complex formation and directed site-specific photoadduct formation by the conjugated psoralen [75]. The formation of photoadduct was confirmed by in vitro assays and mutagenesis in the targeted gene was assayed using SV 40 based episomal shuttle vector assay. The photo adducts directed by PNAs conjugated to psoralen induced mutations at frequencies in the range of 0.46 % (6.5-fold above the background). For intracellular gene targeting in the episomal shuttle vector, 0.13 % of psoralen-PNA induced mutation frequency was achieved (3.5-fold higher than the background). In this study it has been demonstrated that most of the induced mutations were deletions and single-base-pair substitutions at or adjacent to the targeted PNA-binding and photoadduct-formation sites [75]. This set of study further support the development of PNAs as tools for gene-targeting applications.

Above, different types of binding modes of PNA to dsDNA have been discussed. In addition to bis PNAs, pcPNAs were used for intracellular gene targeting at mixed sequence sites. It has been well reported in literature that due to steric hindrance, pcPNAs are unable to form pcPNA–pcPNA duplexes but can bind to complementary DNA sequences by Watson–Crick pairing via double duplex-invasion complex formation. It has been demonstrated that psoralen-conjugated pcPNAs can deliver site-specific photoadducts and mediate targeted gene modification within both episomal and chromosomal DNA in mammalian cells without possessing any off-target effects [76]. Psoralen-pcPNA mutations were single-base substitutions and deletions found at the predicted pcPNA-binding sites. No mutations were induced by the individual pcPNAs alone nor did complementary PNA pairs of the same sequence cause mutations [76].

PNAs combined with donor DNAs have the potential to induce gene correction based on triplex-induced homologous recombination mechanism and have been tested in several disease models including beta-thalassemia [69]. Generally, splice-site mutations in the β-globin gene lead to aberrant transcripts and decreased functional β-globin, causing beta-thalassemia. It was shown that bis-PNAs when co-transfected with recombinant donor DNA fragments containing the corrected mutation site, can promote single base-pair modification at the start of the second intron of the β-globin gene, the site of a common thalassemia-associated mutation. Green fluorescent protein-β-globin fusion gene was used to detect the restoration of proper splicing of transcripts. It was also shown that recombination frequencies can be enhanced when the PNAs/DNA were used with the lysomotropic agent, chloroquine [69].

Later on, Lonkar and coworkers also demonstrated that pcPNAs, when co-transfected with donor DNA fragments, can promote single base pair modification at the start of the second intron of the β-globin gene [68]. Gene editing was detected by examining the restoration of proper splicing of transcripts produced from a green fluorescent protein-beta globin fusion gene. In addition it was also found that pcPNAs stimulate recombination in human fibroblast cells dependent on the nucleotide excision repair factor, XPA. These results signify that pcPNAs can be used as tools for site-specific gene modification in mammalian cells without sequence restriction.

In addition to β-globin gene, genome modification has been also demonstrated in the CCR5 gene. CCR5 encodes a chemokine receptor required for HIV-1 entry into human cells, and individuals carrying mutations in this gene are resistant to HIV-1 infection. Schleifman et al. has shown that transfection of human cells with bisPNAs/donor DNA targeted to the CCR5 gene, can introduce stop codons mimicking the naturally occurring CCR5-delta32 mutation, produced 2.46 % targeted gene modification [67]. In a series of experiments, CCR5 modification was confirmed at the DNA, RNA, and protein levels [67]. It was also shown that introduction of stop codon confer resistance to infection with HIV-1. This work underscores that PNA induced genome modification can be used as a therapeutic strategy for CCR5 knockout in HIV-1-infected individuals.

Further Rogers et al. have shown that through conjugation of a triplex-forming peptide nucleic acid (PNA) to the transport peptide, antennapedia (Antp), successful in vivo chromosomal genomic modification of hematopoietic progenitor cells can be achieved [77]. In addition, hematopoietic progenitor cells still possessed the differentiation capabilities even after gene modification. This strategy obviates the use of transfection-based protocols for PNA/donor DNA.

McNeer et al. proposed new methods for intracellular delivery of PNA/donor DNA molecules for genome editing by using poly(lactic-co-glycolic acid) (PLGA)-based nanoparticles [78]. PLGA is an FDA-approved biocompatible polymer and is used clinically for delivery of drugs for numerous indications including the treatment of prostate cancer (Lupron and Trelstar). The previous work has shown that PLGA nanoparticles can be used for intracellular delivery of nucleic acid polymers and oligomers, including plasmid DNA and siRNAs for gene-silencing studies [79]. PLGA nanoparticles are formulated by using double-emulsion solvent-evaporation technique. PLGA nanoparticles encapsulate tcPNA (Fig. 5), DNA (DNA was neutralized using spermidine as a counter ion), or both tcPNA and DNA (in which the lysines conjugated to the PNA both on C and N termini served as the counter ion for the DNA). Further, treatment of human CD34+ HSCs with nanoparticles containing nucleic acids in dosages of 0.25–2 mg PLGA/mL was performed. It has been found that cells treated with nanoparticles show greater cell recovery and viability as compared to nucleofection protocols. Nanoparticle treatment also led to much higher rates of recombination, corresponding to at least a 60-fold increase in modified and viable cells.

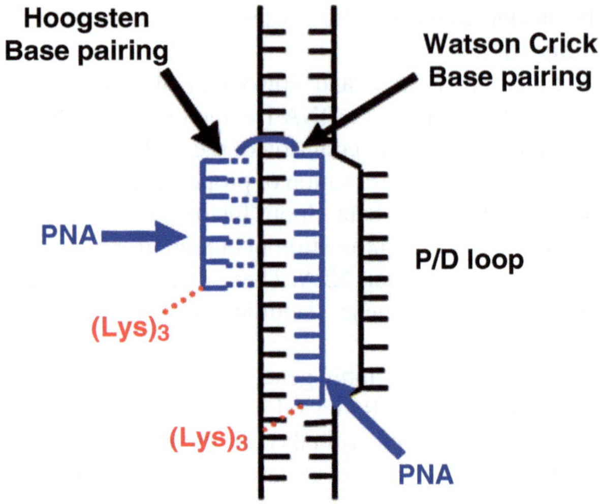

Fig. 5 Lysine conjugated tail clamp PNA binding to dsDNA

It has been successfully shown that PNA/donor DNA delivered by nanoparticle-based approach caused site-specific gene editing of human cells in vivo in hematopoietic stem cell-engrafted NOD-scid mice. Intravenous injection of particles containing PNAs/DNAs produced modification of the human CCR5 gene in hematolymphoid cells in the mice. Deep-sequencing results confirmed in vivo modification of the CCR5 gene at frequencies in the range of 0.1–0.5 % in hematopoietic cells in the spleen and bone marrow. At the same time, off-target modification in the homologous CCR2 gene was two orders of magnitude lower. Application of nanotechnology in the site-specific gene editing by using PNA offers numerous advantages. Firstly, this approach provides a framework for development of a flexible system for direct in vivo genomic modification of HSCs. Second; this work suggests a versatile method for targeted drug delivery to human hematopoietic populations.

tcPNAs targeting always requires homopurine or homopyrimidine rich regions for effective targeting.

In another strategy we have shown that new generation chemically modified gamma PNAs (γPNAs) can be used to target mixed sequence genomic DNA and induce genome modification [80]. In γPNAs, stereogenic chiral center has been induced at the gamma position of PNA backbone (Fig. 6) [81–83]. The presence of chiral center at gamma positions results into preorganized conformation of PNAs which further boost its binding affinity with complementary sequences containing DNA or RNA [84,85].

In order to demonstrate proof of concept, a transgenic mouse model containing a β-globin/EGFP fusion gene, consisting of intron 2 of human β-globin inserted within the GFP coding regions have been employed [86,87]. The intron region contains the IVS2-654 (C → T) mutation which is a common cause of thalassemia in individuals of Southeast Asian heritage. This mutation leads to a cryptic splice site that causes incorrect splicing of the intron and prevents expression. Using nanopar-

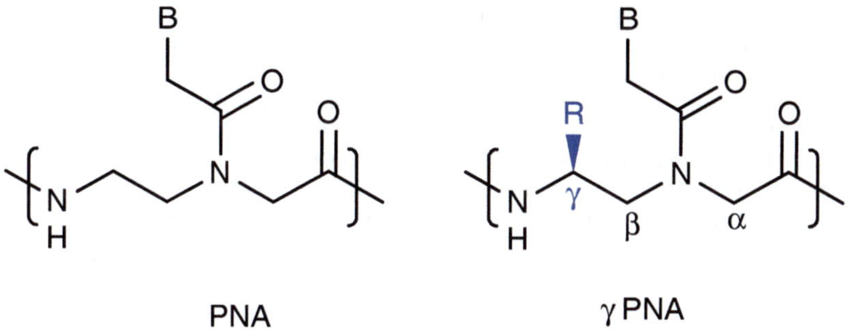

Fig. 6 Chemical structure of PNAs and γPNAs

ticulate based delivery systems we have shown that single-stranded γPNAs designed to target genomic DNA site near IVS2-654 mutation along with donor DNA, induced site-specific gene editing at frequencies of 0.8 % in mouse bone marrow cells treated ex vivo and 0.1 % in vivo via IV injection, without detectable toxicity. Moving one step ahead we have shown that γPNAs provide a new tool for induced gene editing based on Watson-Crick recognition without sequence restriction.

Summary

Overall TFOs and PNAs have played an important role in genome modification by implying different strategies. Promising results have been attained both in ex vivo as well as in vivo studies. By utilizing information at chemistry/biology interface, researchers have come up with many strategies to increase the gene editing frequency. Though PNAs possess promising chemical properties in comparison to TFOs, nonetheless, lot of work still need to be done in order to attain gene-editing frequencies, which are clinically relevant.

References

1. Felsenfeld G, Davies DR, Rich A. Formation of a three-stranded polynucleotide molecule. J Am Chem Soc. 1957;79:2023–4.
2. Francois JC, Saison-Behmoaras T, Thuong NT, Helene C. Inhibition of restriction endonuclease cleavage via triple helix formation by homopyrimidine oligonucleotides. Biochemistry. 1989;28(25):9617–9.
3. Hoogsteen K. The structure of crystals containing a hydrogen-bonded complex of 1-methylthymine and 9-methyladenine. Acta Crystallogr. 1959;12:822–3.
4. Moser HE, Dervan PB. Sequence-specific cleavage of double helical DNA by triple helix formation. Science. 1987;238(4827):645–50.

5. Letai AG, Palladino MA, Fromm E, Rizzo V, Fresco JR. Specificity in formation of triple-stranded nucleic acid helical complexes: studies with agarose-linked polyribonucleotide affinity columns. Biochemistry. 1988;27(26):9108–12.
6. Beal PA, Dervan PB. Second structural motif for recognition of DNA by oligonucleotide-directed triple-helix formation. Science. 1991;251(4999):1360–3.
7. Durland RH, Kessler DJ, Gunnell S, Duvic M, Hogan ME, Pettitt BM. Binding of triple helix forming oligonucleotides to sites in gene promoters. Biochemistry. 1991;30(38):9246–55.
8. Arnott S, Selsing E. Structures for the polynucleotide complexes poly dA.poly dT and poly dT.poly dA.poly dT. J Mol Biol. 1974;88(2):509–21.
9. Radhakrishnan I. De, l. S. C.; Patel, D. J., Nuclear magnetic resonance structural studies of intramolecular purine · purine · pyrimidine DNA triplexes in solution. Base triple pairing alignments and strand direction. J Mol Biol. 1991;221(4):1403–18.
10. Francois JC, Saison-Behmoaras T, Helene C. Sequence-specific recognition of the major groove of DNA by oligodeoxynucleotides via triple helix formation. Footprinting studies. Nucleic Acids Res. 1988;16(24):11431–40.
11. Hartman DA, Kuo SR, Broker TR, Chow LT, Wells RD. Intermolecular triplex formation distorts the DNA duplex in the regulatory region of human papillomavirus type-11. J Biol Chem. 1992;267(8):5488–94.
12. Pilch DS, Brousseau R, Shafer RH. Thermodynamics of triple helix formation: spectrophotometric studies on the d(A)10 · 2d(T)10 and d(C+3T4C+3) · d(G3A4G3) · d(C3T4C3) triple helixes. Nucleic Acids Res. 1990;18(19):5743–50.
13. Lee JS, Johnson DA, Morgan AR. Complexes formed by (pyrimidine)n. (purine)n DNAs on lowering the pH are three-stranded. Nucleic Acids Res. 1979;6(9):3073–91.
14. Cheng AJ, Van DMW. Monovalent cation effects on intermolecular purine-purine-pyrimidine triple-helix formation. Nucleic Acids Res. 1993;21(24):5630–5.
15. Milligan JF, Krawczyk SH, Wadwani S, Matteucci MD. An anti-parallel triple helix motif with oligodeoxynucleotides containing 2′-deoxyguanosine and 7-deaza-2′-deoxyxanthosine. Nucleic Acids Res. 1993;21(2):327–33.
16. Olivas WM, Maher III LJ. Overcoming potassium-mediated triplex inhibition. Nucleic Acids Res. 1995;23(11):1936–41.
17. Rao TS, Durland RH, Seth DM, Myrick MA, Bodepudi V, Revankar GR. Incorporation of 2′-deoxy-6-thioguanosine into G-rich oligodeoxyribonucleotides inhibits G-tetrad formation and facilitates triplex formation. Biochemistry. 1995;34(3):765–72.
18. Cheng A-J, Van DMW. Oligodeoxyribonucleotide length and sequence effects on intermolecular purine-purine-pyrimidine triple-helix formation. Nucleic Acids Res. 1994;22(22):4742–7.
19. Brown PM, Fox KR. Nucleosome core particles inhibit DNA triple helix formation. Biochem J. 1996;319(2):607–11.
20. Lee JS, Woodsworth ML, Latimer LJP, Morgan AR. Poly(pyrimidine) · poly(purine) synthetic DNAs containing 5-methylcytosine form stable triplexes at neutral pH. Nucleic Acids Res. 1984;12(16):6603–14.
21. Singleton SF, Dervan PB. Influence of pH on the equilibrium association constants for oligodeoxyribonucleotide-directed triple helix formation at single DNA sites. Biochemistry. 1992;31(45):10995–1003.
22. Miller PS, Bhan P, Cushman CD, Trapane TL. Recognition of a guanine-cytosine base pair by 8-oxoadenine. Biochemistry. 1992;31(29):6788–93.
23. Jetter MC, Hobbs FW. 7,8-Dihydro-8-oxoadenine as a replacement for cytosine in the third strand of triple helixes. Triplex formation without hypochromicity. Biochemistry. 1993;32(13):3249–54.
24. Berressem R, Engels JW. 6-Oxocytidine--a novel protonated C-base analog for stable triple helix formation. Nucleic Acids Res. 1995;23(17):3465–72.
25. Krawczyk SH, Milligan JF, Wadwani S, Moulds C, Froehler BC, Matteucci MD. Oligonucleotide-mediated triple helix formation using an N3-protonated deoxycytidine

analog exhibiting pH-independent binding within the physiological range. Proc Natl Acad Sci U S A. 1992;89(9):3761–4.

26. Ishibashi T, Yamakawa H, Wang Q, Tsukahara S, Takai K, Maruyama T, Takaku H. Triple helix formation with oligodeoxyribonucleotides containing 8-oxo-2′-deoxyadenosine and 2′-modified nucleoside derivatives. Nucleic Acids Symp Ser. 1995;34:127–8 (Twentysecond Symposium on Nucleic Acids Chemistry, 1995).
27. Soliva R, Garcia RG, Bias JR, Eritjal R, Asensio JL, Gonzalez C, Luque FJ, Orozco M. DNA-triplex stabilizing properties of 8-aminoguanine. Nucleic Acids Res. 2000;28(22):4531–9.
28. Froehler BC, Wadwani S, Terhorst TJ, Gerrard SR. Oligodeoxyribonucleotides containing C-5 propyne analogs of 2′-deoxyuridine and 2′-deoxycytidine. Tetrahedron Lett. 1992;33(37): 5307–10.
29. Bijapur J, Keppler MD, Bergqvist S, Brown T, Fox KR. 5-(1-propargylamino)-2′-deoxyuridine (UP): a novel thymidine analogue for generating DNA triplexes with increased stability. Nucleic Acids Res. 1999;27(8):1802–9.
30. Sollogoub M, Darby RAJ, Cuenoud B, Brown T, Fox KR. Stable DNA triple helix formation using oligonucleotides containing 2′-aminoethoxy,5-propargylamino-U. Biochemistry. 2002;41(23):7224–31.
31. Osborne SD, Powers VEC, Rusling DA, Lack O, Fox KR, Brown T. Selectivity and affinity of triplex-forming oligonucleotides containing 2′-aminoethoxy-5-(3-aminoprop-1-ynyl)uridine for recognizing AT base pairs in duplex DNA. Nucleic Acids Res. 2004;32(15):4439–47.
32. Rusling DA, Rachwal PA, Brown T, Fox KR. The stability of triplex DNA is affected by the stability of the underlying duplex. Biophys Chem. 2009;145(2-3):105–10.
33. Inoue H, Hayase Y, Imura A, Iwai S, Miura K, Ohtsuka E. Synthesis and hybridization studies on two complementary nona(2′-O-methyl)ribonucleotides. Nucleic Acids Res. 1987;15(15): 6131–48.
34. Shimizu M, Konishi A, Shimada Y, Inoue H, Ohtsuka E. Oligo(2′-O-methyl)ribonucleotides. Effective probes for duplex DNA. FEBS Lett. 1992;302(2):155–8.
35. Blommers MJJ, Natt F, Jahnke W, Cuenoud B. Dual recognition of double-stranded DNA by 2′-aminoethoxy-modified oligonucleotides: the solution structure of an intramolecular triplex obtained by NMR spectroscopy. Biochemistry. 1998;37(51):17714–25.
36. Cuenoud B, Casset F, Huesken D, Natt F, Wolf RM, Altmann K-H, Martin P, Moser HE. Dual recognition of double-stranded DNA by 2′-aminoethoxy-modified oligonucleotides. Angew Chem Int Ed. 1998;37(9):1288–91.
37. Puri N, Majumdar A, Cuenoud B, Natt F, Martin P, Boyd A, Miller PS, Seidman MM. Targeted gene knockout by 2′-O-aminoethyl modified triplex forming oligonucleotides. J Biol Chem. 2001;276(31):28991–8.
38. Xodo L, Alunni-Fabbroni M, Manzini G, Quadrifoglio F. Pyrimidine phosphorothioate oligonucleotides form triple-stranded helices and promote transcription inhibition. Nucleic Acids Res. 1994;22(16):3322–30.
39. Escude C, Giovannangeli C, Sun J-S, Lloyd DH, Chen J-K, Gryaznov SM, Garestier T, Helene C. Stable triple helixes formed by oligonucleotide N3′ → P5′ phosphoramidates inhibit transcription elongation. Proc Natl Acad Sci U S A. 1996;93(9):4365–9.
40. Lacroix L, Arimondo PB, Takasugi M, Helene C, Mergny J-L. Pyrimidine morpholino oligonucleotides form a stable triple helix in the absence of magnesium ions. Biochem Biophys Res Commun. 2000;270(2):363–9.
41. Michel T, Debart F, Vasseur J-J, Geinguenaud F, Taillandier E. FTIR and UV spectroscopy studies of triplex formation between α-oligonucleotides with non-ionic phosphoramidate linkages and DNA targets. J Biomol Struct Dyn. 2003;21(3):435–45.
42. Nielsen PE, Egholm M, Berg RH, Buchardt O. Sequence-selective recognition of DNA by strand displacement with a thymine-substituted polyamide. Science. 1991;254(5037):1497–500.
43. Demidov VV, Potaman VN, Frank-Kamenetskii MD, Egholm M, Buchard O, Sonnichsen SH, Nielsen PE. Stability of peptide nucleic acids in human serum and cellular extracts. Biochem Pharmacol. 1994;48(6):1310–3.

44. Nielsen PE. Targeted gene repair facilitated by peptide nucleic acids (PNA). Chembiochem. 2010;11(15):2073–6.
45. Havre PA, Glazer PM. Targeted mutagenesis of simian virus 40 DNA mediated by a triple helix-forming oligonucleotide. J Virol. 1993;67(12):7324–31.
46. Havre PA, Gunther EJ, Gasparro FP, Glazer PM. Targeted mutagenesis of DNA using triple helix-forming oligonucleotides linked to psoralen. Proc Natl Acad Sci U S A. 1993;90(16):7879–83.
47. Gasparro FP, Havre PA, Olack GA, Gunther EJ, Glazer PM. Site-specific targeting of psoralen photoadducts with a triple helix-forming oligonucleotide: characterization of psoralen monoadduct and crosslink formation. Nucleic Acids Res. 1994;22(14):2845–52.
48. Sandor Z, Bredberg A. Repair of triple helix directed psoralen adducts in human cells. Nucleic Acids Res. 1994;22(11):2051–6.
49. Takasugi M, Guendouz A, Chassignol M, Decout JL, Lhomme J, Thuong NT, Helene C. Sequence-specific photo-induced cross-linking of the two strands of double-helical DNA by a psoralen covalently linked to a triple helix-forming oligonucleotide. Proc Natl Acad Sci U S A. 1991;88(13):5602–6.
50. Sancar A, Tang M-S. Nucleotide excision repair. Photochem Photobiol. 1993;57(5):905–21.
51. Wood RD, Aboussekhra A, Biggerstaff M, Jones CJ, O'Donovan A, Shivji MK, Szymkowski DE. Nucleotide excision repair of DNA by mammalian cell extracts and purified proteins. Cold Spring Harb Symp Quant Biol. 1993;58:625–32.
52. Wang G, Levy DD, Seidman MM, Glazer PM. Targeted mutagenesis in mammalian cells mediated by intracellular triple helix formation. Mol Cell Biol. 1995;15(3):1759–68.
53. Seidman MM, Glazer PM. The potential for gene repair via triple helix formation. J Clin Invest. 2003;112(4):487–94.
54. Majumdar A, Khorlin A, Dyatkina N, Lin FLM, Powell J, Liu J, Fei Z, Khripine Y, Watanabe KA, George J, Glazer PM, Seidman MM. Targeted gene knockout mediated by triple helix forming oligonucleotides. Nat Genet. 1998;20(2):212–4.
55. Vasquez KM, Wang G, Havre PA, Glazer PM. Chromosomal mutations induced by triplex-forming oligonucleotides in mammalian cells. Nucleic Acids Res. 1999;27(4):1176–81.
56. Vasquez KM, Marburger K, Intody Z, Wilson JH. Manipulating the mammalian genome by homologous recombination. Proc Natl Acad Sci U S A. 2001;98(15):8403–10.
57. Vos JMH, Hanawalt PC. DNA interstrand cross-links promote chromosomal integration of a selected gene in human cells. Mol Cell Biol. 1989;9(7):2897–905.
58. Bhattacharyya NP, Maher VM, McCormick JJ. Intrachromosomal homologous recombination in human cells which differ in nucleotide excision-repair capacity. Mutat Res. 1990;234(1):31–41.
59. Tsujimura T, Maher VM, Godwin AR, Liskay RM, McCormick JJ. Frequency of intrachromosomal homologous recombination induced by UV radiation in normally repairing and excision repair-deficient human cells. Proc Natl Acad Sci U S A. 1990;87(4):1566–70.
60. Faruqi AF, Seidman MM, Segal DJ, Carroll D, Glazer PM. Recombination induced by triple-helix-targeted DNA damage in mammalian cells. Mol Cell Biol. 1996;16(12):6820–8.
61. Faruqi AF, Datta HJ, Carroll D, Seidman MM, Glazer PM. Triple-helix formation induces recombination in mammalian cells via a nucleotide excision repair-dependent pathway. Mol Cell Biol. 2000;20(3):990–1000.
62. Luo Z, Macris MA, Faruqi AF, Glazer PM. High-frequency intrachromosomal gene conversion induced by triplex-forming oligonucleotides microinjected into mouse cells. Proc Natl Acad Sci U S A. 2000;97(16):9003–8.
63. Vasquez KM, Christensen J, Li L, Finch RA, Glazer PM. Human XPA and RPA DNA repair proteins participate in specific recognition of triplex-induced helical distortions. Proc Natl Acad Sci U S A. 2002;99(9):5848–53.
64. Knauert MP, Kalish JM, Hegan DC, Glazer PM. Triplex-stimulated intermolecular recombination at a single-copy genomic target. Mol Ther. 2006;14(3):392–400.
65. Chan PP, Lin M, Faruqi AF, Powell J, Seidman MM, Glazer PM. Targeted correction of an episomal gene in mammalian cells by a short DNA fragment tethered to a triplex-forming oligonucleotide. J Biol Chem. 1999;274(17):11541–8.

66. Majumdar A, Muniandy PA, Liu J, Liu J-L, Liu S-T, Cuenoud B, Seidman MM. Targeted gene knock in and sequence modulation mediated by a psoralen-linked triplex-forming oligonucleotide. J Biol Chem. 2008;283(17):11244–52.
67. Schleifman EB, Bindra R, Leif J, del Campo J, Rogers FA, Uchil P, Kutsch O, Shultz LD, Kumar P, Greiner DL, Glazer PM. Targeted disruption of the ccr5 gene in human hematopoietic stem cells stimulated by peptide nucleic acids. Chem Biol. 2011;18(9):1189–98.
68. Lonkar P, Kim K-H, Kuan JY, Chin JY, Rogers FA, Knauert MP, Kole R, Nielsen PE, Glazer PM. Targeted correction of a thalassemia-associated β-globin mutation induced by pseudo-complementary peptide nucleic acids. Nucleic Acids Res. 2009;37(11):3635–44.
69. Chin JY, Kuan JY, Lonkar PS, Krause DS, Seidman MM, Peterson KR, Nielsen PE, Kole R, Glazer PM. Correction of a splice-site mutation in the β-globin gene stimulated by triplex-forming peptide nucleic acids. Proc Natl Acad Sci U S A. 2008;105(36):13514–9. S13514/1-S13514/7.
70. Egholm M, Christensen L, Dueholm KL, Buchardt O, Coull J, Nielsen PE. Efficient pH-independent sequence-specific DNA binding by pseudoisocytosine-containing bis-PNA. Nucleic Acids Res. 1995;23(2):217–22.
71. Nielsen PE, Egholm M, Buchardt O. Evidence for (PNA)2/DNA triplex structure upon binding of PNA to dsDNA by strand displacement. J Mol Recognit. 1994;7(3):165–70.
72. Faruoi AF, Egholm M, Glazer PM. Peptide nucleic acid-targeted mutagenesis of a chromosomal gene in mouse cells. Proc Natl Acad Sci U S A. 1998;95(4):1398–403.
73. Rogers FA, Vasquez KM, Egholm M, Glazer PM. Site-directed recombination via bifunctional PNA-DNA conjugates. Proc Natl Acad Sci U S A. 2002;99(26):16695–700.
74. Wang G, Xu X, Pace B, Dean DA, Glazer PM, Chan P, Goodman SR, Shokolenko I. Peptide nucleic acid (PNA) binding-mediated induction of human γ-globin gene expression. Nucleic Acids Res. 1999;27(13):2806–13.
75. Kim K-H, Nielsen PE, Glazer PM. Site-Specific Gene Modification by PNAs Conjugated to Psoralen. Biochemistry. 2006;45(1):314–23.
76. Kim K-H, Nielsen PE, Glazer PM. Site-directed gene mutation at mixed sequence targets by psoralen-conjugated pseudo-complementary peptide nucleic acids. Nucleic Acids Res. 2007;35(22):7604–13.
77. Rogers FA, Lin SS, Hegan DC, Krause DS, Glazer PM. Targeted Gene Modification of Hematopoietic Progenitor Cells in Mice Following Systemic Administration of a PNA-peptide Conjugate. Mol Ther. 2012;20(1):109–18.
78. McNeer NA, Chin JY, Schleifman EB, Fields RJ, Glazer PM, Saltzman WM. Nanoparticles Deliver Triplex-forming PNAs for Site-specific Genomic Recombination in CD34+ Human Hematopoietic Progenitors. Mol Ther. 2011;19(1):172–80.
79. Blum JS, Saltzman WM. High loading efficiency and tunable release of plasmid DNA encapsulated in submicron particles fabricated from PLGA conjugated with poly-L-lysine. J Control Release. 2008;129(1):66–72.
80. Bahal R, Quijano E, McNeer NA, Liu Y, Bhunia DC, Lopez-Giraldez F, Fields RJ, Saltzman WM, Ly DH, Glazer PM. Single-Stranded γPNAs for In Vivo Site-Specific Genome Editing via Watson-Crick Recognition. Curr Gene Ther. 2014;14(5):331–42.
81. Rapireddy S, Bahal R, Ly DH. Strand Invasion of Mixed-Sequence, Double-Helical B-DNA by γ-Peptide Nucleic Acids Containing G-Clamp Nucleobases under Physiological Conditions. Biochemistry. 2011;50(19):3913–8.
82. Dragulescu-Andrasi A, Rapireddy S, Frezza BM, Gayathri C, Gil RR, Ly DH. A Simple γ-Backbone Modification Preorganizes Peptide Nucleic Acid into a Helical Structure. J Am Chem Soc. 2006;128(31):10258–67.
83. He G, Rapireddy S, Bahal R, Sahu B, Ly DH. Strand Invasion of Extended, Mixed-Sequence B-DNA by γPNAs. J Am Chem Soc. 2009;131(34):12088–90.
84. Sahu B, Sacui I, Rapireddy S, Zanotti KJ, Bahal R, Armitage BA, Ly DH. Synthesis and characterization of conformationally preorganized, (R)-diethylene glycol-containing γ-peptide nucleic acids with superior hybridization properties and water solubility. J Org Chem. 2011;76(14):5614–27.

85. Bahal R, Sahu B, Rapireddy S, Lee C-M, Ly DH. Sequence-unrestricted, Watson-Crick recognition of double helical B-DNA by (R)-MiniPEG-γPNAs. Chembiochem. 2012;13(1):56–60.
86. Svasti S, Suwanmanee T, Fucharoen S, Moulton HM, Nelson MH, Maeda N, Smithies O, Kole R. RNA repair restores hemoglobin expression in IVS2-654 thalassemic mice. Proc Natl Acad Sci U S A. 2009;106(4):1205–10.
87. Sazani P, Gemignani F, Kang S-H, Maier MA, Manoharan M, Persmark M, Bortner D, Kole R. Systemically delivered antisense oligomers upregulate gene expression in mouse tissues. Nat Biotechnol. 2002;20(12):1228–33.

Genome Editing by Aptamer-Guided Gene Targeting (AGT)

Patrick Ruff and Francesca Storici

Abstract DNA aptamers are sequences of DNA that because of their unique secondary structure are capable of binding to a specific target. Aptamer technology has only recently been applied to gene correction. The effectiveness of using aptamers for gene targeting comes from their versatility, as aptamers can be used in conjunction with currently existing genome modification systems. Here we describe how DNA aptamers can be exploited to increase donor DNA availability, and thus promote the transfer of genetic information from a donor DNA molecule to a desired chromosomal locus. Although still in its infancy compared to other more well-characterized systems, aptamer-guided gene targeting (AGT) offers a new direction to the field of genetic engineering.

Keywords Aptamer-guided gene targeting (AGT) • Aptamer • I-SceI • Gene correction • Genetic engineering • DNA oligonucleotides • Donor DNA • Yeast • Human cells

Introduction

DNA double-strand breaks (DSBs) are a form of DNA damage, which, if not repaired properly, can cause mutations, genome rearrangements, or even be lethal to the cell. The cell must repair this damage, even if repair leads to mutation at the site of the damage. Taking advantage of this aspect of cellular repair machinery, site-specific "homing" endonucleases can be used as tools to introduce genome modifications. The process requires the endonuclease to generate the DSB as well as a donor or targeting DNA sequence that can be used as a template to repair the

P. Ruff, Ph.D.
Department of Microbiology and Immunology, Columbia University Medical Center, New York, NY, USA

F. Storici, Ph.D. (✉)
School of Biology, Georgia Institute of Technology, 310 Ferst Drive, Atlanta, GA 30332, USA
e-mail: storici@gatech.edu

T. Cathomen et al. (eds.), *Genome Editing*, Advances in Experimental Medicine and Biology 895, DOI 10.1007/978-1-4939-3509-3_7

DSB. Generation of a DSB at the desired locus increases gene targeting frequency several orders of magnitude and has been shown to be effective in bacteria [1], yeast [2], plants [3], fruit flies [4], mice [5], human embryonic stem cells [6], and many other cell types. Since naturally found homing endonucleases target only a specific recognition sequence, it became important for researchers to develop a way of making DSBs at other desired loci. Several strategies have been adopted including modifying naturally occurring endonucleases such as Cre [7] or Flp [8], generating modular DNA-binding proteins fused to the non-specific cleavage domain from FokI (ZFNs [9, 10] and later TALENs [11, 12]), and the recently discovered RNA-guided clustered regularly interspaced short palindromic repeat (CRISPR)/Cas9 system [13]. The advances made in generating site-specific DSBs modularly are quite impressive and these authors would refer you to other chapters in this book for understanding better how they work and their potential off-target effects.

Despite the advances in generating the site-specific DSBs needed for highly efficient gene targeting, there has been less focus on the other essential component for gene targeting: the donor DNA necessary to make the desired modification. To address the problem of donor DNA availability, we developed a novel gene targeting approach, aptamer-guided gene targeting, AGT, in which we bound the homing endonuclease I-SceI by a DNA aptamer fused to the donor DNA of choice, to target the donor DNA to a desired genetic locus located next to an I-SceI cut site [14]. DNA aptamers are sequences of DNA that are able to bind to a specific target with high affinity because of their unique secondary structure [15]. The AGT approach increases the efficiency of gene targeting by guiding an exogenous donor DNA into the vicinity of the site targeted for genetic modification. By tethering the exogenous donor DNA to I-SceI, which recognizes an 18-bp sequence and generates a DSB [16], we achieved targeted delivery of exogenous donor DNA to the site of the I-SceI DSB in different genomic locations in yeast and human cells, facilitating gene correction by the donor DNA [14]. DNA aptamers were selected for binding to the I-SceI protein using a variant of capillary electrophoresis systematic evolution of ligands by exponential enrichment (CE-SELEX) called "Non-SELEX" [17]. By utilizing DNA oligodeoxyribonucleotides (oligos) that contained the I-SceI aptamer sequence as well as homology to repair the I-SceI DSB and correct a target gene, we were able to increase gene targeting frequencies up to 32-fold over a non-binding control in yeast and up to 16-fold over a non-binding control in human cells [14]. Our strategy offers a novel way to increase gene targeting efficiency and represents the first study to use aptamers in the context of gene correction. The general model for how AGT works is represented in Fig. 1.

Fig. 1 (continued) or in the nucleus. (**c**) I-SceI drives the bifunctional oligo to the targeted locus containing the I-SceI site, and (**d**) generates a DSB at the I-SceI site. (**e**) Resection of the 5′ ends of the DSB gives rise to single-stranded 3′ DNA tails. (**f**) The 3′ tail of the bifunctional oligo anneals to its complementary DNA sequence on the targeted DNA, and after the non-homologous sequence is clipped, (**g**) DNA synthesis proceeds on the template sequence. (**h**) After unwinding of the bifunctional oligo, a second annealing step occurs between the extended 3′ end and the other 3′ end generated from the DSB. (**i**) Further processing, gap-filling DNA synthesis, and subsequent ligation complete repair and modification of the target locus. This Figure is reproduced from Ruff et al., 2014 [14]

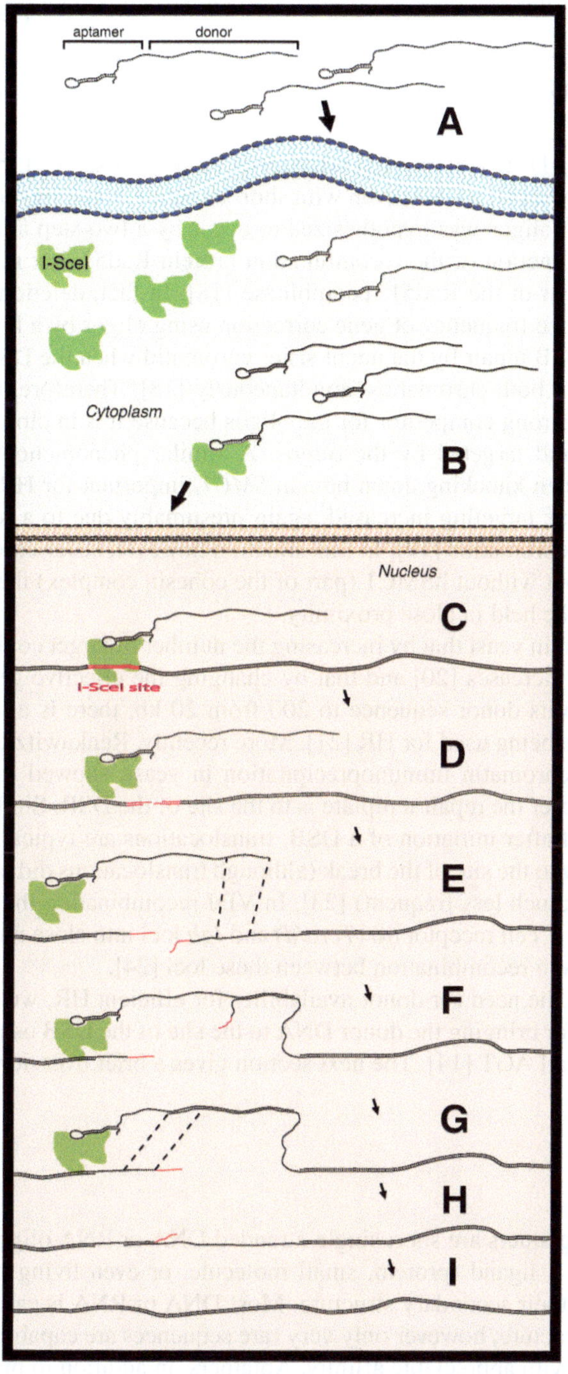

Fig. 1 Aptamer-guided gene targeting model. (**a**) Bifunctional targeting oligos containing the A7 aptamer at the 5′ end along with a region of homology to restore the function of a defective gene of interest are transformed/transfected into the cell. The I-SceI endonuclease is produced from the Fig. 1 (continued) chromosome (yeast cells) or from a transfected expression vector (human cells). (**b**) The A7 aptamer then binds to the I-SceI protein, either in the cytoplasm (shown here)

AGT Origins

Using an I-SceI-DSB system in yeast, Storici et al. [18] showed that DNA oligos could efficiently repair a DSB, even with short targeting homology to the break site. Repair by these oligos was hypothesized to occur by a two-step annealing process reliant on the function of the recombination protein Rad52, but not on the strand invasion function of the Rad51 recombinase [18]. In fact, deletion of the *RAD51* gene increases the frequency of gene correction using oligos by a few fold, because it suppresses DSB repair by the uncut sister chromatid when the DSB occurs in G2 cells and not on both chromatids simultaneously [18]. Therefore, the uncut sister chromatid is a strong competitor for the oligos because it is in close vicinity to the broken chromatid targeted by the oligos. A similar phenomenon was shown in human cells when knocking down human *SMC1*, important for HR between sister chromatids, gene targeting increased, again presumably due to a reduction in HR between sister chromatids [19]. In this situation however, sister chromatid HR was still possible, but without hSMC1 (part of the cohesin complex) the sister chromatids would not be held in close proximity.

It was shown in yeast that by increasing the number of target copies gene correction efficiency increases [20] and that by changing the effective genomic distance from a DSB to its donor sequence to 200 from 20 kb, there is an increase in the donor sequence being used for HR [21]. More recently, Renkawitz et al. [22], using time-resolved chromatin immunoprecipitation in yeast, showed that HR is more efficient the closer the repair template is to the site of the DSB. Similarly, in mice it was shown that after initiation of a DSB, translocations are typically found within close proximity to the site of the break (although translocations did occur at distance sites, this was much less frequent) [23]. In VDJ recombination the RAG recombinase brings the T cell receptor αδ (*Tcra/d*) and *Igh* loci into close nuclear proximity and contributes to recombination between these loci [24].

Based upon the need for donor availability for efficient HR, we came up with a novel method for bringing the donor DNA to the site of the DSB using DNA aptamers which we call AGT [14]. The next section gives a brief overview of aptamers.

Aptamers

Nucleic acid aptamers are short single-stranded DNA or RNA oligos that are capable of binding a ligand (protein, small molecule, or even living cells) with high affinity due to their secondary structure. Most DNA or RNA is capable of forming a secondary structure, however only very rare sequences are capable of binding to a specific target with appreciable affinity. Aptamers, in addition to binding with high affinity, also bind with high specificity, such as for an aptamer selected to bind theophylline but not caffeine, the two molecules differing only by a single methyl group [25]. Aptamers are sometimes referred to as artificial antibodies, but aptamers have several advantages over antibodies, including ease and low cost of production which does not involve animals. Aptamers are less immunogenic than antibodies and are already being used as a therapeutic for humans [26].

Aptamers are obtained by rigorous selection, in which aptamers are "evolved" from pools of random DNA or RNA, leaving few (if any) sequences capable of binding the target out of a high number (usually 10^{14} or more) of starting sequences [27]. The random library is typically flanked by fixed primer regions such that each oligo in the pool contains the sequence 5′-primer1-N_{20-60}-primer2(reverse complement)-3′, where N is a random base [28]. The primers are used to amplify the library after selection by PCR. The process to generate aptamers by in vitro selection was developed by the Szostak [29] and Gold [30] groups independently in 1990 and the process has become known as systematic evolution of ligands by exponential enrichment (SELEX). The SELEX procedure involves the use of the random library of DNA/RNA sequences being incubated with the target, followed by a partitioning step to remove unbound sequences, then followed by an elution step to recover the binding sequences, and then an amplification step to generate a library of sequences enriched for binding (see Fig. 2 for a schematic). Over the years, several variants of

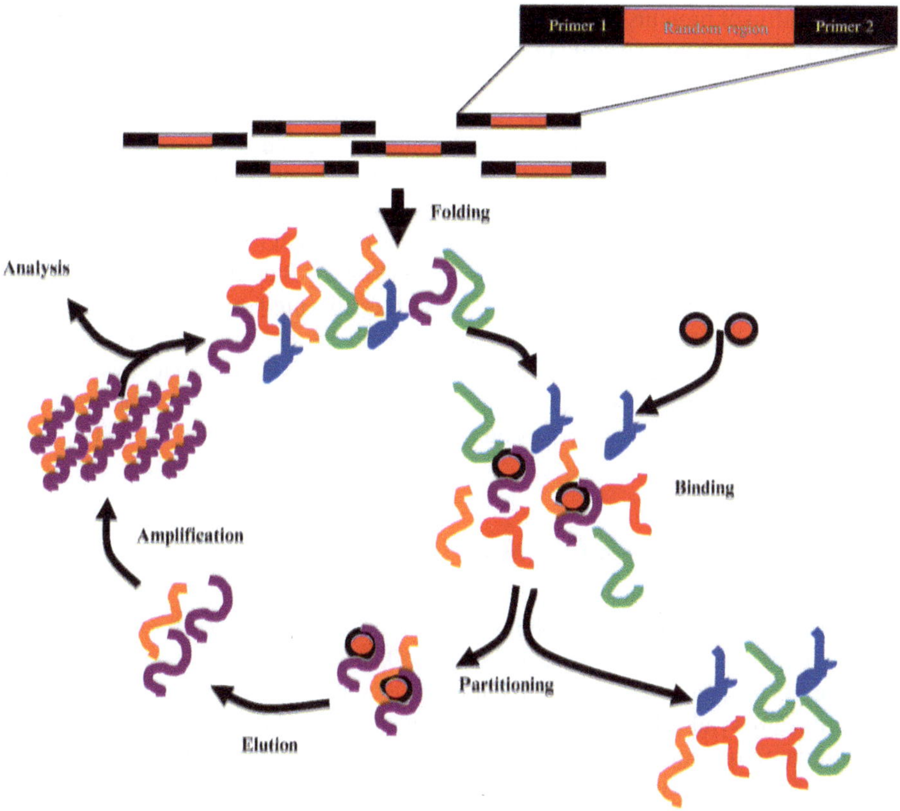

Fig. 2 The Systematic Evolution of Ligands by EXponential enrichment (SELEX) procedure. The ligand of interest, shown as a *red dot*, is incubated with a random library of ssDNA molecules containing a central random sequence flanked by known sequences of primers to be used in PCR amplification. Sequences that form a suitable secondary structure, shown as the *purple* and *orange* DNA sequences, bind to the ligand. These sequences are partitioned from the weaker binders and are eluted. After amplification of these binders, the process is repeated

SELEX have arisen. One variant of SELEX using capillary electrophoresis (CE) allows for SELEX to be performed in a much shorter amount of time due to much more efficient partitioning and the prevention of aptamers binding to the ligand support (the ligand flows freely in buffer). In as little as one round of selection [31], and almost always less than five, strong binding highly specific aptamers may be selected, as opposed to traditional SELEX which typically takes ten or more rounds of selection. CE-SELEX generated aptamers can have nM and even pM level disassociation constants [32]. The efficiency and ease of generating aptamers using CE-SELEX led us to use this methodology when selecting for an aptamer to our chosen site-specific endonuclease I-SceI.

The Site-Specific Endonuclease I-SceI

DNA binding proteins exist in all forms of life, but despite their prevalence there are only a handful of proteins evolved that are capable of binding to and cleaving double-stranded DNA in a site-specific manner. Those restriction endonucleases capable of achieving site-specific DNA DSBs are known as "homing" endonucleases, and they have high specificity due to a long recognition sequence (12–40 bp) [33]. Homing endonucleases have been studied since the late 1970s, and one of the first homing endonucleases studied was originally called "Omega" which later became known as I-SceI [34]. The I-SceI endonuclease's natural function is to recognize a nonsymmetrical 18-bp sequence in yeast mitochondria of 5′ TAG GGA TAA CAG GGT AAT 3′ on an intron-less allele and generate a DNA DSB at that location, propagating the intron containing allele and overwriting the previously intron-less allele through homologous recombination and gene conversion [35]. Since its discovery, I-SceI has been used and continues to be used in almost every model system from bacteria to human cells to model DSB damage and repair. Due to the widespread use of I-SceI, we chose to use I-SceI as our site-specific endonuclease for AGT.

Single-Stranded DNA Oligos as Donor DNA

Synthetic single-stranded DNA oligos are short sequences of DNA typically 90 nt or less that are often used in genome editing. Oligos are used to modify a specific sequence in the genome by containing homology to the targeted sequence. Oligos can be chemically synthesized quickly and cheaply and can achieve efficient gene editing at a similar frequency to donors with longer homology lengths, including donor plasmids or PCR products [36]. Gene correction by oligos can be obtained even with homology to the target locus as low as 30 nucleotides [14, 37]. Oligos have been successfully used for efficient gene targeting in many cell systems [38, 39]. For AGT, we used bipartite oligos in which half of each molecule was used to bind to I-SceI (the aptamer region) and the other half of each molecule was used as a template to repair the DSB by homologous recombination and modify the chosen target site [14].

Characterization of AGT

After designing the AGT system and performing selection of aptamers to I-SceI, a few sequences were identified that bind to I-SceI in vitro. Of these, two of the stronger binders, named aptamer A4 and aptamer A7, were first tested in yeast *Saccharomyces cerevisiae*. The first locus tested was the yeast *trp5* gene, in which a cassette was integrated and gene targeting was measured as the number of Trp^+ transformants that had deleted the cassette and repaired the sequence of the gene. Schemes of the integrated cassettes for the *trp5* locus are shown in Fig. 3A. The first step in the characterization of AGT was to determine at which part of the DNA donor oligo the aptamer sequence should be positioned. Initial testing revealed that the 5′ end of the targeting oligo was much more efficient at gene targeting compared to the 3′ end, presumably because having the homology region of the donor sequence at the 3′ end facilitates the homology search to the target locus.

The next step in characterizing AGT was to test the importance of the primer regions for binding by the aptamer sequence. As discussed above, in aptamer selection, each random DNA sequence is flanked by two fixed primer regions for amplification by PCR (Fig. 2) and often these primer sequences are not necessary for the aptamer to bind to its target [40–42]. Based on this, we removed the primer regions from our selected I-SceI aptamers and at the same time increased the homology length of the targeting DNA. Again, using the *trp5* locus, the aptamers A7 and A4 were tested as oligos A7.TRP5.54 and A4.TRP5.54 with the aptamers being at the 5′ end of the oligos followed directly by 54 bases of homology to *trp5*. Additionally, a control sequence of the same length as the aptamer region but that was not shown to bind to I-SceI, the oligo C.TRP5.54, was used. The weaker binding aptamer A4 showed no significant difference to the non-binding control, whereas the stronger binding aptamer A7 showed a statistically significant seven-fold increase in gene targeting relative to the control [14]. In situations in which there was no I-SceI site but I-SceI was still expressed, or there was an I-SceI site but no I-SceI expression, or with glucose repression of the *GAL1-10* promoter there was no significant difference between the A7.TRP5.54 or the C.TRP5.54 oligos [14].

After seeing that the A7 aptamer could increase gene targeting, it was hypothesized that the aptamer might work even better if the homology length was shorter, as a longer homology sequence might interfere with the aptamer binding to I-SceI. Alternatively, the stimulatory effect provided by the aptamer might be exacerbated using shorter DNA donors. To test whether the I-SceI aptamer was effective on shorter DNA donors, the homology length of the oligos was shortened to 40 bases instead of 54 giving the A7.TRP5.40 and C.TRP5.40 oligos. While the overall level of repair was lower for the shorter oligos, the A7 aptamer even further stimulated gene targeting to a 9.2-fold significant difference over the control (Fig. 3b). It is possible that the aptamer, being at the 5′ end of the molecule, may not always be fully present in the oligos if the oligos are not purified. Indeed, non-purified 100-mer oligos synthesized at a coupling efficiency of 99.5 % contain ~60 % full-length product, with the other 40 % being 5′ truncated oligos [43]. This also means that the A7.TRP5.40 oligonucleotide might be more effective at gene targeting because,

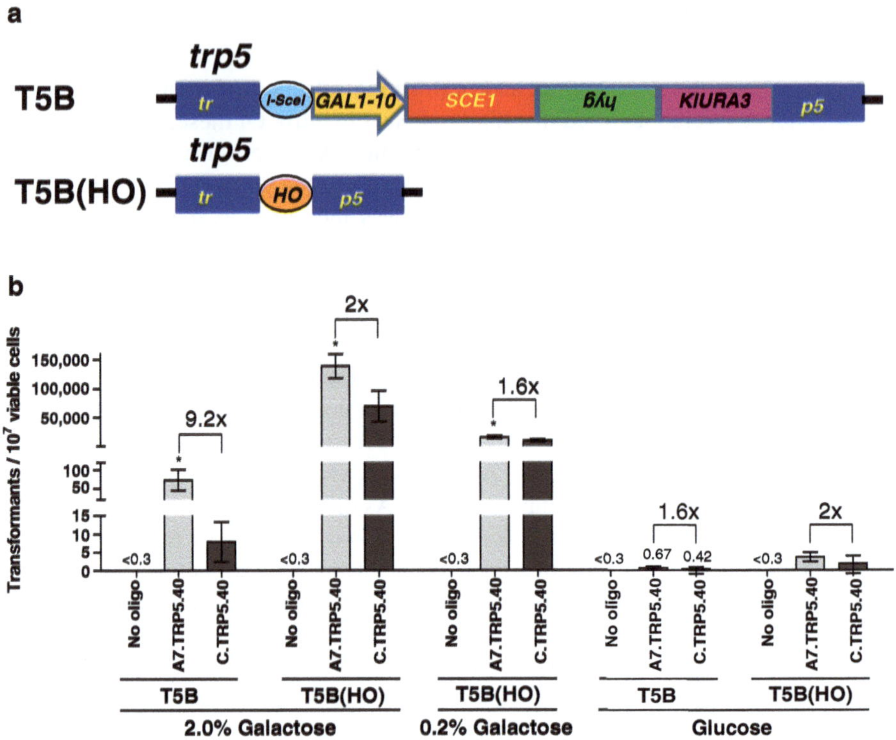

Fig. 3 AGT stimulates gene targeting in yeast. (**a**) Scheme of targeted yeast loci. The FRO-155/156 strain, shown as T5B, contains the I-SceI break site (*blue ellipse*), and a cassette with the I-SceI gene *SCE1* under the galactose inducible *GAL1-10* promoter, the hygromycin resistance gene *hyg*, as well as the counterselectable *K.l. URA3* gene in a construct that has been inserted into the *TRP5* gene. Strain HK-225/226, shown as T5B(HO), contains the HO break site (*orange ellipse* with HO) inserted into the *TRP5* gene. (**b**) Frequency of gene correction in yeast using aptamer-containing oligos shown in *light gray* coloring and non-binding control oligos in *dark gray* coloring (X axis), measured by the number of transformants per 10^7 viable cells (Y axis), with no oligo controls for all strains averaged in. Targeting occurred at the *trp5* locus in the I-SceI-containing strain (T5B) or in the HO-containing strain (T5B(HO)) grown on galactose for the induction of I-SceI or HO, or on glucose for the repression of I-SceI or HO. Bars correspond to the mean value and error bars represent 95 % confidence intervals. *Asterisks* denote statistical significant difference between the aptamer-containing oligo and the corresponding non-binding control (* for $p<0.05$), and the fold change in the gene correction frequency is indicated. Part of this Figure is reproduced from Ruff et al., 2014 [14]

being shorter, less sequences are truncated at the 5′ aptamer end. To test this theory, polyacrylamide gel electrophoresis (PAGE) purified oligos were used, testing both the oligos with 54 bases of homology as well as those with 40 bases of homology. For both sets of PAGE-purified oligos, there were statistically significant increases in gene targeting (27-fold for the purified A7.TRP5.54 over the purified C.TRP5.54 and 30-fold for the purified A7.TRP5.40 over the purified C.TRP5.40) [14]. This result suggests the importance of the aptamer region being intact for efficient gene targeting; however the length of the homology may also have an effect since the shorter oligos had a greater increase in gene targeting.

AGT was tested not only at the yeast chromosomal *trp5* locus, but also at the *ade2* and *leu2* genomic loci in yeast, and at the DsRed2 (a red fluorescent protein) gene locus carried on a plasmid in human embryonic kidney (HEK-293) cells. At every locus tested there was a significant increase in gene targeting using the A7 aptamer containing oligos over the control oligos each with 54 bases of homology to the targeted gene (Fig. 3B, Fig. 4A and [14]). Not all aptamer-containing oligos stimulated gene targeting to the same extent. For the DsRed2 locus, the A.Red.40 oligo (the A7 aptamer-containing oligo with 40 bases of homology) had a 6.2-fold increase in gene targeting and the A.Red.30 oligo had a 16-fold increase in gene targeting (Fig. 4A). The variability of AGT at the different loci tested could be due to the different secondary structure each oligo forms. In order to bind I-SceI, the aptamer region must assume a specific secondary structure and it is possible that the long ssDNA within the homology region interferes with this structure. For further analysis of the aptamer secondary structure, see Ruff et al. [14].

In order to further validate the specificity of the I-SceI aptamer in AGT to stimulate gene targeting only in the presence of I-SceI, controls were done in which a DSB was generated either by a nuclease different from I-SceI (experiment using yeast cells) or was generated in vitro by I-SceI (experiment using HEK-293 cells). In yeast, the homing endonuclease HO was used in place of I-SceI to generate a DSB at *trp5* (Fig. 3A). HO is much more efficient than I-SceI in generating a DSB and stimulates gene targeting to higher levels [44]. However, gene correction frequency by the aptamer-containing donor oligo A7.TRP5.40 was much higher than that obtained using the control oligo C.TRP5.40 only when the DSB was induced by I-SceI in the cells (Fig. 3B). Similarly, in human cells the plasmid containing the DsRed2 gene was predigested with I-SceI in vitro overnight. By not expressing I-SceI inside the HEK-293 cells there was no significant difference between the A.Red.40 and A.Red.30 oligos compared to the controls and only a 1.8-fold increase with the A.Red.54 oligo (Fig. 4B). These data demonstrate the potential of AGT to increase donor DNA availability in a specific manner during the gene targeting process.

The Utility of AGT for Gene Targeting

The concept of having a sequence of DNA that is both capable of binding to a target while at the same time capable of repairing a targeted gene is advantageous for gene correction. AGT uses DNA aptamers, which themselves are a relatively new discovery with aptamer selection only being done since the early 1990s [29, 30]. Aptamers have been utilized as biosensors [45, 46] and as therapeutics [47] but not in the context of gene targeting. Our aptamer binds I-*Sce*I and is targeted to a specific

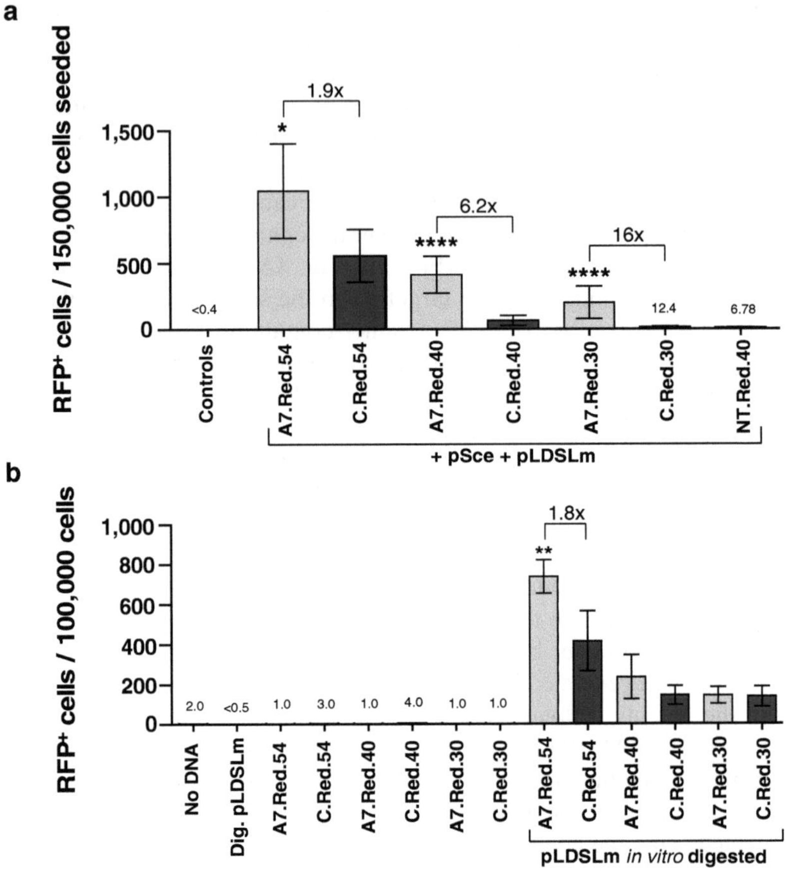

Fig. 4 AGT stimulates gene targeting in humans cells. Transient transfections of HEK-293 cells using an I-SceI expression vector (pSce), a vector (pLDSLm) that contained the DsRed2 gene disrupted with two stop codons and the I-SceI site, and oligos. Negative controls were the HEK-293 cells alone (no DNA, only transfection reagent alone), pSce alone, pLDSLm alone, and the individual oligos alone. (**a**) Hand counts of each transfection were done in HEK-293 cells in lieu of flow cytometry which was over reporting the number of background RFP+ cells for the shorter oligos (see also [14]). The different samples are shown on the X axis and the number of RFP+ cells per 150,000 cells seeded is shown on the Y axis. Negative controls did not show any RFP+ cells. (**b**) Flow cytometry analysis of transfections of the in vitro digested pLDSLm vector. The different samples are shown on the X axis and the number of RFP+ cells per 100,000 cells is shown on the Y axis. Negative controls were the cells alone (no DNA), the digested vector alone, and the individual oligos alone. Transfections with both the digested vector and an oligo are bracketed. Bars correspond to the mean value and error bars represent 95 % confidence intervals. Asterisks denote statistical significant difference between the aptamer-containing oligo and the corresponding non-binding control (* for $p<0.05$, ** for $p<0.01$, and **** for $p<0.0001$), and the fold change in the gene correction frequency is indicated. This Figure is reproduced from Ruff et al., 2014 [14]

genomic DNA site. This represents a novel gene targeting strategy that lays the foundation for other similar systems to come.

Future Directions for AGT

The AGT system described here provides a proof-of-principle using aptamers for gene targeting, and based upon this initial work further studies to elucidate, validate, and improve the system should follow. For example, there is the limitation we found with the aptamer system whereby different single-stranded homology regions may disrupt the aptamer region's secondary structure to inhibit binding, reducing the efficiency of AGT. Secondly, the aptamer binding affinity may be below optimum. Additionally, the current design for AGT is specific for the I-SceI endonuclease.

There are many potential ways to improve the AGT system. For example, to reduce annealing between the aptamer region and the homology region of the DNA oligos, the donor region could be present as a double-stranded instead of single-stranded sequence. Alternatively, a linker region could be designed between the aptamer and the homology region. The linker could be a DNA sequence unlikely to be bound by either the aptamer or the homology region, or a sulfide linker [48] or a carbon linker [49] between the two nucleic acids. Moreover, the aptamer itself could be improved for enhanced binding to its target. The binding affinity of the A7 aptamer for I-SceI was measured at a modest 3.16 μM, which was approximately 15-fold higher than that of the A4 aptamer (52.49 μM) that had no effect on gene targeting. Although the I-SceI aptamer A7 was selected from a large pool of oligos, it is possible that the strongest binder to I-SceI is yet to be selected. Assuming the aptamer binding affinity could be improved, modifications to the A7 aptamer sequence might lead to a further increase in gene targeting efficiency. From the standpoint of the DNA targeting molecule and oligo stability, which could be a limitation to gene correction, this could be enhanced by changing the chemistry of the oligos. Oligo modifications, such as the use of phosphorothioate linkages, could improve the stability of the donor molecules which could improve gene targeting frequencies. While the proof of AGT was obtained using an aptamer specific to the I-SceI enzyme, it may be possible to obtain an aptamer for FokI (ZFNs and TALENs) or an aptamer to Cas9 (CRISPRs) that could stimulate gene targeting. There is also the potential that AGT might increase gene targeting for aptamer targets not limited to endonucleases but theoretically other DNA binding proteins or proteins involved in recombination. We anticipate that selection of strong binding aptamers to tether the DNA binding factors of choice for use in AGT may become a powerful strategy to augment donor DNA availability and improve genome editing.

Acknowledgment We acknowledge support from the Georgia Cancer Coalition Grant R9028, the NIH Grant R21EB9228, and the Georgia Tech Fund for Innovation in Research and Education GT-FIRE-1021763.

References

1. Nussbaum A, Shalit M, Cohen A. Restriction-stimulated homologous recombination of plasmids by the RecE pathway of Escherichia coli. Genetics. 1992;130(1):37–49.
2. Storici F, Durham CL, Gordenin DA, Resnick MA. Chromosomal site-specific double-strand breaks are efficiently targeted for repair by oligonucleotides in yeast. Proc Natl Acad Sci U S A. 2003;100(25):14994–9.
3. Puchta H, Dujon B, Hohn B. Homologous recombination in plant cells is enhanced by in vivo induction of double strand breaks into DNA by a site-specific endonuclease. Nucleic Acids Res. 1993;21(22):5034–40.
4. Banga SS, Boyd JB. Oligonucleotide-directed site-specific mutagenesis in Drosophila melanogaster. Proc Natl Acad Sci U S A. 1992;89(5):1735–9.
5. Rouet P, Smih F, Jasin M. Expression of a site-specific endonuclease stimulates homologous recombination in mammalian cells. Proc Natl Acad Sci U S A. 1994;91(13):6064–8.
6. Smih F, Rouet P, Romanienko PJ, Jasin M. Double-strand breaks at the target locus stimulate gene targeting in embryonic stem cells. Nucleic Acids Res. 1995;23(24):5012–9.
7. Santoro SW, Schultz PG. Directed evolution of the site specificity of Cre recombinase. Proc Natl Acad Sci U S A. 2002;99(7):4185–90.
8. Voziyanov Y, Konieczka JH, Stewart AF, Jayaram M. Stepwise manipulation of DNA specificity in Flp recombinase: progressively adapting Flp to individual and combinatorial mutations in its target site. J Mol Biol. 2003;326(1):65–76.
9. Palpant NJ, Dudzinski D. Zinc finger nucleases: looking toward translation. Gene Ther. 2013;20(2):121–7.
10. Kim YG, Cha J, Chandrasegaran S. Hybrid restriction enzymes: zinc finger fusions to Fok I cleavage domain. Proc Natl Acad Sci U S A. 1996;93(3):1156–60.
11. Carlson DF, Fahrenkrug SC, Hackett PB. Targeting DNA With Fingers and TALENs. Mol Ther Nucleic Acids. 2012;1, e3.
12. Moscou MJ, Bogdanove AJ. A simple cipher governs DNA recognition by TAL effectors. Science. 2009;326(5959):1501.
13. Cong L, Ran FA, Cox D, Lin S, Barretto R, Habib N, et al. Multiplex genome engineering using CRISPR/Cas systems. Science. 2013;339(6121):819–23.
14. Ruff P, Koh KD, Keskin H, Pai RB, Storici F. Aptamer-guided gene targeting in yeast and human cells. Nucleic Acids Res. 2014;42(7), e61.
15. Ni X, Castanares M, Mukherjee A, Lupold SE. Nucleic acid aptamers: clinical applications and promising new horizons. Curr Med Chem. 2011;18(27):4206–14.
16. Niu Y, Tenney K, Li H, Gimble FS. Engineering variants of the I-SceI homing endonuclease with strand-specific and site-specific DNA-nicking activity. J Mol Biol. 2008;382(1):188–202.
17. Berezovski M, Musheev M, Drabovich A, Krylov SN. Non-SELEX selection of aptamers. J Am Chem Soc. 2006;128(5):1410–1.
18. Storici F, Snipe JR, Chan GK, Gordenin DA, Resnick MA. Conservative repair of a chromosomal double-strand break by single-strand DNA through two steps of annealing. Mol Cell Biol. 2006;26(20):7645–57.
19. Potts PR, Porteus MH, Yu H. Human SMC5/6 complex promotes sister chromatid homologous recombination by recruiting the SMC1/3 cohesin complex to double-strand breaks. EMBO J. 2006;25(14):3377–88.
20. Wilson JH, Leung WY, Bosco G, Dieu D, Haber JE. The frequency of gene targeting in yeast depends on the number of target copies. Proc Natl Acad Sci U S A. 1994;91(1):177–81.
21. Coic E, Martin J, Ryu T, Tay SY, Kondev J, Haber JE. Dynamics of homology searching during gene conversion in Saccharomyces cerevisiae revealed by donor competition. Genetics. 2011;189(4):1225–33.
22. Renkawitz J, Lademann CA, Kalocsay M, Jentsch S. Monitoring homology search during DNA double-strand break repair in vivo. Mol Cell. 2013;50(2):261–72.

23. Roukos V, Voss TC, Schmidt CK, Lee S, Wangsa D, Misteli T. Spatial dynamics of chromosome translocations in living cells. Science. 2013;341(6146):660–4.
24. Rocha PP, Chaumeil J, Skok JA. Molecular biology. Finding the right partner in a 3D genome. Science. 2013;342(6164):1333–4.
25. Jenison RD, Gill SC, Pardi A, Polisky B. High-resolution molecular discrimination by RNA. Science. 1994;263(5152):1425–9.
26. Singerman LJ, Masonson H, Patel M, Adamis AP, Buggage R, Cunningham E, et al. Pegaptanib sodium for neovascular age-related macular degeneration: third-year safety results of the VEGF Inhibition Study in Ocular Neovascularisation (VISION) trial. Br J Ophthalmol. 2008;92(12):1606–11.
27. Hall B, Micheletti JM, Satya P, Ogle K, Pollard J, Ellington AD. Design, synthesis, and amplification of DNA pools for in vitro selection. Curr Protoc Mol Biol. 2009;Chapter 24:Unit 24.2.
28. Sabeti PC, Unrau PJ, Bartel DP. Accessing rare activities from random RNA sequences: the importance of the length of molecules in the starting pool. Chem Biol. 1997;4(10):767–74.
29. Ellington AD, Szostak JW. In vitro selection of RNA molecules that bind specific ligands. Nature. 1990;346(6287):818–22.
30. Tuerk C, Gold L. Systematic evolution of ligands by exponential enrichment: RNA ligands to bacteriophage T4 DNA polymerase. Science. 1990;249(4968):505–10.
31. Berezovski M, Drabovich A, Krylova SM, Musheev M, Okhonin V, Petrov A, et al. Nonequilibrium capillary electrophoresis of equilibrium mixtures: a universal tool for development of aptamers. J Am Chem Soc. 2005;127(9):3165–71.
32. Mosing RK, Mendonsa SD, Bowser MT. Capillary electrophoresis-SELEX selection of aptamers with affinity for HIV-1 reverse transcriptase. Anal Chem. 2005;77(19):6107–12.
33. Belfort M, Roberts RJ. Homing endonucleases: keeping the house in order. Nucleic Acids Res. 1997;25(17):3379–88.
34. Stoddard BL. Homing endonucleases: from microbial genetic invaders to reagents for targeted DNA modification. Structure. 2011;19(1):7–15.
35. Colleaux L, D'Auriol L, Galibert F, Dujon B. Recognition and cleavage site of the intron-encoded omega transposase. Proc Natl Acad Sci U S A. 1988;85(16):6022–6.
36. Wang Z, Zhou ZJ, Liu DP, Huang JD. Double-stranded break can be repaired by single-stranded oligonucleotides via the ATM/ATR pathway in mammalian cells. Oligonucleotides. 2008;18(1):21–32.
37. Chen F, Pruett-Miller SM, Huang Y, Gjoka M, Duda K, Taunton J, et al. High-frequency genome editing using ssDNA oligonucleotides with zinc-finger nucleases. Nat Methods. 2011;8(9):753–5.
38. Olsen PA, Randol M, Krauss S. Implications of cell cycle progression on functional sequence correction by short single-stranded DNA oligonucleotides. Gene Ther. 2005;12(6):546–51.
39. Papaioannou I, Simons JP, Owen JS. Oligonucleotide-directed gene-editing technology: mechanisms and future prospects. Expert Opin Biol Ther. 2012;12(3):329–42.
40. Cowperthwaite MC, Ellington AD. Bioinformatic analysis of the contribution of primer sequences to aptamer structures. J Mol Evol. 2008;67(1):95–102.
41. Pan W, Clawson GA. The shorter the better: reducing fixed primer regions of oligonucleotide libraries for aptamer selection. Molecules. 2009;14(4):1353–69.
42. Legiewicz M, Lozupone C, Knight R, Yarus M. Size, constant sequences, and optimal selection. RNA. 2005;11(11):1701–9.
43. Stafford P, Brun M. Three methods for optimization of cross-laboratory and cross-platform microarray expression data. Nucleic Acids Res. 2007;35(10), e72.
44. Plessis A, Perrin A, Haber JE, Dujon B. Site-specific recombination determined by I-SceI, a mitochondrial group I intron-encoded endonuclease expressed in the yeast nucleus. Genetics. 1992;130(3):451–60.
45. Cheng AK, Ge B, Yu HZ. Aptamer-based biosensors for label-free voltammetric detection of lysozyme. Anal Chem. 2007;79(14):5158–64.
46. Stojanovic MN, Kolpashchikov DM. Modular aptameric sensors. J Am Chem Soc. 2004;126(30):9266–70.

47. Barbas AS, Mi J, Clary BM, White RR. Aptamer applications for targeted cancer therapy. Future Oncol. 2010;6(7):1117–26.
48. Chu TC, Twu KY, Ellington AD, Levy M. Aptamer mediated siRNA delivery. Nucleic Acids Res. 2006;34(10), e73.
49. Zhou J, Neff CP, Swiderski P, Li H, Smith DD, Aboellail T, et al. Functional in vivo delivery of multiplexed anti-HIV-1 siRNAs via a chemically synthesized aptamer with a sticky bridge. Mol Ther. 2013;21(1):192–200.

Stimulation of AAV Gene Editing via DSB Repair

Angela M. Mitchell, Rachel Moser, Richard Jude Samulski, and Matthew Louis Hirsch

Abstract Recent advancements in mammalian genome editing technologies have demonstrated precise genetic manipulations at the chromosomal level at efficiencies relevant for disease therapy. In fact, zinc-finger nucleases (ZFNs) that induce deletions in the HIV CCR5 receptor in patient T cells ex vivo have demonstrated promise upon treated cell infusion in the clinic. In these applications, adenoviral delivery vectors were employed however; there is growing popularity for the use of adeno-associated virus (AAV) gene delivery which has been used in over 100 clinical trials without any vector-related toxicity. This review chapter summarizes the development of AAV for clinical gene therapy, the early observations of AAV gene targeting, and the current status of AAV vectors for gene editing via site specific DNA double strand break repair. In addition, the remaining obstacles towards the combination of AAV vectorology and site-specific endonucleases for genetic engineering are discussed.

Keywords AAV • Gene therapy • Zinc-finger nuclease • Cas9/CRISPR • TALEN • DNA repair • Homologous recombination • Gene editing • Gene targeting

---

A.M. Mitchell, Ph.D.
Gene Therapy Center, University of North Carolina at Chapel Hill, Chapel Hill, NC, USA
e-mail: angelamm@princeton.edu

R. Moser, B.A.
Department of Ophthalmology, University of North Carolina at Chapel Hill, Chapel Hill, NC, USA
e-mail: rmoser@med.unc.edu

R.J. Samulski, Ph.D.
Department of Pharmacology, Gene Therapy Center, University of North Carolina at Chapel Hill, Chapel Hill, NC, USA
e-mail: rjs@med.unc.edu

M.L. Hirsch, Ph.D. (✉)
Gene Therapy Center, University of North Carolina at Chapel Hill, Chapel Hill, NC, USA

Department of Ophthalmology, University of North Carolina at Chapel Hill, Chapel Hill, NC, USA
e-mail: mhirsch@email.unc.edu

T. Cathomen et al. (eds.), *Genome Editing*, Advances in Experimental Medicine and Biology 895, DOI 10.1007/978-1-4939-3509-3_8

The Vectorization of Adeno-Associated Virus

Adeno-associated virus (AAV) was discovered in 1965 as contaminating viral particles in adenoviral stocks and was determined to be a new species of virus, which depends on adenovirus to complete its replication cycle [1]. AAV is a member of the parvovirus family and so consists of a single-stranded DNA genome contained within a protein capsid and is further classified as a dependovirus as it cannot replicate without the presence of a helper virus [2]. AAV infection is common and most people are seropositive for AAV by a very young age [3]; however, no pathogenicity has been conclusively linked to AAV.

The wild type AAV genome is 4.7 kb in length and encodes at least two genes, *Rep* and *Cap*, flanked by inverted terminal repeats (ITRs), which are the only viral elements required in *cis* for pack aging of the genome [4]. The *Rep* gene encodes the four non-structural Rep proteins required for replication and packaging of the AAV genome [5, 6], while the *Cap* gene encodes the VP1, VP2, and VP3 proteins that make up the icosahedral capsid as well as the assembly activating protein (AAP) necessary for assembly of the capsid and packaging of the genome [7]. Numerous AAV serotypes have been discovered that have between 49 and 99 % identity in their capsid amino acid sequence [8] and differ in several aspects of their biology including their tropism and transduction kinetics [9, 10]. For use as gene therapy vectors, the viral genes are removed from the genome and replaced with a transgene cassette [4]. The production and characterization of recombinant AAV (rAAV) vectors have been thoroughly reviewed previously [11].

AAV begins its transduction pathway by binding to a primary glycan receptor, such as heparan sulfate, and then to a secondary receptor, such as a growth factor receptor or integrin. The specific receptors bound by AAV depend on the serotype and the cell type being transduced (more thoroughly reviewed in [12]). AAV then enters the cell through receptor-mediated endocytosis [13] and traffics through endolysosomal pathways to the perinuclear microtubule-organizing center [14]. The virions are retained at this region both pre- and post-endosomal escape facilitated by the phospholipase domain present in the N-terminal region of VP1 [14–16]. After endosomal escape, intact AAV particles travel to the nucleus where uncoating of the genome occurs [17]. As AAV is a single-stranded virus, second-strand DNA synthesis must occur before the genome can be transcriptionally active, and the virus depends on cellular polymerases to synthesize the second strand [18]. In the absence of a helper virus, the wild-type genome is transcriptionally repressed by the actions of the Rep proteins, allowing the virus to remain latent until a helper-virus is present to facilitate the completion of AAV's lifecycle [19]. For rAAV, the use of constitutively active promoters allows for the expression of the transgene cassette following second-strand synthesis.

Episomal rAAV genomes form monomer episomes and concatemers that show long-term persistence, although they do not replicate [20–22]. In addition, AAV2 genomes can site specifically integrate into the AAVS1 site on chromosome 19 in a Rep dependent manner [23, 24]. However, in the absence of Rep, only a low level (less than 0.5 %) of illegitimate integration occurs [25–27]. This integration often involves interaction between the ITR and the host DNA leading to a small deletion of the host DNA and an insertion of the vector genome [28]. Although the integration has been shown to occur throughout the genome, current studies suggest that genes, transcriptional initiation site, ribosomal DNA repeats, CpG islands, and palindromic sequences are hotspots for rAAV integration [27, 28]. Despite this low level of integration, many studies have found no oncogenic effect of rAAV [25, 26], while others have found low levels of oncogenesis to be transgene and promoter specific [25–27].

Through understanding of the rAAV transduction pathway, several strategies have been developed to utilize the biology of rAAV to create vectors that are more efficient, the most effective of which is the development of self-complementary rAAV [29, 30]. The development of self-complementary rAAV began with the observation that rAAV transduction was significantly enhanced by co-infection with adenovirus even though the known helper virus functions of adenovirus (e.g. promoter activation) were not a part of rAAV's transduction [18]. The increase in transduction observed with adenovirus co-infection was, in part, due to increased second-strand DNA synthesis caused by the E4Orf6 protein [18], suggesting that second-strand synthesis is a rate-limiting step in rAAV transduction and that, if rates of second-strand synthesis could be increased, then transduction would increase.

To avoid the rate-limiting step of second-strand synthesis, vectors were first designed to less than half the packaging capacity of rAAV, allowing single-stranded replication dimers of the transgene cassette to be packaged [29]. When released from the capsid, these complementary dimers undergo intrastrand base pairing to form transcriptionally active, duplexed single-stranded DNA. This process was made more efficient by deleting the sequence that is nicked by Rep (the D element containing the Rep nicking stem) on one ITR, thereby forcing the formation of a self-complementary dimer [31]. Self-complementary vectors lead to both faster transduction kinetics and to greater transgene expression both in vivo and in vitro [31] and are useful in gene editing applications [32], and show promise in clinical gene therapy.

AAV as a Clinical Gene Therapy Vector

Due to many advantageous properties, AAV has been developed as a vector (rAAV) for clinical gene delivery. The lack of known pathogenicity associated with AAV and AAV's helper dependence both suggest the safety of vectors derived from AAV. In addition, AAV is less inflammatory than many viruses decreasing risk of immune related complications and immune responses to the transgene delivered [33]. Furthermore, all viral genes are removed from rAAV, decreasing the likelihood of immune responses to viral gene products. AAV can also infect both dividing and non-dividing cells and can lead to long-term expression of a transgene [3, 34, 35]. Moreover, transgene cassettes flanked by AAV2 ITRs can be packaged into the capsids of a wide variety of AAV serotypes, through a process known as transencapsidation [36, 37], allowing for the production of vectors with highly variable tropism and distinct immune profiles. The transduction potential of various rAAV serotypes has been studied in detail, allowing for the selection of the appropriate serotype for different applications [9, 10, 38]. Due to these properties, rAAV has been used for gene delivery in over 100 clinical trials (http://www.abedia.com/wiley), which have repeatedly demonstrated the safety of rAAV-mediated gene delivery [39]. The diseases targeted by these trials have included monogenetic disease, such as hemophilia [40, 41], central nervous system diseases [42], and heart disease [43] illustrating the versatility of AAV gene therapy applications.

Recently, several lines of investigation have demonstrated the potential of rAAV as a clinical vector by demonstrating increasing success in meeting efficacy goals. In fact, the first commercial gene therapy product licensed for use in Europe, Glybera® utilizes rAAV to deliver the lipoprotein lipase (LPL) gene to treat lipoprotein lipase deficiency (LPLD) [44]. LPLD is an orphan disease caused by loss of function mutations to LPL or other genes involved in clearing triglycerides from circulation, leading to high serum triglyceride levels and to acute and chronic pancreatitis [44]. Glybera® consists of the AAV1 capsid, which is muscle-tropic, carrying a naturally occurring gain of function mutant of LPL (LPL^{S447X}) and is administered by intramuscular injection [44]. Although the population with LPLD is small, clinical trials testing the safety and effectiveness of Glybera® demonstrated no safety concerns. Furthermore, it was determined that treatment resulted in decreased incidence of pancreatitis, decreased severity of pancreatitis, and decreased hospitalization from pancreatitis, exemplifying the effectiveness of the approach [44–47]. Based on these clinical results, Glybera® was approved for commercial use in Europe in 2012.

Another target that has demonstrated a great deal of success with rAAV mediated gene therapy is retinal gene therapy. Ocular diseases, in general, make very promising targets for gene therapy due to the small size of the eye [48], its immunoprivileged status [49], the ease of vector administration, and well-established techniques for evaluating outcomes (e.g. tomography and electroretinography) [50]. In addition, several rAAV serotypes preferentially transduce different retinal cell types, allowing the delivery to be customized for the specific therapeutic application [51].

Of the retinal diseases, rAAV-mediated gene therapy for Leber's congenital amaurosis (LCA) has generated a great deal of interest due to its success in improving vision in clinical trials. LCA causes early onset retinal degeneration and is the most prevalent cause of childhood blindness [51]. Although LCA can be caused by mutation in several genes, clinical trials utilizing rAAV have focused on LCA caused by loss of function mutations in *RPE65* [52–57]. RPE65 converts the all-*trans*-retinal formed from photoreceptor signaling to 11-*cis*-retinal allowing for repeated signaling of the photoreceptors [58]. The gene therapy approach for LCA consists of subretinal delivery of a rAAV2 vector carrying a functional copy of *RPE65*. These vectors have been used to treat 30 patients who have demonstrated lasting improvements in visual fields, nystagmus, dark-adapted perimetry, and mobility in low light [52–57].

Based on the great success of these trials, eight more clinical trials have been initiated to treat retinal disease with rAAV, including a phase III trial targeting LCA (http://www.abedia.com/wiley). These results, as well as those with LPLD, demonstrate the strong promise of rAAV-mediated gene therapy. Due to this promise, AAV studies have expanded to investigate not only gene addition strategies, but also strategies for correction of disease mutations via double strand break repair.

AAV Vectors and Gene Editing

In mammalian cells, the ability to precisely alter the human genome is hampered by the inefficiency of homologous recombination, which is approximated at one event in a million in dividing cells. Around the turn of this century, it was observed that AAV vector genomes, following transduction of dividing human cells, stimulated HR 1000-fold compared to the efficiency using a plasmid repair substrate [59, 60]. Characterization of this AAV gene editing event demonstrated that particular vector modifications could be used to further enhance the AAV-mediated HR including: (1) increasing the homology between the AAV vector genome and the target loci, (2) centering the non-homologous or modified sequence within the vector genome, and (3) a direct relationship was observed between the induced HR event and time [61]. In addition, the nature of the targeted mutation influences AAV-mediated HR. It was reported that AAV vector induced deletion via HR occurred at a much higher frequency (14-fold) than a targeted insertion [60]. However, controversy on this point exists as that same group, and others, have observed that the nature of the actual mutation (deletion vs. insertion vs. point mutations) does not always reflect a particular bias in HR [60, 61]. Another factor influencing the efficiency of AAV gene editing is the cell cycle kinetics. In general, AAV transduction occurs more efficiently in cells that enter S-phase [62]. For instance, an early observation reported that cellular division is necessary for AAV gene targeting [63] and consistently no targeting was observed in skeletal muscle fibers. It was also demonstrated that cells enriched at the G1/S boundary at the time of infection resulted in increased gene targeting [63]. The use of genotoxic stress agents that increase AAV vector transduction, however, did not stimulate AAV gene editing suggesting that chromosomal damage and a DNA repair milieu is independent of the ability of AAV genomes to function in HR [61, 64]. Although it remains unknown why AAV genomes enhance HR, speculations include the single-strand nature of AAV genomes, the induced DNA damage response upon transduction, and/or the localization of AAV genomes within distinct nuclear compartments including the nucleolus [17, 64–67].

Regarding the efficiencies of AAV-mediated gene editing, reports over the last decade have been quite variable with frequencies ranging from undetectable to 1/1000 despite early reports suggesting that 1 % of cells in the absence of selection [59, 63]. In fact, recent reports demonstrate that without a selection scheme, such as the increased fitness of corrected cells [68], AAV gene targeting remains too low for most applications. Therefore, a trend has been observed towards the use of AAV genome in conjunction with site-specific endonucleases to achieve increased targeted repair [69, 70].

Stimulation of AAV Gene Editing via Site-Specific DNA Damage

Towards higher efficiencies of targeted homologous recombination with the convenience of efficient AAV transduction, two groups reported stimulation of AAV gene editing by a site-specific nuclease in 2003 [69, 70]. To do so, defective reporters containing the I-SceI site were integrated into the human chromosome and used as targets for AAV-mediated gene editing. These instances relied on an AAV co-transduction strategy in which the homologous repair substrate and the I-SceI gene were delivered separately. As had been observed in previous transfection experiments, reporter correction indicative of homologous repair was dramatically enhanced upwards of 100-fold in the absence of a selective agent. Quantitation of site-specific I-SceI mediated genome processing via Southern blotting suggested that approximately one in five breaks were "corrected" by rAAV under the tested conditions [69]. Despite the potential promiscuity of I-SceI at non-target sites, AAV vector integration was not enhanced by the nuclease [69, 70]. The collective data of these cell culture reports using the gold standard nuclease demonstrate that stimulation of AAV-mediated HR by specific DSBs results in efficiencies relevant to the treatment of particular genetic diseases. Furthermore, these gene correction efficiencies could be significantly increased nearly tenfold using the self-complementary AAV vector genome for I-SceI delivery, highlighting the need for efficient vector transduction [32].

Following the encouraging demonstrations of AAV-mediated HR via induced DSB repair, the use of endonuclease platforms amenable to programmable DNA targeting was required. At that time, ZFNs were the most widely explored format and, remarkably, the initial report of AAV-ZFN mediated HR was demonstrated in a mouse model of hemophilia B [71]. In particular, hemophilia B represents a particularly well-suited model for AAV-mediated HR as a particular point mutation in the factor 9 (F9) gene is over-represented in patients, AAV transduction of the liver following IV administration is excellent, in some instances liver cells can undergo cellular replication, and only a low level of circulating F9 is necessary to significantly improve the bleeding phenotype. Furthermore, liver-directed AAV-F9 gene addition strategies were already showing promise in the clinic [40]. In an attempt to expand the general utility of the approach to the most abundant mutations in the F9 patient population, a strategy was employed to precisely insert exons 2–8 of the F9 cDNA thereby restoring F9 production [71]. Regarding vectorization of the necessary components for the event, the individual reagents were packaged in separate AAV8 capsids: (1) ZFN1, (2) ZFN2, and (3) the repair substrate encoding F9 exons

2–8 flanked by homologous arms to endogenous F9 sequence. Therefore, each particular cell requires successful transduction by three separate vectors to allow the ZFN-mediated repair event to occur. Further complications included the inherent ability of the liver cells to mediate HR, in general, as cellular replication in the liver is minimal in a non-induced state. Despite these concerns, it was demonstrated that co-transduction of AAV8-ZFNs induced the cleavage of the targeted locus and stimulated HR in the presence of the AAV repair substrate delivered by an additional AAV vector [71]. Following triple vector administration, the treated hemophilic mice demonstrated circulating levels of F9 at levels sufficient to nearly correct the coagulation deficiency. Importantly, both WT and F9 deficient mice treated with the nuclease and repair AAV vector cocktail demonstrated no observed toxicity and were well tolerated over time [71].

In the year to follow (2012), additional reports in cell cultures expanded the applications of multiple AAV particles for nuclease mediated gene editing [72, 73]. In particular, AAV-ZFN induced targeted gene disruption and, of significant interest, proviral genome deletion was reported at levels of 30 % and 12 % of cells treated, respectively [73]. In contrast, homology mediated repair occurred at lower rates (1–5 %), which may reflect the efficiency of NHEJ compared to HR and/or the requirement for transduction of an additional AAV repair vector [72, 73]. To address the later deficiency, a single promoter platform relying on the T2A ribosome skipping sequence to connect two ZFNs was developed such that all three components (ZFN1, ZFN2, and repair sequence) are encoded by a genome competent for single AAV particle packaging (<5 kb) [74]. Importantly the single AAV-ZFN1-ZFN2-repair vector format ensures that all transduced cells have the necessary factors for the desired modification. This result is consistent with observations that capsid and genome enhancements that increase transduction, in addition to pharmacological agents, increase AAV gene editing [32, 72, 74–76].

In the most recent years, it appears that AAV vector development for nuclease induced gene editing reflects the advancements in the design of these proteins. For instance, the same group that demonstrated F9 correction in the hemophilia B mouse model, employed a similar AAV vector strategy in adult mice, however, now using obligate heterodimeric ZFNs which decreased off-target nuclease induced mutations [76]. That particular report is note-worthy as it implies that the HR repair event occurs within non-dividing cells, although this was not rigorously demonstrated [76]. Although, the designer nuclease field has expanded to encompass TALENs, homing endonucleases, and the Cas9/CRISPR system, the application of these newer technologies for AAV-mediated induced gene editing is currently lacking. This is likely due to the larger TALEN and Cas9/guide strand sequence requirements, which are not readily accommodated by the limited packaging capacity of the AAV capsid (<5 kb). However, characterization of smaller Cas9 variants from multiple species, as well as the compact TALEN format, offer smaller tools that can be packaged in a single intact AAV particle for efficient delivery to relatively non-permissive cells (Cong et al. 2014 ASGCT abstract) [77, 78].

Challenges for AAV Gene Editing via DSB Repair

Currently, AAV vectors are the most clinically accepted and efficient gene delivery strategy for mammalian cell delivery. However, the attributes that have made them popular for gene addition strategies also represent deficiencies concerning AAV vector approaches to gene editing. For instance, in the LCA ocular trial, gene expression and phenotypic correction has been observed out several years following a single vector administration. In the case of a potentially mutagenic nuclease, AAV's remarkable propensity for long-term episomal expression is not desired, and this highlights the importance of temporal regulation of endonuclease production when using this delivery context. Another apparent disadvantage of current AAV approaches, in particular for ex vivo applications, is its inability to efficiently transduce particular types of stem cells. For instance, despite over 30 years of AAV vector optimization experiments, only a handful of reports demonstrate transduction of hematopoietic stems and we have noted AAV ITR-mediated toxicity in human embryonic stem cells [79–82]. Therefore, depending on the cellular response to AAV vector transduction, it remains likely that particular cell types and states of pluripotency will remain on the periphery of the AAV-mediate gene editing following site-specific DNA breaks. At the level of the vector genome, although several creative strategies have been characterized that allow oversized or large gene AAV transduction, their efficiencies are remarkably reduced compared to single particle AAV transduction potentially compromising their clinical utilization [83–88]. Therefore, the collective design of transgene regulation, an expanded capsid reservoir, a better understanding of the AAV vector biology in clinically important cell types, and smaller nuclease formats in general likely represent the next wave of AAV vectors for gene editing applications via DSB repair.

References

1. Atchison RW, Casto BC, Hammon WM. Adenovirus-associated defective virus particles. Science. 1965;149(3685):754–6.
2. Fields BN, Knipe DM, Howley PM. Ovid technologies Inc. Fields' virology. Philadelphia: Wolters Kluwer Health/Lippincott Williams & Wilkins; 2007. Available from: http://eresources.lib.unc.edu/external_db/external_database_auth.html?A=P%7CF=N%7CID=162631%7CREL=HSL%7CSO=HSL%7CURL=http://libproxy.lib.unc.edu/login?url=http://gateway.ovid.com/ovidweb.cgi?T=JS&NEWS=N&PAGE=booktext&D=books&AN=00139921/5th_Edition/.
3. Berns KI, Pinkerton TC, Thomas GF, Hoggan MD. Detection of adeno-associated virus (AAV)-specific nucleotide sequences in DNA isolated from latently infected Detroit 6 cells. Virology. 1975;68(2):556–60.
4. Xiao X, Xiao W, Li J, Samulski RJ. A novel 165-base-pair terminal repeat sequence is the sole cis requirement for the adeno-associated virus life cycle. J Virol. 1997;71(2):941–8.
5. King JA, Dubielzig R, Grimm D, Kleinschmidt JA. DNA helicase-mediated packaging of adeno-associated virus type 2 genomes into preformed capsids. EMBO J. 2001;20(12): 3282–91.

6. Kyostio SR, Owens RA, Weitzman MD, Antoni BA, Chejanovsky N, Carter BJ. Analysis of adeno-associated virus (AAV) wild-type and mutant Rep proteins for their abilities to negatively regulate AAV p5 and p19 mRNA levels. J Virol. 1994;68(5):2947–57.
7. Sonntag F, Schmidt K. Kleinschmidt JrA. A viral assembly factor promotes AAV2 capsid formation in the nucleolus. Proc Natl Acad Sci. 2010;107(22):10220–5.
8. Gao G, Vandenberghe LH, Alvira MR, Lu Y, Calcedo R, Zhou X, et al. Clades of adeno-associated viruses are widely disseminated in human tissues. J Virol. 2004;78(12):6381–8.
9. Ellis B, Hirsch M, Barker J, Connelly J, Steininger R, Porteus M. A survey of ex vivo/in vitro transduction efficiency of mammalian primary cells and cell lines with Nine natural adeno-associated virus (AAV1-9) and one engineered adeno-associated virus serotype. Virol J. 2013;10(1):74. doi:10.1186/1743-422X-10-74.
10. Zincarelli C, Soltys S, Rengo G, Rabinowitz JE. Analysis of AAV serotypes 1-9 mediated gene expression and tropism in mice after systemic injection. Mol Ther. 2008;16(6):1073–80.
11. Grieger JC, Choi VW, Samulski RJ. Production and characterization of adeno-associated viral vectors. Nat Protoc. 2006;1(3):1412–28. PubMed Epub 2007/04/05. eng.
12. Asokan A, Schaffer DV, Samulski RJ. The AAV vector toolkit: poised at the clinical crossroads. Mol Ther. 2012;20(4):699–708. PubMed Epub 2012/01/26. eng.
13. Bartlett JS, Wilcher R, Samulski RJ. Infectious entry pathway of adeno-associated virus and adeno-associated virus vectors. J Virol. 2000;74(6):2777–85.
14. Xiao PJ, Samulski RJ. Cytoplasmic trafficking, endosomal escape, and perinuclear accumulation of adeno-associated virus type 2 particles are facilitated by microtubule network. J Virol. 2012;86(19):10462–73. PubMed Epub 2012/07/20. eng.
15. Popa-Wagner R, Porwal M, Kann M, Reuss M, Weimer M, Florin L, et al. Impact of VP1-specific protein sequence motifs on adeno-associated virus type 2 intracellular trafficking and nuclear entry. J Virol. 2012;86(17):9163–74.
16. Rabinowitz JE, Samulski RJ. Building a better vector: the manipulation of AAV virions. Virology. 2000;278(2):301–8.
17. Johnson JS, Samulski RJ. Enhancement of adeno-associated virus infection by mobilizing capsids into and out of the nucleolus. J Virol. 2009;83(6):2632–44. PubMed Pubmed Central PMCID: 2648275, Epub 2008/12/26. eng.
18. Ferrari FK, Samulski T, Shenk T, Samulski RJ. Second-strand synthesis is a rate-limiting step for efficient transduction by recombinant adeno-associated virus vectors. J Virol. 1996;70(5):3227–34.
19. Beaton A, Palumbo P, Berns KI. Expression from the adeno-associated virus p5 and p19 promoters is negatively regulated in trans by the rep protein. J Virol. 1989;63(10):4450–4. PubMed Epub 1989/10/01. eng.
20. Miao CH, Snyder RO, Schowalter DB, Patijn GA, Donahue B, Winther B, et al. The kinetics of rAAV integration in the liver. Nat Genet. 1998;19(1):13–5. PubMed Epub 1998/05/20. eng.
21. Cheung AK, Hoggan MD, Hauswirth WW, Berns KI. Integration of the adeno-associated virus genome into cellular DNA in latently infected human Detroit 6 cells. J Virol. 1980;33(2):739–48.
22. Hirsch ML, Storici F, Li C, Choi VW, Samulski RJ. AAV recombineering with single strand oligonucleotides. PLoS One. 2009;4(11), e7705. PubMed Pubmed Central PMCID: 2765622, Epub 2009/11/06. eng.
23. Kotin RM, Siniscalco M, Samulski RJ, Zhu XD, Hunter L, Laughlin CA, et al. Site-specific integration by adeno-associated virus. Proc Natl Acad Sci U S A. 1990;87(6):2211–5.
24. Samulski RJ, Zhu X, Xiao X, Brook JD, Housman DE, Epstein N, et al. Targeted integration of adeno-associated virus (AAV) into human chromosome 19. EMBO J. 1991;10(12):3941–50. PubMed Epub 1991/12/01. eng.
25. Deyle DR, Russell DW. Adeno-associated virus vector integration. Curr Opin Mol Ther. 2009;11(4):442–7. PubMed Epub 2009/08/04. eng.
26. McCarty DM, Young Jr SM, Samulski RJ. Integration of adeno-associated virus (AAV) and recombinant AAV vectors. Annu Rev Genet. 2004;38:819–45. PubMed Epub 2004/12/01. eng.

27. Rosas LE, Grieves JL, Zaraspe K, La Perle KMD, Fu H, McCarty DM. Patterns of scAAV vector insertion associated with oncogenic events in a mouse model for genotoxicity. Mol Ther. 2012;20(11):2098–110.
28. Miller DG, Trobridge GD, Petek LM, Jacobs MA, Kaul R, Russell DW. Large-scale analysis of adeno-associated virus vector integration sites in normal human cells. J Virol. 2005;79(17):11434–42.
29. McCarty DM, Monahan PE, Samulski RJ. Self-complementary recombinant adeno-associated virus (scAAV) vectors promote efficient transduction independently of DNA synthesis. Gene Ther. 2001;8(16):1248–54. PubMed Epub 2001/08/18. eng.
30. Wang Z, Ma HI, Li J, Sun L, Zhang J, Xiao X. Rapid and highly efficient transduction by double-stranded adeno-associated virus vectors in vitro and in vivo. Gene Ther. 2003;10(26):2105–11. PubMed Epub 2003/11/20. eng.
31. McCarty DM, Fu H, Monahan PE, Toulson CE, Naik P, Samulski RJ. Adeno-associated virus terminal repeat (TR) mutant generates self-complementary vectors to overcome the rate-limiting step to transduction in vivo. Gene Ther. 2003;10(26):2112–8. PubMed Epub 2003/11/20. eng.
32. Hirsch ML, Green L, Porteus MH, Samulski RJ. Self-complementary AAV mediates gene targeting and enhances endonuclease delivery for double-strand break repair. Gene Ther. 2010;17(9):1175–80. PubMed Pubmed Central PMCID: 3152950, Epub 2010/05/14. eng.
33. Mueller C, Flotte TR. Clinical gene therapy using recombinant adeno-associated virus vectors. Gene Ther. 2008;15(11):858–63.
34. Podsakoff G, Wong KK, Chatterjee S. Efficient gene transfer into nondividing cells by adeno-associated virus-based vectors. J Virol. 1994;68(9):5656–66.
35. Xiao X, Li J, Samulski RJ. Efficient long-term gene transfer into muscle tissue of immunocompetent mice by adeno-associated virus vector. J Virol. 1996;70(11):8098–108.
36. Rabinowitz JE, Bowles DE, Faust SM, Ledford JG, Cunningham SE, Samulski RJ. Cross-dressing the virion: the transcapsidation of adeno-associated virus serotypes functionally defines subgroups. J Virol. 2004;78(9):4421–32. PubMed Epub 2004/04/14. eng.
37. Rabinowitz JE, Rolling F, Li C, Conrath H, Xiao W, Xiao X, et al. Cross-packaging of a single adeno-associated virus (AAV) type 2 vector genome into multiple AAV serotypes enables transduction with broad specificity. J Virol. 2002;76(2):791–801. PubMed Epub 2001/12/26. eng.
38. Ellis BL, Hirsch ML, Barker JC, Connelly JP, Steininger 3rd RJ, Porteus MH. A survey of ex vivo/in vitro transduction efficiency of mammalian primary cells and cell lines with Nine natural adeno-associated virus (AAV1-9) and one engineered adeno-associated virus serotype. Virol J. 2013;10:74. PubMed Pubmed Central PMCID: 3607841.
39. Mitchell AM, Nicolson SC, Warischalk JK, Samulski RJ. AAV's anatomy: roadmap for optimizing vectors for translational success. Curr Gene Ther. 2010;10(5):319–40. PubMed Epub 2010/08/18. eng.
40. Nathwani AC, Tuddenham EG, Rangarajan S, Rosales C, McIntosh J, Linch DC, et al. Adenovirus-associated virus vector-mediated gene transfer in hemophilia B. N Engl J Med. 2011;365(25):2357–65. PubMed Pubmed Central PMCID: 3265081, Epub 2011/12/14. eng.
41. Manno CS, Pierce GF, Arruda VR, Glader B, Ragni M, Rasko JJ, et al. Successful transduction of liver in hemophilia by AAV-Factor IX and limitations imposed by the host immune response. Nat Med. 2006;12(3):342–7. PubMed Epub 2006/02/14. eng.
42. Gray SJ. Gene therapy and neurodevelopmental disorders. Neuropharmacology. 2013;68:136–42. PubMed Epub 2012/07/04. eng.
43. Zouein FA, Booz GW. AAV-mediated gene therapy for heart failure: enhancing contractility and calcium handling. F1000Prime Rep. 2013;5:27. PubMed Epub 2013/08/24. eng.
44. Gaudet D, Methot J, Kastelein J. Gene therapy for lipoprotein lipase deficiency. Curr Opin Lipidol. 2012;23(4):310–20. PubMed Epub 2012/06/14. eng.
45. Carpentier AC, Frisch F, Labbe SM, Gagnon R, de Wal J, Greentree S, et al. Effect of alipogene tiparvovec (AAV1-LPL(S447X)) on postprandial chylomicron metabolism in lipoprotein

lipase-deficient patients. J Clin Endocrinol Metab. 2012;97(5):1635–44. PubMed Epub 2012/03/23. eng.
46. Gaudet D, Methot J, Dery S, Brisson D, Essiembre C, Tremblay G, et al. Efficacy and long-term safety of alipogene tiparvovec (AAV1-LPLS447X) gene therapy for lipoprotein lipase deficiency: an open-label trial. Gene Ther. 2013;20(4):361–9. PubMed Epub 2012/06/22. eng.
47. Stroes ES, Nierman MC, Meulenberg JJ, Franssen R, Twisk J, Henny CP, et al. Intramuscular administration of AAV1-lipoprotein lipase S447X lowers triglycerides in lipoprotein lipase-deficient patients. Arterioscler Thromb Vasc Biol. 2008;28(12):2303–4. PubMed Epub 2008/09/20. eng.
48. Ali RR, Reichel MB, Thrasher AJ, Levinsky RJ, Kinnon C, Kanuga N, et al. Gene transfer into the mouse retina mediated by an adeno-associated viral vector. Hum Mol Genet. 1996;5(5):591–4. PubMed Epub 1996/05/01. eng.
49. Barker SE, Broderick CA, Robbie SJ, Duran Y, Natkunarajah M, Buch P, et al. Subretinal delivery of adeno-associated virus serotype 2 results in minimal immune responses that allow repeat vector administration in immunocompetent mice. J Gene Med. 2009;11(6):486–97.
50. Jacobson SG, Cideciyan AV, Ratnakaram R, Heon E, Schwartz SB, Roman AJ, et al. Gene therapy for leber congenital amaurosis caused by RPE65 mutations: safety and efficacy in 15 children and adults followed up to 3 years. Arch Ophthalmol. 2012;130(1):9–24. PubMed Epub 2011/09/14. eng.
51. McClements ME, MacLaren RE. Gene therapy for retinal disease. Transl Res. 2013;161(4):241–54. PubMed Epub 2013/01/12. eng.
52. Testa F, Maguire AM, Rossi S, Pierce EA, Melillo P, Marshall K, et al. Three-year follow-up after unilateral subretinal delivery of adeno-associated virus in patients with Leber congenital Amaurosis type 2. Ophthalmology. 2013;120(6):1283–91. PubMed Pubmed Central PMCID: 3674112.
53. Bainbridge JW, Smith AJ, Barker SS, Robbie S, Henderson R, Balaggan K, et al. Effect of gene therapy on visual function in Leber's congenital amaurosis. N Engl J Med. 2008;358(21):2231–9. PubMed Epub 2008/04/29. eng.
54. Hauswirth WW, Aleman TS, Kaushal S, Cideciyan AV, Schwartz SB, Wang L, et al. Treatment of leber congenital amaurosis due to RPE65 mutations by ocular subretinal injection of adeno-associated virus gene vector: short-term results of a phase I trial. Hum Gene Ther. 2008;19(10):979–90. PubMed Epub 2008/09/09. eng.
55. Maguire AM, High KA, Auricchio A, Wright JF, Pierce EA, Testa F, et al. Age-dependent effects of RPE65 gene therapy for Leber's congenital amaurosis: a phase 1 dose-escalation trial. Lancet. 2009;374(9701):1597–605. PubMed Epub 2009/10/27. eng.
56. Maguire AM, Simonelli F, Pierce EA, Pugh Jr EN, Mingozzi F, Bennicelli J, et al. Safety and efficacy of gene transfer for Leber's congenital amaurosis. N Engl J Med. 2008;358(21):2240–8. PubMed Epub 2008/04/29. eng.
57. Bennett J, Ashtari M, Wellman J, Marshall KA, Cyckowski LL, Chung DC, et al. AAV2 gene therapy readministration in three adults with congenital blindness. Sci Transl Med. 2012;4(120):120ra15.
58. Cai X, Conley SM, Naash MI. RPE65: role in the visual cycle, human retinal disease, and gene therapy. Ophthalmic Genet. 2009;30(2):57–62. PubMed Epub 2009/04/18. eng.
59. Russell DW, Hirata RK. Human gene targeting by viral vectors. Nat Genet. 1998;18(4):325–30. PubMed Pubmed Central PMCID: 3010411, Epub 1998/04/16. eng.
60. Russell DW, Hirata RK. Human gene targeting favors insertions over deletions. Hum Gene Ther. 2008;19(9):907–14. PubMed Pubmed Central PMCID: 2940567.
61. Hirata RK, Russell DW. Design and packaging of adeno-associated virus gene targeting vectors. J Virol. 2000;74(10):4612–20. PubMed Pubmed Central PMCID: 111981, Epub 2000/04/25. eng.
62. Russell DW, Miller AD, Alexander IE. Adeno-associated virus vectors preferentially transduce cells in S phase. Proc Natl Acad Sci U S A. 1994;91(19):8915–9. PubMed Pubmed Central PMCID: 44717.

63. Trobridge G, Hirata RK, Russell DW. Gene targeting by adeno-associated virus vectors is cell-cycle dependent. Hum Gene Ther. 2005;16(4):522–6. PubMed.
64. Liu X, Yan Z, Luo M, Zak R, Li Z, Driskell RR, et al. Targeted correction of single-base-pair mutations with adeno-associated virus vectors under nonselective conditions. J Virol. 2004;78(8):4165–75. PubMed Pubmed Central PMCID: 374254.
65. Cervelli T, Palacios JA, Zentilin L, Mano M, Schwartz RA, Weitzman MD, et al. Processing of recombinant AAV genomes occurs in specific nuclear structures that overlap with foci of DNA-damage-response proteins. J Cell Sci. 2008;121(Pt 3):349–57. PubMed.
66. Inagaki K, Ma C, Storm TA, Kay MA, Nakai H. The role of DNA-PKcs and artemis in opening viral DNA hairpin termini in various tissues in mice. J Virol. 2007;81(20):11304–21. PubMed Pubmed Central PMCID: 2045570, Epub 2007/08/10. eng.
67. Schwartz RA, Carson CT, Schuberth C, Weitzman MD. Adeno-associated virus replication induces a DNA damage response coordinated by DNA-dependent protein kinase. J Virol. 2009;83(12):6269–78. PubMed Pubmed Central PMCID: 2687378.
68. Paulk NK, Loza LM, Finegold MJ, Grompe M. AAV-mediated gene targeting is significantly enhanced by transient inhibition of nonhomologous end joining or the proteasome in vivo. Hum Gene Ther. 2012;23(6):658–65. PubMed Pubmed Central PMCID: 3392621.
69. Miller DG, Petek LM, Russell DW. Human gene targeting by adeno-associated virus vectors is enhanced by DNA double-strand breaks. Mol Cell Biol. 2003;23(10):3550–7. PubMed Pubmed Central PMCID: 164770.
70. Porteus MH, Baltimore D. Chimeric nucleases stimulate gene targeting in human cells. Science. 2003;300(5620):763. PubMed Epub 2003/05/06. eng.
71. Li H, Haurigot V, Doyon Y, Li T, Wong SY, Bhagwat AS, et al. In vivo genome editing restores haemostasis in a mouse model of haemophilia. Nature. 2011;475(7355):217–21. PubMed Pubmed Central PMCID: 3152293.
72. Asuri P, Bartel MA, Vazin T, Jang JH, Wong TB, Schaffer DV. Directed evolution of adeno-associated virus for enhanced gene delivery and gene targeting in human pluripotent stem cells. Mol Ther. 2012;20(2):329–38. PubMed Pubmed Central PMCID: 3277219.
73. Handel EM, Gellhaus K, Khan K, Bednarski C, Cornu TI, Muller-Lerch F, et al. Versatile and efficient genome editing in human cells by combining zinc-finger nucleases with adeno-associated viral vectors. Hum Gene Ther. 2012;23(3):321–9. PubMed Pubmed Central PMCID: 3300077.
74. Ellis BL, Hirsch ML, Porter SN, Samulski RJ, Porteus MH. Zinc-finger nuclease-mediated gene correction using single AAV vector transduction and enhancement by Food and Drug Administration-approved drugs. Gene Ther. 2013;20(1):35–42. PubMed Epub 2012/01/20. Eng.
75. Rahman SH, Bobis-Wozowicz S, Chatterjee D, Gellhaus K, Pars K, Heilbronn R, et al. The nontoxic cell cycle modulator indirubin augments transduction of adeno-associated viral vectors and zinc-finger nuclease-mediated gene targeting. Hum Gene Ther. 2013;24(1):67–77. PubMed Pubmed Central PMCID: 3555098.
76. Anguela XM, Sharma R, Doyon Y, Miller JC, Li H, Haurigot V, et al. Robust ZFN-mediated genome editing in adult hemophilic mice. Blood. 2013;122(19):3283–7. PubMed Pubmed Central PMCID: 3821724.
77. Beurdeley M, Bietz F, Li J, Thomas S, Stoddard T, Juillerat A, et al. Compact designer TALENs for efficient genome engineering. Nat Commun. 2013;4:1762. PubMed Pubmed Central PMCID: 3644105.
78. Esvelt KM, Mali P, Braff JL, Moosburner M, Yaung SJ, Church GM. Orthogonal Cas9 proteins for RNA-guided gene regulation and editing. Nat Methods. 2013;10(11):1116–21. PubMed Pubmed Central PMCID: 3844869.
79. Hirsch ML, Fagan BM, Dumitru R, Bower JJ, Yadav S, Porteus MH, et al. Viral single-strand DNA induces p53-dependent apoptosis in human embryonic stem cells. PLoS One. 2011;6(11), e27520. PubMed Pubmed Central PMCID: 3219675, Epub 2011/11/25. eng.

80. Song L, Li X, Jayandharan GR, Wang Y, Aslanidi GV, Ling C, et al. High-efficiency transduction of primary human hematopoietic stem cells and erythroid lineage-restricted expression by optimized AAV6 serotype vectors in vitro and in a murine xenograft model in vivo. PLoS One. 2013;8(3), e58757. PubMed Pubmed Central PMCID: 3597592.
81. Maina N, Han Z, Li X, Hu Z, Zhong L, Bischof D, et al. Recombinant self-complementary adeno-associated virus serotype vector-mediated hematopoietic stem cell transduction and lineage-restricted, long-term transgene expression in a murine serial bone marrow transplantation model. Hum Gene Ther. 2008;19(4):376–83. PubMed.
82. Srivastava A. Hematopoietic stem cell transduction by recombinant adeno-associated virus vectors: problems and solutions. Hum Gene Ther. 2005;16(7):792–8. PubMed.
83. Duan D, Yue Y, Engelhardt JF. Expanding AAV packaging capacity with trans-splicing or overlapping vectors: a quantitative comparison. Mol Ther. 2001;4(4):383–91. PubMed Epub 2001/10/11. eng.
84. Hirsch ML, Agbandje-McKenna M, Samulski RJ. Little vector, big gene transduction: fragmented genome reassembly of adeno-associated virus. Mol Ther. 2010;18(1):6–8. PubMed Pubmed Central PMCID: 2839225, Epub 2010/01/06. eng.
85. Hirsch ML, Li C, Bellon I, Yin C, Chavala S, Pryadkina M, et al. Oversized AAV transduction is mediated via a DNA-PKcs independent, Rad51C-dependent repair pathway. Mol Ther. 2013;13. PubMed.
86. Lai Y, Yue Y, Duan D. Evidence for the failure of adeno-associated virus serotype 5 to package a viral genome>or=8.2 kb. Mol Ther. 2010;18(1):75–9. PubMed Pubmed Central PMCID: 2839223, Epub 2009/11/12. eng.
87. Wu Z, Yang H, Colosi P. Effect of genome size on AAV vector packaging. Mol Ther. 2010;18(1):80–6. PubMed Pubmed Central PMCID: 2839202, Epub 2009/11/12. eng.
88. Halbert CL, Allen JM, Miller AD. Efficient mouse airway transduction following recombination between AAV vectors carrying parts of a larger gene. Nat Biotechnol. 2002;20(7): 697–701. PubMed Epub 2002/06/29. eng.

Engineered Nucleases and Trinucleotide Repeat Diseases

John H. Wilson, Christopher Moye, and David Mittelman

Abstract Tandem repeats are consecutively repeated sets of nucleotides that mutate by the addition or loss of one or more repeat units. Tandem repeats can modulate gene function and heritable traits in a number of species including human. Mutations at a subset of repeats, most of which are trinucleotide repeats, trigger devastating human neurological and skeletal disorders. In particular, at least a dozen neurological disorders share a common etiology—the expansion of a CAG repeat tract from less than 30 units, to pathogenic lengths of up to hundreds and sometimes thousands of copies. These repeats are attractive targets for therapy because they underlie several disorders, but at the same time they are challenging to attack. Here we describe studies from our lab and others that have exploited engineered nucleases to target or disrupt disease-causing CAG repeats. We discuss the current challenges with this therapeutic approach and highlight the use of next-generation sequencing as a powerful tool to measure repeat mutation and the efficacy of nuclease-mediated genome editing.

Keywords Triplet repeat disorders • Microsatellite repeats • Engineered nucleases • Next generation sequencing

Abbreviations

APRT	Adenine phosphoribosyltransferase
CRISPR	Clustered regulatory interspaced short palindromic repeats
DM1	Myotonic dystrophy Type I
DSBs	Double strand breaks

J.H. Wilson, Ph.D. (✉) • C. Moye, Ph.D.
Verna and Marrs McLean Biochemistry and Molecular Biology, Baylor College of Medicine, One Baylor Plaza, Houston, TX 77030, USA
e-mail: jwilson@bcm.edu

D. Mittelman
Department of Biological Sciences, Virginia Tech, Blacksburg, VA, USA

Virginia Bioinformatics Institute, Blacksburg, VA, USA

T. Cathomen et al. (eds.), *Genome Editing*, Advances in Experimental Medicine and Biology 895, DOI 10.1007/978-1-4939-3509-3_9

FRAXA	Fragile X syndrome
FRDA	Friedreich ataxia
FXTAS	Fragile X-associated tremor and ataxia syndrome
GFP	Green fluorescent protein
HD	Huntington disease
HPRT	Hypoxanthine-guanine phosphoribosyltransferase
PCR	Polymerase chain reaction
PolyQ	Polyglutamine
SCA	Spinocerebellar ataxia
TALENs	Transcription activator-like effector nucleases
TC-NER	Transcription-coupled nucleotide excision repair
TNRs	Trinucleotide repeats
3′UTRs	3′ Untranslated region
5′UTRs	5′ Untranslated region
ZFNs	Zinc-finger nucleases
ZFNickase	Zinc-finger nickase

Introduction

The human genome is rife with simple sequence repeats termed microsatellites. Arranged head-to-tail, these tandem repeats constitute roughly 3 % of the human genome, and are common in all genomes [1]. Microsatellite repeats are the most mutable elements in the genome; they gain repeat units (expand) or lose them (contract) at rates of 0.01–1 % [2–4]. Their variability would seem to make it unlikely that they would be tolerated in coding regions, yet 17 % of human genes contain such repeats [3]. Indeed, a subset of microsatellite repeats—trinucleotide repeats (TNRs)—are overrepresented in exons, especially in genes for transcription factors and other regulatory proteins, suggesting that TNRs may provide an evolutionary benefit [1, 5–8].

These potential evolutionary benefits, however, come with tangible human costs. In 25 human genes, expansion of a TNR causes a genetic disease, with well-known examples including fragile X syndrome, Friedreich ataxia, myotonic dystrophy, and Huntington disease [9–13]. In fragile X syndrome and Friedreich ataxia, for example, expansion of the repeat interferes with gene expression, while in myotonic dystrophy and Huntington disease, expansion generates toxic products. Identifying the cellular defects responsible for the phenotypes of these diseases has opened up productive avenues for drug development, with tantalizing results in model systems, but it has yielded few treatments for patients, and no cures.

An orthogonal approach to defining the proximate causes of disease pathology is the dissection of mechanisms that drive repeat expansion, with the idea that knowing the molecular details might generate therapeutic targets that could then be exploited to prevent or reduce repeat expansions in the first place [14]. The approach is deceptively simple since the mechanisms that underlie TNR instability have proven extremely diverse. Studies in model organisms have identified a broad spectrum of

DNA transactions, including replication, recombination, DNA repair, and transcription that can contribute to TNR instability [1, 3, 15, 16]. In fact, it's hard to identify a process that affects DNA that does not also modulate TNR instability. Adding to this complexity, studies in mice have revealed that different mechanisms of TNR instability predominate in different tissues [17–21]. Finding suitable therapeutic targets in this mechanistic soup will be a challenge.

A third option would be to develop reagents that promote contraction of the repeat tract, a potential cure for the patient so long as the existing cellular damage is not too great. At first glance, such an approach would also seem to require deep knowledge of instability mechanisms; however, recent advances in custom-designed DNA nucleases offer a potential end-run around the common biological pathways of TNR instability. A double-strand break (DSB) introduced into a repeat tract can stimulate TNR contraction as a natural consequence of break repair [22, 23]. Our current toolkit of engineered nucleases includes zinc-finger nucleases (ZFNs), transcription activator-like effector nucleases (TALENs), and clustered regulatory interspaced short palindromic repeats (CRISPR)/Cas nucleases, any of which could, in principle, be used to cleave TNRs. The challenge will be to introduce DSBs, which are themselves dangerous lesions, at specific sites with high efficiency, yet with minimal collateral damage.

Engineered nucleases are transformative tools for genome manipulation, but can they be transformed into reliable reagents for gene surgery in humans? To place the therapeutic challenges in context, we begin with an overview of repeated sequences in the human genome, the inherent mechanisms of instability for TNRs, and how TNR expansions cause disease.

Microsatellite Repeats and TNR Instability

Repetitive DNA sequences are present in all genomes to a greater or lesser extent, but they constitute a surprisingly high fraction of the human genome, accounting for roughly half our DNA [1]. Tandem repeats form a subcategory of such sequences, in which the repeat units are arrayed end to end like boxcars in a train. If the repeat unit is ten or more nucleotides, the repeat tract is referred to as a minisatellite repeat; if the unit is less than ten nucleotides, the tract is commonly known as a microsatellite repeat [3]. Disease-causing TNRs fall squarely into the microsatellite category. Microsatellites are described by their overall length, which can range from just a few units to well over a thousand, and the purity of the sequence, which is the percentage of nucleotides in the tract that match the repeat unit. The tendency for a TNR to expand or contract depends primarily on the overall length of the repeat tract and secondarily on the purity of the repeat tract and the composition of the repeat unit.

TNRs mutate at rates 100- to 1000-fold higher than unique sequences and almost exclusively by altering the number of repeat units in the repeat tract [2, 3]. Thus, TNR mutations are predictable: they change the length of the repeat tract, a property that makes them readily reversible, as well, and allows different tract lengths to be efficiently tested for evolutionary fitness [7, 8, 24]. The properties of the repeat tract, for the most part, determine the rates of mutation: long repeat tracts mutate more rapidly that short ones, pure tracts more rapidly than impure ones [4]. Genomic context, however, also plays a role since similar TNRs at different locations in the genome mutate at different rates. Although it is not yet clear, it seems likely that these context effects speak to the mechanisms of TNR instability, reflecting the location of nearby origins of replication, promoters, chromatin structure, or epigenetic status, among other possibilities [1].

Studies in bacteria, yeast, flies, mammalian cells, and mice have shown that the whole gamut of DNA transactions—replication, recombination, DNA repair, and transcription—can contribute to the instability of microsatellite repeats [1, 3, 15, 16]. By exposing single strands of DNA, these processes allow strands to misalign within the repeat tract, leading to a variety of alternative DNA structures. For example, slippage of the primer strand relative to the template strand during replication could increase or decrease the length of the repeat tract (Fig. 1A) [1]. Similarly, in yeast and human cells, breaks in CAG tracts trigger TNR instability by engaging the single-strand annealing pathway of recombination, which repairs the break by annealing CAG repeat units that flank the break (Fig. 1B) [22, 25]. Finally, and perhaps most surprising, transcription destabilizes TNRs [15, 26–33]. In human cells, transcription-induced TNR instability is stimulated by R-loop formation, requires the MutSβ recognition component of mismatch repair, and depends on the complete pathway of transcription-coupled nucleotide excision repair (TC-NER) (Fig. 1C) [34, 35].

The true diversity of mechanisms of TNR instability is likely to be much greater than suggested in Fig. 1, as indicated by the multitude of factors and processes that influence TNR instability in model organisms and in cell-free systems [36]. Consider just those that affect CAG TNRs: replicative polymerases [37], flap structure-specific endonuclease [25], replication factors [37–42], translesion synthesis [43], supercoiling and topoisomerases [44, 45], helicases [46–48], mismatch repair proteins [27, 49–53], components of base-excision repair [19, 20, 54], nucleotide excision repair [21, 33, 35, 55], single-strand break repair [45], double-strand break repair and homologous recombination [22, 25, 56–59], transcription [27, 33, 60], convergent transcription [31, 32], R-loops [34, 61], E3 ubiquitin ligases and the proteasome [35], CpG methylation [17, 62], histone deacetylases [63], Hsp90 chaperone [8, 58], and DNA-damage checkpoint pathways [31, 42, 64]. Understanding how these processes interconnect is a daunting challenge, but one that must be considered when contemplating therapeutic interventions aimed at shrinking the repeat tract.

TNRs and Human Disease

Expansions of TNRs at 25 loci in the human genome cause devastating neurological, muscular, or developmental diseases [9–13]. Disease-associated expanded repeats are found in the transcribed portions of the gene, including exons, introns,

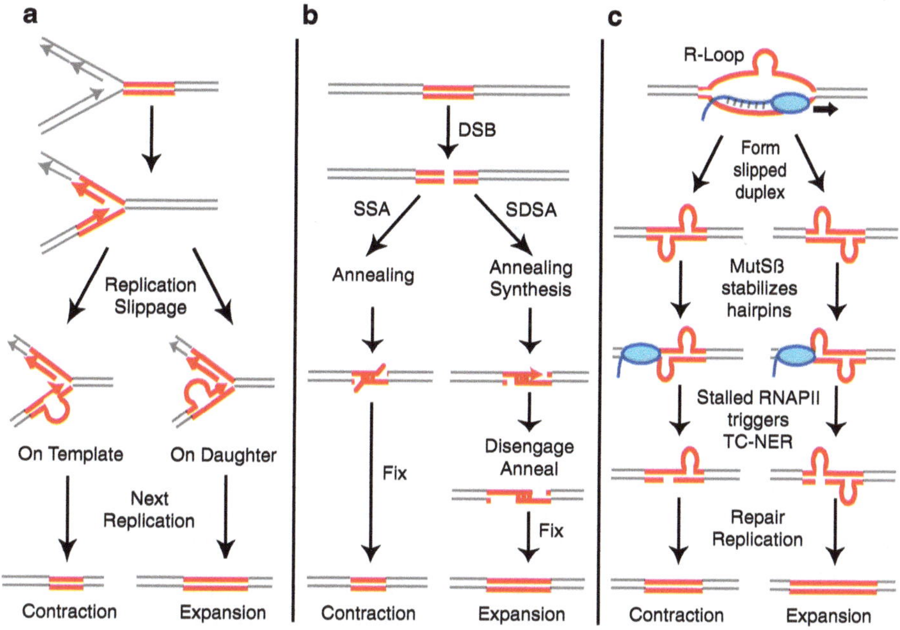

Fig. 1 Mechanisms of microsatellite repeat instability. (**a**) Replication slippage. Mispairing of the daughter strand with its template will loop out a stretch of repeat units. Subsequent replication will generate a contraction or an expansion. (**b**) Homologous recombination. DSBs will expose single-strands that can pair directly by single-strand annealing (SSA), leading to contraction. If one of the paired strands were extended by DNA synthesis, the extended strand could disengage and re-pair to generate an expanded repeat in a version of the synthesis-dependent strand-annealing (SDSA) pathway [119]. (**c**) Transcription. R-loop formation behind RNA polymerase would allow hairpins to form in the nontemplate strand, which would lead to a slipped duplex when the R-loop was resolved. The binding of MutSβ to CAG and CTG hairpins might block RNA polymerase, which is an initiating signal for TC-NER [49, 120, 121]. Repair and subsequent replication would generate expansions and contractions. Adapted from Fig. 1 in Chatterjee, N., Santillan, B.A., and Wilson, J.H. (2013) Microsatellite Repeats: Canaries in the Coalmine. In *Stress-Induced Mutagenesis*, D. Mittelman, ed, Springer Publishing Co, New York, NY, pp. 119–150, with permission

5′UTRs, and 3′UTRs (Fig. 2). Some expanded repeats interfere with gene function, causing recessive, loss-of-function phenotypes [9, 36]. For example, in fragile X syndrome (FRAXA), expansion of CGG in the 5′UTR becomes hypermethylated and inactivates the promoter [65]. Similarly, in Friedreich ataxia (FRDA), expansion of GAA in an intron interferes with transcription elongation [66–68]. More commonly, however, expanded TNR alleles are associated with toxic protein or RNA product, giving rise to dominant, gain-of-function phenotypes.

The dominant gain-of-function phenotype of Huntington disease (HD) and several spinocerebellar ataxias (SCAs) is due to "protein toxicity" caused by exonic CAG repeats that encode polyglutamine (polyQ). In each case, the extended polyQ segment is thought to alter the properties of the mutated protein, making it toxic and

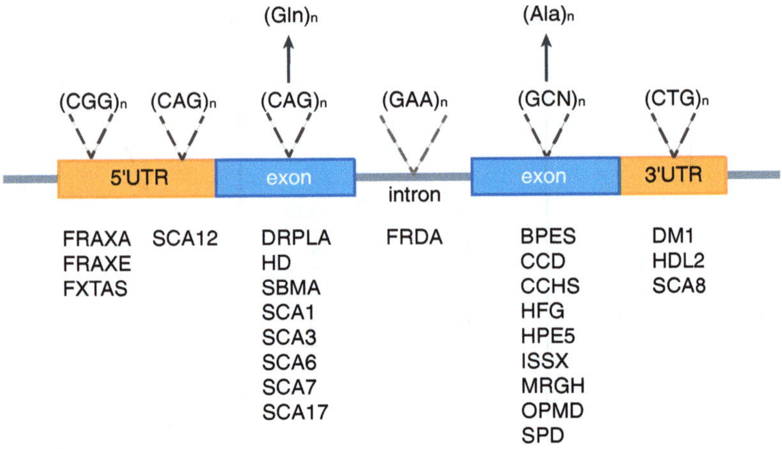

Fig. 2 Human diseases caused by TNRs. Diseases associated with specific TNRs are shown below the repeat units. *BPES* blepharophimosis, ptosis, and epicanthus inversus; *CCD* cleidocranial dysplasia, *CCHS* congenital central hypoventilation syndrome, *DM1* myotonic dystrophy type 1, *DRPLA* dentatorubral pallidoluysian atrophy, *FRAXA* fragile X syndrome, *FRAXE* fragile X mental retardation associated with FRAXE site, *FRDA* Friedreich ataxia, *FXTAS* fragile X tremor and ataxia syndrome, *HD* Huntington disease, *HDL2* Huntington-disease-like 2, *HFG* hand-foot-genital syndrome, *HPE5* holoprosencephaly 5, *ISSX* X-linked infantile spasm syndrome, *MRGH* mental retardation with isolated growth hormone deficiency, *OPMD* oculopharyngeal muscular dystrophy, *SBMA* spinal and bulbar muscular atrophy, *SCA* spinocerebellar ataxia types 1, 3, 6, 7, 8, 12, and 17; *SPD* synpolydactyly. Adapted from Fig. 3 in Chatterjee, N., Santillan, B.A., and Wilson, J.H. (2013) Microsatellite Repeats: Canaries in the Coalmine. In *Stress-Induced Mutagenesis*, D. Mittelman, ed, Springer Publishing Co, New York, NY, pp. 119-150, with permission

a trigger for disease [9]. The phenomenon of "RNA toxicity" was first identified at the myotonic dystrophy type I (DM1) locus (CTG repeat in the 3′UTR) [69, 70]. The expanded repeats in these DM RNAs bind the alternative splicing factor muscleblind, leading to the aberrant splicing of a number of other transcripts, which is the basis for the dominant phenotype of these diseases [69]. Interestingly, the CGG repeats at the fragile X locus, at an intermediate level of expansion, from 55 to 200 repeats, cause a gain-of-function disease, fragile X-associated tremor and ataxia syndrome (FXTAS), due to overexpression of the RNA, which is toxic to the cell [71, 72]. A recent observation that convergent transcription through CAG repeat tracts causes cell death raises the possibility that "DNA toxicity" may contribute to cell death in some of these diseases [31, 73], especially given the unexpectedly high frequency of antisense transcription in the human genome [74, 75].

Disease-associated TNRs include only CNG and GAA repeats. In contrast to most other possible triplets, these TNRs readily form non-B-DNA structures, including hairpins and slipped duplexes (CAG and CGG), G-quartets (CGG), and triplexes (GAA) [76, 77]. These TNRs are thought to be unstable because of their tendency to form non-B-DNA secondary structures when they become transiently single-stranded during replication, repair, recombination, or transcription. The propensity to form troublesome secondary structures increases with tract length to the point that TNR instability approaches 100 % in some cases [78, 79]. But the instability observed for the same repeat tract varies dramatically in different tissues.

Disease-associated TNRs typically expand and contract at high rates in the germline, usually with a bias toward expansion in one or the other parent, depending on the disease. Patients with HD or any of several SCAs (all caused by CAG repeats) typically show expansion bias in the paternal germline. By contrast, patients with FRAXA (CGG repeat), FRDA (GAA repeat) or DM1 (CTG repeat) display a maternal bias for expansion [11]. This bias toward expansion underlies the phenomenon of anticipation, the clinical observation that disease symptoms become more severe from one generation to the next. Though disease phenotypes always arise from TNR expansion, TNR instability is not always biased toward expansion; for example, in patients with FRDA and FRAXA, there is a marked contraction bias in the paternal germline [11, 80]. These parent-of-origin effects are not understood.

TNR instability also occurs in somatic tissues as affected individuals age, especially in the brain, accelerating the onset of neuronal dysfunction and death [79, 81, 82]. TNR diseases caused by CAG and CTG display similar patterns of instability [83]. Microsatellites in genomes of white blood cells and cardiac muscle cells tend to be stable, for example, while those in liver and kidney cells display an intermediate level of instability. Similarly, in the brain, repeats in the striatum are typically very unstable, while those in the cerebral cortex and hippocampus are moderately unstable, and those in the cerebellum are fairly stable [83]. By contrast, the GAA repeats in FRDA patients are very unstable in blood and cerebellum, which distinguishes them from CAG repeats [84]. In addition, GAA repeat instability is biased toward contractions in all tissues; however, in the dorsal root ganglia—whose degeneration gives rise to the disease symptoms—there is a notably higher frequency of large expansions [81, 84]. In FRAXA patients (CGG repeats), long repeats are commonly methylated at their CpG sequences and display minor instability [80, 85], while their rare, unmethylated counterparts are very unstable [86, 87].

The extreme variation in repeat instability across tissues and diseases forms one of the most puzzling features of TNR instability [83]. It suggests that TNR instability involves multiple mechanisms that depend on the repeat sequence, the genomic context, the type of tissue, and developmental status. Proliferating cells in the male germline, for example, may expand TNRs via replication slippage [16], but in terminally differentiated brain neurons, which do not replicate their DNA, expansion likely involves DNA repair processes linked to transcription (Fig. 1) [15, 21, 88]. In mouse models of CAG repeat diseases, genetic experiments reinforce the idea of multiple mechanisms. The recognition components of mismatch repair (Msh2 and Msh3, which form the MutSβ complex) affect instability in the male and female germline and in a variety of somatic tissues; the major maintenance DNA methyltransferase, Dnmt1, affects repeat instability in the germline, but not in somatic tissues [17]; DNA ligase 1 alters repeat instability in the female germline, but has no effect in the male germline [18]; the glycosylase Ogg1 selectively changes instability only in somatic tissues [19, 20]; and the key nucleotide excision repair component Xpa selectively affects instability in neuronal tissues [21].

Understanding the tissue specificity of TNR instability and the networks of proteins that control TNR instability will be a challenge, to say the least. The key question, however, is can we bypass all this complexity using a therapeutic strategy that directly attacks the offending TNR. One such strategy is to use engineered nucleases to introduce DNA strand breaks directly into the repeat locus.

Engineered Nucleases and TNR Contraction

Among the available engineered nucleases—ZFNs, TALENs, and CRISPR/Cas nucleases—only ZFNs and TALENs have thus far been tested for cleavage of TNRs in cells. In the first report, Mittelman et al. constructed a CAG-specific ZFN from two components: zfGCA and zfGCT, which recognize the CAG and CTG strands, respectively (Fig. 3A) [22]. Optimal binding of the individual zinc-finger recognition domains places the two components seven nucleotides apart, slightly farther than the preferred six-nucleotide separation, but acceptable for cleavage [89, 90]. Digestions with the purified components showed that the mixture of zfGCA and zfGCT efficiently cleaved repeat-containing plasmid DNA, as expected; however, the individual components also cleaved the DNA by themselves, presumably after homodimerization to activate the FokI cleavage domain. This result was not entirely unexpected given the cross-reactivity of GCA fingers for GCT sequences, and vice versa [91, 92], and the permissiveness of in vitro systems [93].

To assess ZFN-mediated cleavage of CAG repeats in mammalian cells, two selective assays were used to detect CAG repeat contraction. In both assays—one based on the APRT gene in hamster cells, the other on the HPRT minigene in human cells—the selectable gene was inactive due to a CAG_{95} tract in an intron, which behaves like an extra exon, causing aberrant splicing and blocking gene expression [27, 94]. Contraction of the repeat to less than about 40 units, however, allows sufficient gene expression for cells to survive selection. Treatment with zfGCA + zfGCT increased the frequency of surviving cells some 15-fold above background to an overall frequency of ~0.01 % [22]. Although treatment with zfGCA had no effect, the zfGCT component, by itself, was about 70 % as effective as the complete nuclease, confirming the in vitro results for this component.

These studies also examined the effect of tract length on the efficiency with which ZFNs generated $APRT^+$ or $HPRT^+$ colonies. Reduction in tract length by about 30 % (from CAG_{95} to CAG_{61} or CAG_{68}) caused a 2–3-fold drop in surviving cells [22]. This preference for longer repeat tracts has also been observed in a second study, which showed that ZFNs cleave CAG_{102} tracts efficiently, but do not destabilize CAG_{12} tracts in the same cells [95]. Steep length dependence suggests that the long repeat tracts typical of TNR diseases may be preferred targets over other repeats in the genome, which are uniformly shorter.

Previous analysis of the CAG tracts in surviving $APRT^+$ and $HRPT^+$ colonies had shown that they consisted almost entirely of simple contractions to less than 40 repeat units [27, 35, 94]. Surprisingly, analysis of the repeat tracts in colonies

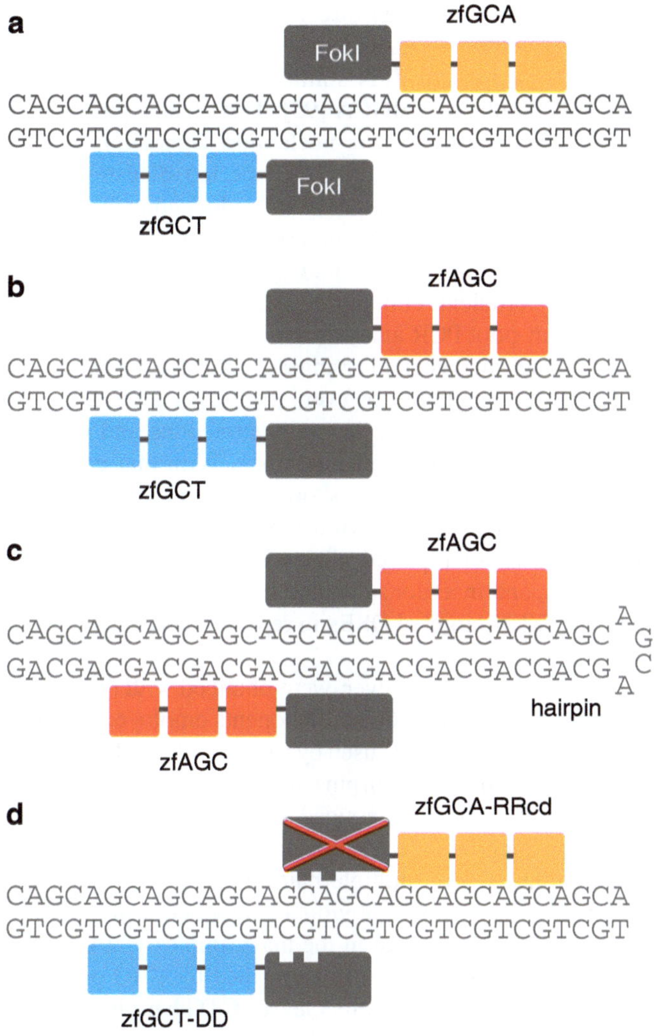

Fig. 3 ZFNs targeted to CAG TNRs. (**a**) Alignment of zfGCA and zfGCT on a CAG TNR. Each ZFN consists of three zinc-finger domains linked to a FokI DNA cleavage domain [22]. (**b**) Alignment of zfAGC and zfGCT on a CAG TNR [95]. (**c**) Alignment of zfAGC on a CAG hairpin [95]. (**d**) A ZFNickase. Both components were mutated to make them obligate heterodimers. One component was further mutated to render it cleavage dead

arising after ZFN treatment showed that only 55 % contained the expected contractions; the remainder displayed deletions or insertions at the CAG tract that prevented it from being included in the mRNA. Insertions and deletions support the idea that ZFN introduce DSBs, but they raise a potential concern for the goal of surgically shrinking TNRs, since DSBs can cause events that extend beyond the borders of the repeat tract.

Results in another study of CAG-directed ZFNs offer a potentially powerful approach to selectively shrinking CAG repeat tracts in a way that avoids DSBs. Liu et al. designed a two-component ZFN similar to the one discussed above, except that they used zfAGC in place of zfGCA [95]. This change positions the two components of the ZFN at the optimal six-nucleotide separation, which should give more efficient cleavage (Fig. 3B) [89, 90]. As expected, digestion with a mixture of zfAGC and zfGCT efficiently cleaved a PCR product containing a CAG_{102} repeat tract. Moreover, transfection of the two components into human cells containing a CAG_{102} tract (two transfections at day 0 and day 3, with analysis at day 6) revealed that cleavage was very efficient, eliminating virtually all the CAG_{102} alleles [95].

Liu et al. used small-pool PCR analysis to detect changes at the CAG_{102} tract. In principle, this technique can detect all changes—contractions and expansions, deletions and insertions. In practice, however, there is a PCR bias toward shorter sequences with fewer CAG repeats; hence, contractions and deletions are more readily detected than expansions and insertions. With that caveat, the products generated by ZFN cleavage are mostly shorter than the parental CAG_{102} tract and some products were isolated and shown to be contractions [95]. An analysis that defined the full spectrum of repaired alleles in ZFN-treated cells—contractions, expansions, deletions, and insertions—would be extremely valuable for evaluating the viability of using ZFN-targeted DSBs in repeat tracts as a therapeutic approach to TNR diseases.

The most surprising results, however, were obtained with the individual components [95]. In contrast to the individual components employed by Mittelman et al., neither the zfGCT and nor the zfAGC used by Liu et al. cleaved duplex DNA. Instead, they cleaved the CTG and CAG hairpins formed by the repeats (Fig. 3C). This hairpin-cleavage activity was first noticed in PCR amplification products of a CAG_{102} repeat tract, which include a set of higher bands, previously characterized as slipped-strand duplexes with multiple hairpins in out-of-register regions of the repeat [96]. Incubation with either individual component caused the hairpin bands to disappear with no obvious change in the main, duplex DNA band. These inferences with confirmed with defined hairpins formed by CAG or CTG strands, and extended to show that zfGCT specifically cleaves CTG hairpins and zfAGC specifically cleaves CAG hairpins.

These results are remarkable from a structural standpoint. Normally, each zinc-finger domain binds to three adjacent bases in the major groove of the DNA; however, CTG and CAG hairpins present major grooves with mismatched bases (T:T and A:A, respectively) every third nucleotide. Evidently, these mismatches minimally alter the ability of homodimeric zfGCT and homodimeric zfAGC to bind and cleave TNR hairpins [95].

Hairpin-specific reagents offer the possibility of probing the mechanism of TNR instability. Liu et al. showed the individual components, zfGCT and zfAGC, when transfected into cells, caused a reduction in the CAG_{102} band, albeit to a lower extent that transfection with the mixture of components [95]. In contrast to the results with the mixture of zfGCT and zfAGC, the individual components generated products that were all shorter than the parental CAG_{102} tract. The authors interpreted this result as evidence that hairpins form in cells and suggest that it is because of the adjacent origin of replication built into their initial constructs. They buttressed this speculation by showing that CAG_{102} tracts in serum-starved cells, which do not replicate their DNA, were not sensitive to the individual components, yet were fully cleaved by the mixture [95].

Liu et al. did not characterize the short products generated to determine whether they were due solely to repeat contractions, but that is a reasonable possibility. Since cleavage of a hairpin delivers damage to just one strand—presumably generating a repeat tract with a gap in one strand—treatment with the individual components may avoid DSBs, which are the source of deletions and insertions. From a therapeutic standpoint, hairpin-specific ZFNs offer two distinct advantages over ZFNs that cleave the repeat tract itself. First, they produce contractions exclusively, unlike ZFN-induced DSBs, which also generate expansions that can increase deleterious effects in cells, perhaps exacerbating the disease phenotype in patients. Second, hairpin-specific ZFNs are naturally targeted to longer repeat tracts—those more likely associated with disease—because only longer repeats can form stable hairpins [97, 98]. One disadvantage is that hairpin-specific ZFNs seem to require replication to generate the substrate for cleavage [95], which may render them unsuited for use in terminally differentiated cells, which do not replicate their DNA. In such cells, however, transcription through repeat tracts may generate the necessary hairpin substrates [21, 27, 31, 35]. Clearly, the effects of hairpin-directed ZFNs will need to be tested in vivo to define their properties.

Moye et al. have taken a different approach—developing zinc-finger nickases (ZFNickases)—to try to focus damage onto one strand and avoid the potential problems related to DSBs [99]. In general, building ZFNickases requires that the FokI domain of one component be rendered incapable of cleavage, so that when the two components are paired, only one component can cut its DNA strand. In addition to introducing such cleavage-dead (cd) mutations, designing ZFNickases for CAG repeat tracts required another modification to eliminate cleavage by homodimers of the individual component [22]. By introducing mutations into the FokI dimerization domain, Moye et al. generated the so-called "RR" (D483R) and "DD" (R487D) variants of zfGCT and zfGCA, which do not homodimerize (Fig. 3D) [100].

To test the activity of the various components, Moye et al. used a supercoiled 582-bp minicircle that contained a CAG_{54} repeat tract (Fig. 4a). Each of the zinc-finger components was transcribed and translated in vitro from plasmid DNA and tested by incubation with supercoiled minicircle DNA. The individual unmodified components (wild type) yielded a mixture of nicked circular and linear products (Fig. 4b, lanes 1 and 2) [99]. The presence of linear products rules out the possibility that zfGCT or zfGCA is specific for CTG or CAG hairpins, which would yield exclusively nicked circles, since the damage would be confined to one strand. The different behavior of zfGCT observed by Moye et al. and Liu et al. may reflect differences in the actual zinc-finger domains used to construct the zinc-finger components.

As expected, the individual modified components—zfGCA-RR, zfGCT-DD, zfGCA-DD, and zfGCT-RR—did not cut the supercoiled minicircle (Fig. 4b, lanes

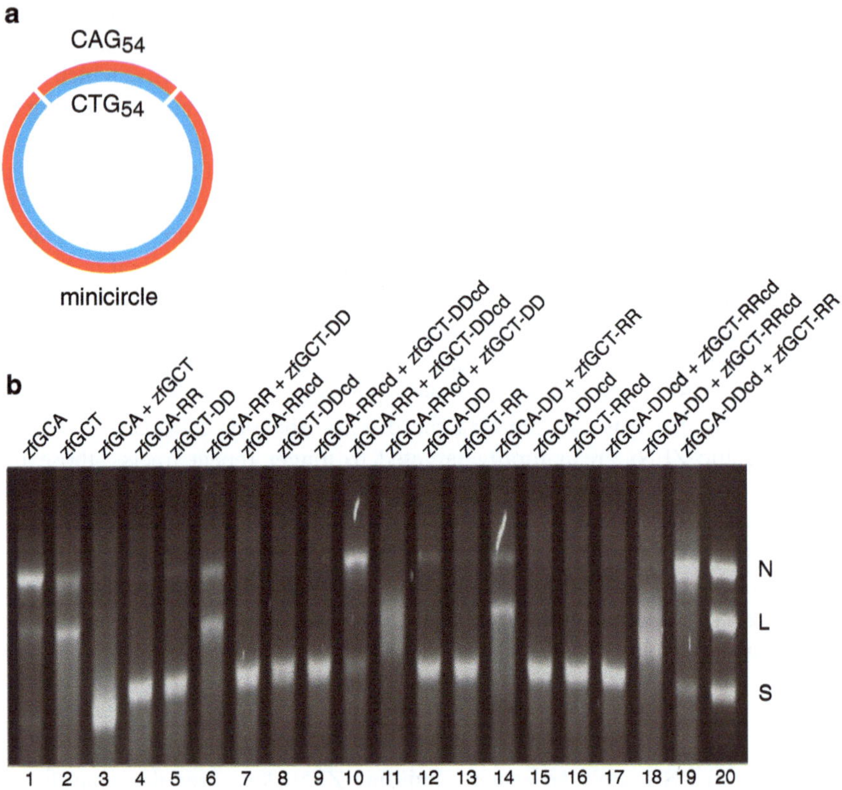

Fig. 4 In vitro assay to test ZFNs and ZFNickases. (**a**) *Minicircle* containing a CAG_{54} TNR. (**b**) Cleavage by various ZFN components. Cleavage by various ZFN components. In vitro transcribed and translated ZFN components were incubated with supercoiled minicircles for 60 min and the products were separated by electrophoresis on agarose gels. *Nicked circles* (N), *linears* (*L*), and *supercoils* (*S*). Trimmed linears migrate just below the *supercoiled minicircle*; *gapped circles* migrate as a broad band just below the *linear minicircle*

4, 5, 12, and 13). Appropriate combinations of modified components (zfGCA-RR with zfGCT-DD and zfGCA-DD with zfGCT-RR) cleaved the minicircles to mixtures of linears and nicked circles (lanes 6 and 14). The modified components cleaved the minicircle at about one third the rate of the wild type components, which cleaved the minicircles to shortened linears with most of the CAG repeat tract removed (Fig. 4b, lane 3) [99]. Decreased rates of cleavage have also been noted in other obligate heterodimeric ZFNs [101].

To generate ZFNickases, the obligate heterodimer mutations with combined with a mutation in the FokI active site (D450A). As expected, neither the individual cleavage-dead components nor mixtures of cleavage-dead components cleaved the minicircles (Fig. 4b, lanes 7, 8, 9, 15, 16, and 17). By contrast, the mixtures of zfGCA-RR with zfGCT-DDcd (lane 10) and zfGCA-DDcd with zfGCT-RR (lane 19), converted supercoils nearly completely to nicked circles. Surprisingly, the seemingly equivalent mixtures of zfGCA-RRcd with zfGCT-DD (lane 11) and zfGCA-DD with zfGCT-RRcd (lane 18) generated a novel band that proved to be a gapped circle with much of the repeat tract removed from one strand [99]. These results correlate with the rate of nicking, which is 3–4-fold more rapid in mixtures that produce gaps as compared to those that make nicks [99]. This result suggests that the RR modification decreases the activity of the FokI cleavage domain.

Moye et al. addressed the key question of whether ZFN nickases destabilize CAG repeats in cells using a GFP-based assay [102] analogous to the HPRT and APRT assays used by Mittelman et al. [22, 99]. The main difference is that cells with shorter CAG repeat tracts can be identified by cell sorting instead of selection. Treatment with the wild type ZFN (zfGCA plus zfGCT) increased GFP+ cells 9.3-fold increase above background. Treatment with the more active ZFNickase (zfGCA-DD plus zfGCT-RRcd) increased GFP+ cells 3.5-fold, whereas treatment with the less active ZFNickase (zfGCA-DDcd plus zfGCT-RR) gave no increase above background [99]. Notably, all the analyzed colonies from the more active ZFNickase all contained contractions of the repeat. These results suggest that ZFNickases, like hairpin-specific ZFNs, may be reasonable alternatives to ZFNs that introduce DSBs into CAG repeat tracks.

Most recently, a TALEN has been demonstrated to be an effective cleavage reagent for a chromosomal CAG repeat tract in yeast [23]. The TALEN was designed with one component anchored to the unique sequences flanking the repeat tract and the other directed at the repeat itself. When expressed in cells, this TALEN efficiently cleaved the repeat, producing exclusively contractions. Moreover, contractions were found only in the targeted repeat, not in any of the other CAG repeats in the yeast genome, as expected by the design of the TALEN. Although DSBs produce indels and expansions in human cells [22, 95], TALENs (and CRISPR/Cas nucleases) can also be modified to introduce nicks.

Challenges and Limitations of DNA-Directed Therapy

Disease-causing TNR tracts present a distinct challenge for therapy by engineered nucleases. Whether the disease is dominant or recessive, the ideal therapy would specifically target the expanded allele to shrink the TNR at the genomic level. The key requirement would be to preferentially manipulate the expanded repeat allele. The therapy would not necessarily have to shrink the expanded repeat completely to normal lengths. Even small contractions to the repeat tract would have therapeutic potential, as disease severity and onset are not simply a function of initial length, but also a function of length-dependent somatic instability that develops in affected tissue such as the striatum and cortex [82, 103, 104]. Because the rate of somatic expansions depends on TNR length, reducing the repeat tract even slightly would reduce the frequency of somatic expansion events, which would tend to delay disease onset and decrease its severity. Permanently shrinking an expanded repeat allele would significantly advance therapy for TNR disorders.

A key consideration—unique to TNR diseases—is that the same repeat responsible for disease at one gene locus resides at many other sites in the human genome, in various lengths and purities: in no case, is a disease-causing TNR unique [3]. Thus, strategies for therapeutic attack on an offending TNR must contend with the consequences of friendly fire at other sites in the genome. The most common TNRs in the genome, however, are not long enough to be effective targets. Less than 20 % of perfect CAG repeats in the human genome are longer than 9 units—the minimum cleavage site for zfGCA/zfGCT [22]—and only 6 repeat tracts are longer than 20 units [105]. Although there may relatively few natural targets for these TNR-directed engineered nucleases, it will still be critical to fully characterize mutagenic effects to the genome. Even nucleases designed to unique genomic targets can trigger off-target DNA breaks [106–108].

Next-generation sequencing offers rapid and cost-effective methods for measuring nuclease-mediated changes to target loci, as well a means to quantify genome-wide off-target cleavage events that lead to mutagenic repair. A variety of these newer sequencing platforms are now available, with many more emerging platforms in development. Today, sequencing methods offer different combinations of tradeoffs in cost, speed, throughput, read lengths, error rates, and bias [109]. For TNR analysis, the most important parameter will be read length; if the sequencing read does not span the TNR, it will be hard to accurately genotype the locus. In addition, data analysis methods have improved dramatically in the last few years, enabling accurate detection of point mutations and small indels [110], as well as more complex changes such as microsatellite repeat variation [111–113], and larger structural events [114]. Showcasing the power of next-generation sequencing are recent studies that characterized mutations and overall genomic instability from the recent sequencing and reconstruction of the HeLa genome, which is greatly mutated, and in some genomic locations, shattered [115–117]. Furthermore, some of the greatly expanded triplet repeats, which were traditionally inaccessible even to Sanger sequencing, can now be studied at the sequence level, using long read sequencing platforms such as PacBio [118]. Such platforms offer the possibility of rapidly measuring the frequencies repeat expansion and contraction, while retaining the ability to resolve insertions of foreign sequence and deletions outside the repeat tract.

Summary and Perspectives

Directly attacking the TNRs that cause disease has become a distinct possibility with the development of designer nucleases that can target breaks to a repeat tract. Thus far, only ZFNs have been tested in human cells, and only against CAG repeat tracts, but they offer an efficient way to modify TNRs, with a strong bias toward repeat contraction. At present, studies with engineered nucleases have revealed three potential ways to deal with CAG repeat tracts: by introducing DSBs into the tract, by generating nicks in the CAG strand or the CTG strand, and by cutting the CAG or CTG hairpins that form at the repeats [22, 23, 95, 99]. Although the spectrum of events generated by each type of nuclease is not well enough characterized, it appears that ZFNs that break one strand—ZFNickases and hairpin-specific ZFNs—offer the possibility for shrinking repeat tracts without the added complication of deletions and insertions. Thus, it seems likely that engineered nucleases of one kind or another will be found that can produce the desired beneficial effect—to shrink the repeat.

Unlike unique target sites, however, TNRs are present in multiple copies in the genome, rendering the genome especially vulnerable to the off-target action of TNR-specific engineered nucleases. For CAG-specific ZFNs, the steep length dependence of cleavage will mitigate such off-target effects [22, 95], because the vast majority of CAG repeat tracts are short [105]. In addition, most—about three quarters—of the CAG repeats are located outside of exons [105]; thus, their accidental modification may have little consequence for cell viability or function. Alternatively, TALENs and CRISPR/Cas nucleases can be designed to recognize unique sequences at one edge of the target repeat to focus cleavage selectively at the desired genomic site [23]. This strategy is problematical for ZFNs, because zinc-finger domains that bind tightly and selectively are not available for all DNA triplets. Regardless of the strategy, it will be essential to characterize the genome-wide sensitivity of TNRs to modification by the engineered nuclease, a task well within the capabilities of next-generation sequencing technologies. Only then will it be possible to evaluate the true therapeutic usefulness of nuclease-mediated repeat contraction as an approach to treating TNR diseases.

Our focus in this review has been entirely on engineered nucleases and the strengths and limitations of using them to shrink disease-causing TNRs. We've ignored the bigger challenge: how to use this therapy in humans. Assuming we were able to build an engineered nuclease with all the right properties—efficient cutting at the disease TNR, exclusive bias toward contractions, and minimal cleavage at other sites in the genome—how would we deploy it in humans? Treatment of the germline, sperm, egg, fertilized egg, or early embryo would eliminate the disease in the offspring, and in subsequent generations, but that approach is fraught with almost insurmountable ethical and societal questions. The alternative is to treat the affected individual. Most TNR diseases involve a limited set of cells or tissues. Thus, it should be possible to express the nuclease in the affected cells by a combination of local delivery, tissue- or cell-specific vector targeting, and cell-specific nuclease expression. In principle, such an approach could preserve at-risk cells and eliminate disease symptoms. It would not, of course, alter the germline, which would leave the patient's future children at risk. More serious, though, is the possibility that TNR effects in a secondary population of cells—previously masked by the defects in the primary cell population—would cause a new disease, requiring additional treatment. Clearly, there is much to be done if such a therapy is to be developed, but engineered nucleases are beginning to fulfill their promise.

Acknowledgement The authors thank Drs. Shondra Pruett-Miller and Matthew Porteus for construction of ZFNucleases; Drs. Lynn Zechiedrich, Jonathan Fogg, and Vincent Dion for construction of minicircle DNA; and Drs. Beatriz Santillan, Yi Zhang, and Ramon Roman-Sanchez for help with various assays.This work was supported by grants from the NIH to JHW (GM38219 and EY11731) and DM (NS079926), and by a T-32 Training Grant to CM (EY07001).

References

1. Richard GF, Kerrest A, Dujon B. Comparative genomics and molecular dynamics of DNA repeats in eukaryotes. Microbiol Mol Biol Rev. 2008;72(4):686–727.
2. Rando OJ, Verstrepen KJ. Timescales of genetic and epigenetic inheritance. Cell. 2007;128(4):655–68.
3. Gemayel R, Vinces MD, Legendre M, Verstrepen KJ. Variable tandem repeats accelerate evolution of coding and regulatory sequences. Annu Rev Genet. 2010;44:445–77.
4. Legendre M, Pochet N, Pak T, Verstrepen KJ. Sequence-based estimation of minisatellite and microsatellite repeat variability. Genome Res. 2007;17(12):1787–96.
5. Subramanian S, Mishra RK, Singh L. Genome-wide analysis of microsatellite repeats in humans: their abundance and density in specific genomic regions. Genome Biol. 2003;4(2):R13.
6. Verstrepen KJ, Jansen A, Lewitter F, Fink GR. Intragenic tandem repeats generate functional variability. Nat Genet. 2005;37(9):986–90.
7. Kashi Y, King DG. Simple sequence repeats as advantageous mutators in evolution. Trends Genet. 2006;22(5):253–9.
8. Mittelman D, Wilson JH. Stress, genomes, and evolution. Cell Stress Chaperones. 2010;15(5):463–6.
9. Orr HT, Zoghbi HY. Trinucleotide repeat disorders. Annu Rev Neurosci. 2007;30:575–621.
10. Brouwer JR, Willemsen R, Oostra BA. Microsatellite repeat instability and neurological disease. Bioessays. 2009;31(1):71–83.
11. Pearson CE, Edamura KN, Cleary JD. Repeat instability: mechanisms of dynamic mutations. Nat Rev Genet. 2005;6(10):729–42.
12. Albrecht A, Mundlos S. The other trinucleotide repeat: polyalanine expansion disorders. Curr Opin Genet Dev. 2005;15(3):285–93.
13. La Spada AR, Taylor JP. Repeat expansion disease: progress and puzzles in disease pathogenesis. Nat Rev Genet. 2010;11(4):247–58.
14. Gomes-Pereira M, Monckton DG. Chemical modifiers of unstable expanded simple sequence repeats: what goes up, could come down. Mutat Res. 2006;598(1-2):15–34.
15. Lin Y, Hubert Jr L, Wilson JH. Transcription destabilizes triplet repeats. Mol Carcinog. 2009;48(4):350–61.
16. Mirkin SM. Expandable DNA, repeats and human disease. Nature. 2007;447(7147):932–40.
17. Dion V, Lin Y, Hubert Jr L, Waterland RA, Wilson JH. Dnmt1 deficiency promotes CAG repeat expansion in the mouse germline. Hum Mol Genet. 2008;17(9):1306–17.
18. Tome S, Panigrahi GB, Lopez Castel A, Foiry L, Melton DW, Gourdon G, et al. Maternal germline-specific effect of DNA ligase I on CTG/CAG instability. Hum Mol Genet. 2011;20(11):2131–43.
19. Kovtun IV, Liu Y, Bjoras M, Klungland A, Wilson SH, McMurray CT. OGG1 initiates age-dependent CAG trinucleotide expansion in somatic cells. Nature. 2007;447(7143):447–52.
20. Kovtun IV, Johnson KO, McMurray CT. Cockayne syndrome B protein antagonizes OGG1 in modulating CAG repeat length in vivo. Aging (Albany NY). 2011;3:509–14.
21. Hubert Jr L, Lin Y, Dion V, Wilson JH. Xpa deficiency reduces CAG trinucleotide repeat instability in neuronal tissues in a mouse model of SCA1. Hum Mol Genet. 2011;20(24):4822–30.
22. Mittelman D, Moye C, Morton J, Sykoudis K, Lin Y, Carroll D, et al. Zinc-finger directed double-strand breaks within CAG repeat tracts promote repeat instability in human cells. Proc Natl Acad Sci U S A. 2009;106(24):9607–12.
23. Richard GF, Viterbo D, Khanna V, Mosbach V, Castelain L, Dujon B. Highly specific contractions of a single CAG/CTG trinucleotide repeat by TALEN in yeast. PLoS One. 2014;9(4), e95611.
24. Nithianantharajah J, Hannan AJ. Dynamic mutations as digital genetic modulators of brain development, function and dysfunction. Bioessays. 2007;29(6):525–35.

25. Freudenreich CH, Kantrow SM, Zakian VA. Expansion and length-dependent fragility of CTG repeats in yeast. Science. 1998;279(5352):853–6.
26. Kim N, Jinks-Robertson S. Transcription as a source of genome instability. Nat Rev Genet. 2011;13(3):204–14.
27. Lin Y, Dion V, Wilson JH. Transcription promotes contraction of CAG repeat tracts in human cells. Nat Struct Mol Biol. 2006;13(2):179–80.
28. Kim N, Jinks-Robertson S. Guanine repeat-containing sequences confer transcription-dependent instability in an orientation-specific manner in yeast. DNA Repair (Amst). 2012;10(9):953–60.
29. Wierdl M, Greene CN, Datta A, Jinks-Robertson S, Petes TD. Destabilization of simple repetitive DNA sequences by transcription in yeast. Genetics. 1996;143(2):713–21.
30. Mochmann LH, Wells RD. Transcription influences the types of deletion and expansion products in an orientation-dependent manner from GAC*GTC repeats. Nucleic Acids Res. 2004;32(15):4469–79.
31. Lin Y, Leng M, Wan M, Wilson JH. Convergent transcription through a long CAG tract destabilizes repeats and induces apoptosis. Mol Cell Biol. 2010;30(18):4435–51.
32. Nakamori M, Pearson CE, Thornton CA. Bidirectional transcription stimulates expansion and contraction of expanded (CTG)*(CAG) repeats. Hum Mol Genet. 2011;20(3):580–8.
33. Jung J, Bonini N. CREB-binding protein modulates repeat instability in a Drosophila model for polyQ disease. Science. 2007;315(5820):1857–9.
34. Lin Y, Dent SY, Wilson JH, Wells RD, Napierala M. R loops stimulate genetic instability of CTG.CAG repeats. Proc Natl Acad Sci U S A. 2010;107(2):692–7.
35. Lin Y, Wilson JH. Transcription-induced CAG repeat contraction in human cells is mediated in part by transcription-coupled nucleotide excision repair. Mol Cell Biol. 2007;27(17):6209–17.
36. Lopez Castel A, Cleary JD, Pearson CE. Repeat instability as the basis for human diseases and as a potential target for therapy. Nat Rev Mol Cell Biol. 2010;11(3):165–70.
37. Schweitzer JK, Livingston DM. The effect of DNA replication mutations on CAG tract stability in yeast. Genetics. 1999;152(3):953–63.
38. Rosche W, Jaworski A, Kang S, Kramer S, Larson J, Geidroc D, et al. Single-stranded DNA-binding protein enhances the stability of CTG triplet repeats in Escherichia coli [published erratum J Bacteriol 1996;178(23):7024]. J Bacteriol. 1996;178(16):5042-4.
39. Lopez Castel A, Tomkinson AE, Pearson CE. CTG/CAG repeat instability is modulated by the levels of human DNA ligase I and its interaction with proliferating cell nuclear antigen: a distinction between replication and slipped-DNA repair. J Biol Chem. 2009;284(39):26631–45.
40. Hou C, Chan NL, Gu L, Li GM. Incision-dependent and error-free repair of (CAG)(n)/(CTG)(n) hairpins in human cell extracts. Nat Struct Mol Biol. 2009;16(8):869–75.
41. Pelletier R, Krasilnikova MM, Samadashwily GM, Lahue R, Mirkin SM. Replication and expansion of trinucleotide repeats in yeast. Mol Cell Biol. 2003;23(4):1349–57.
42. Razidlo DF, Lahue RS. Mrc1, Tof1 and Csm3 inhibit CAG.CTG repeat instability by at least two mechanisms. DNA Repair (Amst). 2008;7(4):633–40.
43. Collins NS, Bhattacharyya S, Lahue RS. Rev1 enhances CAG.CTG repeat stability in Saccharomyces cerevisiae. DNA Repair (Amst). 2007;6(1):38–44.
44. Napierala M, Bacolla A, Wells RD. Increased negative superhelical density in vivo enhances the genetic instability of triplet repeat sequences. J Biol Chem. 2005;280(45):37366–76.
45. Hubert Jr L, Lin Y, Dion V, Wilson JH. Topoisomerase 1 and single-strand break repair modulate transcription-induced CAG repeat contraction in human cells. Mol Cell Biol. 2011;31(15):3105–12.
46. Bhattacharyya S, Lahue RS. Saccharomyces cerevisiae Srs2 DNA helicase selectively blocks expansions of trinucleotide repeats. Mol Cell Biol. 2004;24(17):7324-30.
47. Dhar A, Lahue RS. Rapid unwinding of triplet repeat hairpins by Srs2 helicase of Saccharomyces cerevisiae. Nucleic Acids Res. 2008;36(10):3366–73.
48. Kerrest A, Anand RP, Sundararajan R, Bermejo R, Liberi G, Dujon B, et al. SRS2 and SGS1 prevent chromosomal breaks and stabilize triplet repeats by restraining recombination. Nat Struct Mol Biol. 2009;16(2):159–67.

49. Pearson CE, Ewel A, Acharya S, Fishel RA, Sinden RR. Human MSH2 binds to trinucleotide repeat DNA structures associated with neurodegenerative diseases. Hum Mol Genet. 1997;6(7):1117–23.
50. Manley K, Shirley TL, Flaherty L, Messer A. Msh2 deficiency prevents in vivo somatic instability of the CAG repeat in Huntington disease transgenic mice. Nat Genet. 1999;23(4):471–3.
51. Schweitzer JK, Livingston DM. Destabilization of CAG trinucleotide repeat tracts by mismatch repair mutations in yeast. Hum Mol Genet. 1997;6(3):349–55.
52. Jaworski A, Rosche WA, Gellibolian R, Kang S, Shimizu M, Bowater RP, et al. Mismatch repair in Escherichia coli enhances instability of (CTG)n triplet repeats from human hereditary diseases. Proc Natl Acad Sci U S A. 1995;92(24):11019–23.
53. Gomes-Pereira M, Fortune MT, Ingram L, McAbney JP, Monckton DG. Pms2 is a genetic enhancer of trinucleotide CAG{middle dot}CTG repeat somatic mosaicism: implications for the mechanism of triplet repeat expansion. Hum Mol Genet. 2004;13(16):1815-25.
54. Jarem DA, Wilson NR, Delaney S. Structure-dependent DNA damage and repair in a trinucleotide repeat sequence. Biochemistry. 2009;48(28):6655–63.
55. Parniewski P, Bacolla A, Jaworski A, Wells RD. Nucleotide excision repair affects the stability of long transcribed (CTG*CAG) tracts in an orientation-dependent manner in Escherichia coli. Nucleic Acids Res. 1999;27(2):616–23.
56. Richard GF, Dujon B, Haber JE. Double-strand break repair can lead to high frequencies of deletions within short CAG/CTG trinucleotide repeats. Mol Gen Genet. 1999;261(4-5):871–82.
57. Richard GF, Goellner GM, McMurray CT, Haber JE. Recombination-induced CAG trinucleotide repeat expansions in yeast involve the MRE11-RAD50-XRS2 complex. EMBO J. 2000;19(10):2381–90.
58. Mittelman D, Sykoudis K, Hersh M, Lin Y, Wilson JH. Hsp90 modulates CAG repeat instability in human cells. Cell Stress Chaperones. 2010;15(5):753–9.
59. Hebert ML, Spitz LA, Wells RD. DNA double-strand breaks induce deletion of CTG.CAG repeats in an orientation-dependent manner in Escherichia coli. J Mol Biol. 2004;336(3):655–72.
60. Bowater RP, Jaworski A, Larson JE, Parniewski P, Wells RD. Transcription increases the deletion frequency of long CTG.CAG triplet repeats from plasmids in Escherichia coli. Nucleic Acids Res. 1997;25(14):2861–8.
61. Reddy K, Tam M, Bowater RP, Barber M, Tomlinson M, Nichol Edamura K, et al. Determinants of R-loop formation at convergent bidirectionally transcribed trinucleotide repeats. Nucleic Acids Res. 2011;39(5):1749–62.
62. Gorbunova V, Seluanov A, Mittelman D, Wilson JH. Genome-wide demethylation destabilizes CTG.CAG trinucleotide repeats in mammalian cells. Hum Mol Genet. 2004;13(23):2979–89.
63. Debacker K, Frizzell A, Gleeson O, Kirkham-McCarthy L, Mertz T, Lahue RS. Histone deacetylase complexes promote trinucleotide repeat expansions. PLoS Biol. 2012;10(2), e1001257.
64. Lahiri M, Gustafson TL, Majors ER, Freudenreich CH. Expanded CAG repeats activate the DNA damage checkpoint pathway. Mol Cell. 2004 2004/7/23;15(2):287-93.
65. Pieretti M, Zhang FP, Fu YH, Warren ST, Oostra BA, Caskey CT, et al. Absence of expression of the FMR-1 gene in fragile X syndrome. Cell. 1991;66(4):817–22.
66. Bidichandani SI, Ashizawa T, Patel PI. The GAA triplet-repeat expansion in Friedreich ataxia interferes with transcription and may be associated with an unusual DNA structure. Am J Hum Genet. 1998;62(1):111–21.
67. Grabczyk E, Usdin K. The GAA*TTC triplet repeat expanded in Friedreich's ataxia impedes transcription elongation by T7 RNA polymerase in a length and supercoil dependent manner. Nucleic Acids Res. 2000;28(14):2815–22.
68. Soragni E, Herman D, Dent SY, Gottesfeld JM, Wells RD, Napierala M. Long intronic GAA*TTC repeats induce epigenetic changes and reporter gene silencing in a molecular model of Friedreich ataxia. Nucleic Acids Res. 2008;36(19):6056–65.

69. Osborne RJ, Thornton CA. RNA-dominant diseases. Hum Mol Genet. 2006;15 Spec No 2:R162-9.
70. Lee JE, Cooper TA. Pathogenic mechanisms of myotonic dystrophy. Biochem Soc Trans. 2009;37(Pt 6):1281–6.
71. Li Y, Jin P. RNA-mediated neurodegeneration in fragile X-associated tremor/ataxia syndrome. Brain Res. 2012;1462:112–7.
72. Willemsen R, Levenga J, Oostra BA. CGG repeat in the FMR1 gene: size matters. Clin Genet. 2011;80(3):214–25.
73. Lin Y, Wilson JH. Transcription-induced DNA toxicity at trinucleotide repeats: double bubble is trouble. Cell Cycle. 2011;10(4):611–8.
74. Katayama S, Tomaru Y, Kasukawa T, Waki K, Nakanishi M, Nakamura M, et al. Antisense transcription in the mammalian transcriptome. Science. 2005;309(5740):1564–6.
75. He Y, Vogelstein B, Velculescu VE, Papadopoulos N, Kinzler KW. The antisense transcriptomes of human cells. Science. 2008;322(5909):1855–7.
76. Wells RD. Non-B DNA, conformations, mutagenesis and disease. Trends Biochem Sci. 2007;32(6):271–8.
77. Sinden RR, Potaman VN, Oussatcheva EA, Pearson CE, Lyubchenko YL, Shlyakhtenko LS. Triplet repeat DNA structures and human genetic disease: dynamic mutations from dynamic DNA. J Biosci. 2002;27(1 Suppl 1):53–65.
78. Yoon SR, Dubeau L, de Young M, Wexler NS, Arnheim N. Huntington disease expansion mutations in humans can occur before meiosis is completed. Proc Natl Acad Sci U S A. 2003;100(15):8834–8.
79. Shelbourne PF, Keller-McGandy C, Bi WL, Yoon SR, Dubeau L, Veitch NJ, et al. Triplet repeat mutation length gains correlate with cell-type specific vulnerability in Huntington disease brain. Hum Mol Genet. 2007;16(10):1133–42.
80. Reyniers E, Martin JJ, Cras P, Van Marck E, Handig I, Jorens HZ, et al. Postmortem examination of two fragile X brothers with an FMR1 full mutation. Am J Med Genet. 1999;84(3):245–9.
81. De Biase I, Rasmussen A, Endres D, Al-Mahdawi S, Monticelli A, Cocozza S, et al. Progressive GAA expansions in dorsal root ganglia of Friedreich's ataxia patients. Ann Neurol. 2007;61(1):55–60.
82. Swami M, Hendricks AE, Gillis T, Massood T, Mysore J, Myers RH, et al. Somatic expansion of the Huntington's disease CAG repeat in the brain is associated with an earlier age of disease onset. Hum Mol Genet. 2009;18(16):3039–47.
83. Lin Y, Dion V, Wilson JH. Transcription and triplet repeat instability. In: Wells RD, Ashizawa T, editors. Genetic instabilities and neurological diseases. 2nd ed. San Diego: Academic Press; 2006. p. 691-704.
84. De Biase I, Rasmussen A, Monticelli A, Al-Mahdawi S, Pook M, Cocozza S, et al. Somatic instability of the expanded GAA triplet-repeat sequence in Friedreich ataxia progresses throughout life. Genomics. 2007;90(1):1–5.
85. Tassone F, Hagerman RJ, Gane LW, Taylor AK. Strong similarities of the FMR1 mutation in multiple tissues: postmortem studies of a male with a full mutation and a male carrier of a premutation. Am J Med Genet. 1999;84(3):240–4.
86. Taylor AK, Tassone F, Dyer PN, Hersch SM, Harris JB, Greenough WT, et al. Tissue heterogeneity of the FMR1 mutation in a high-functioning male with fragile X syndrome. Am J Med Genet. 1999;84(3):233–9.
87. Wohrle D, Salat U, Glaser D, Mucke J, Meisel-Stosiek M, Schindler D, et al. Unusual mutations in high functioning fragile X males: apparent instability of expanded unmethylated CGG repeats. J Med Genet. 1998;35(2):103–11.
88. Kaytor MD, Burright EN, Duvick LA, Zoghbi HY, Orr HT. Increased trinucleotide repeat instability with advanced maternal age. Hum Mol Genet. 1997;6(12):2135–9.
89. Bibikova M, Carroll D, Segal DJ, Trautman JK, Smith J, Kim YG, et al. Stimulation of homologous recombination through targeted cleavage by chimeric nucleases. Mol Cell Biol. 2001;21(1):289–97.

90. Handel EM, Alwin S, Cathomen T. Expanding or restricting the target site repertoire of zinc-finger nucleases: the inter-domain linker as a major determinant of target site selectivity. Mol Ther. 2009;17(1):104–11.
91. Liu Q, Xia Z, Zhong X, Case CC. Validated zinc finger protein designs for all 16 GNN DNA triplet targets. J Biol Chem. 2002;277(6):3850–6.
92. Segal DJ, Dreier B, Beerli RR, Barbas 3rd CF. Toward controlling gene expression at will: selection and design of zinc finger domains recognizing each of the 5′-GNN-3′ DNA target sequences. Proc Natl Acad Sci U S A. 1999;96(6):2758–63.
93. Carroll D, Morton JJ, Beumer KJ, Segal DJ. Design, construction and in vitro testing of zinc finger nucleases. Nat Protoc. 2006;1(3):1329–41.
94. Gorbunova V, Seluanov A, Dion V, Sandor Z, Meservy JL, Wilson JH. Selectable system for monitoring the instability of CTG/CAG triplet repeats in mammalian cells. Mol Cell Biol. 2003;23(13):4485–93.
95. Liu G, Chen X, Bissler JJ, Sinden RR, Leffak M. Replication-dependent instability at (CTG) × (CAG) repeat hairpins in human cells. Nat Chem Biol. 2010;6(9):652–9.
96. Pearson CE, Wang YH, Griffith JD, Sinden RR. Structural analysis of slipped-strand DNA (S-DNA) formed in (CTG)n. (CAG)n repeats from the myotonic dystrophy locus. Nucleic Acids Res. 1998;26(3):816–23.
97. Petruska J, Arnheim N, Goodman MF. Stability of intrastrand hairpin structures formed by the CAG/CTG class of DNA triplet repeats associated with neurological diseases. Nucleic Acids Res. 1996;24(11):1992–8.
98. McMurray CT. DNA secondary structure: a common and causative factor for expansion in human disease. Proc Natl Acad Sci U S A. 1999;96(5):1823–5.
99. Moye C. Generation of Instability within Trinucleotide Repeats. PhD Thesis, Baylor College of Medicine. 2013.
100. Szczepek M, Brondani V, Buchel J, Serrano L, Segal DJ, Cathomen T. Structure-based redesign of the dimerization interface reduces the toxicity of zinc-finger nucleases. Nat Biotechnol. 2007;25(7):786–93.
101. Miller JC, Holmes MC, Wang J, Guschin DY, Lee YL, Rupniewski I, et al. An improved zinc-finger nuclease architecture for highly specific genome editing. Nat Biotechnol. 2007;25(7):778–85.
102. Santillan BA, Moye C, Mittelman D, Wilson JH. GFP-Based Fluorescence Assay for CAG Repeat Instability in Cultured Human Cells. PLoS One. 2014;9(11), e113952.
103. Dragileva E, Hendricks A, Teed A, Gillis T, Lopez ET, Friedberg EC, et al. Intergenerational and striatal CAG repeat instability in Huntington's disease knock-in mice involve different DNA repair genes. Neurobiol Dis. 2009;33(1):37–47.
104. Wheeler VC, Lebel L-A, Vrbanac V, Teed A, te Riele H, MacDonald ME. Mismatch repair gene Msh2 modifies the timing of early disease in HdhQ111 striatum. Hum Mol Genet. 2003;12(3):273-81.
105. Kozlowski P, de Mezer M, Krzyzosiak WJ. Trinucleotide repeats in human genome and exome. Nucleic Acids Res. 2010;38(12):4027–39.
106. Gabriel R, Lombardo A, Arens A, Miller JC, Genovese P, Kaeppel C, et al. An unbiased genome-wide analysis of zinc-finger nuclease specificity. Nat Biotechnol. 2011;29(9):816–23.
107. Fu Y, Foden JA, Khayter C, Maeder ML, Reyon D, Joung JK, et al. High-frequency off-target mutagenesis induced by CRISPR-Cas nucleases in human cells. Nat Biotechnol. 2013;31(9):822–6.
108. Pattanayak V, Ramirez CL, Joung JK, Liu DR. Revealing off-target cleavage specificities of zinc-finger nucleases by in vitro selection. Nat Methods. 2011;8(9):765–70.
109. Ward RM, Schmieder R, Highnam G, Mittelman D. Big data challenges and opportunities in high-throughput sequencing. Syst Biomed. 2013;1:29–34.
110. DePristo MA, Banks E, Poplin R, Garimella KV, Maguire JR, Hartl C, et al. A framework for variation discovery and genotyping using next-generation DNA sequencing data. Nat Genet. 2011;43(5):491–8.

111. Fondon 3rd JW, Martin A, Richards S, Gibbs RA, Mittelman D. Analysis of microsatellite variation in Drosophila melanogaster with population-scale genome sequencing. PLoS One. 2012;7(3), e33036.
112. Gymrek M, Golan D, Rosset S, Erlich Y. lobSTR: A short tandem repeat profiler for personal genomes. Genome Res. 2012;22(6):1154–62.
113. Highnam G, Franck C, Martin A, Stephens C, Puthige A, Mittelman D. Accurate human microsatellite genotypes from high-throughput resequencing data using informed error profiles. Nucleic Acids Res. 2013;41(1), e32.
114. Rausch T, Zichner T, Schlattl A, Stutz AM, Benes V, Korbel JO. DELLY: structural variant discovery by integrated paired-end and split-read analysis. Bioinformatics. 2012;28(18):i333–i9.
115. Landry JJ, Pyl PT, Rausch T, Zichner T, Tekkedil MM, Stutz AM, et al. The genomic and transcriptomic landscape of a HeLa cell line. G3 (Bethesda). 2013;3(8):1213-24.
116. Mittelman D, Wilson JH. The fractured genome of HeLa cells. Genome Biol. 2013;14(4):111.
117. Adey A, Burton JN, Kitzman JO, Hiatt JB, Lewis AP, Martin BK, et al. The haplotype-resolved genome and epigenome of the aneuploid HeLa cancer cell line. Nature. 2013;500(7461):207–11.
118. Loomis EW, Eid JS, Peluso P, Yin J, Hickey L, Rank D, et al. Sequencing the unsequenceable: expanded CGG-repeat alleles of the fragile X gene. Genome Res. 2013;23(1):121–8.
119. Paques F, Haber JE. Multiple pathways of recombination induced by double-strand breaks in Saccharomyces cerevisiae. Microbiol Mol Biol Rev. 1999;63(2):349–404.
120. Owen BA, Yang Z, Lai M, Gajec M, Badger 2nd JD, Hayes JJ, et al. (CAG)(n)-hairpin DNA binds to Msh2-Msh3 and changes properties of mismatch recognition. Nat Struct Mol Biol. 2005;12(8):663–70.
121. Salinas-Rios V, Belotserkovskii BP, Hanawalt PC. DNA slip-outs cause RNA polymerase II arrest in vitro: potential implications for genetic instability. Nucleic Acids Res. 2011;39(17): 7444–54.

Using Engineered Nucleases to Create HIV-Resistant Cells

George Nicholas Llewellyn, Colin M. Exline, Nathalia Holt, and Paula M. Cannon

Abstract HIV-1/AIDS is often considered a priority disease in the development of genetic and cell based therapies because of the high burden imposed by current treatments, which require life-long adherence to antiretroviral drug regimens. Engineered nucleases have the capability to either disrupt a specific gene, or to promote precise gene edits or additions at the targeted gene. As one application for the gene disruption capabilities of the nucleases, HIV-1 infection provides an exceptional target in the CCR5 gene. This is the most commonly used entry co-receptor through which the virus enters into CD4+ T cells. Importantly, the loss of CCR5 is expected to be well-tolerated, since a relatively high percentage of individuals are naturally homozygous for a defective CCR5 allele. As a result, CCR5 disruption by zinc finger nuclease treatment of autologous T cells was the first-in-man use of engineered nucleases. Future applications to refine this therapy may include disrupting CCR5 in precursor hematopoietic stem cells, the additional disruption of the alternate HIV-1 co-receptor, CXCR4, in T cells, and the addition of other anti-HIV genes at a disrupted CCR5 locus to provide a combinatorial therapy. Finally, the gene disrupting actions of engineered nucleases could also be harnessed to inactivate the integrated HIV-1 genomes that persist in patients' cells despite drug therapy, and which thereby prevent the complete eradication of the virus by drug treatments.

Keywords Homology directed repair (HDR) • Hematopoietic stem cells (HSCs) • Non-homologous end-joining (NHEJ) • TAL effector nuclease (TALENs) • Zinc finger nucleases (ZFNs)

G.N. Llewellyn, Ph.D. • C.M. Exline, Ph.D. • N. Holt, Ph.D. • P.M. Cannon, Ph.D. (✉)
Department of Molecular Microbiology and Immunology, Keck School of Medicine, University of Southern California, 2011 Zonal Avenue, HMR502, Los Angeles, CA 90033, USA
e-mail: gllewell@usc.edu; colinexline@gmail.com; nat@nathaliaholt.com; pcannon@usc.edu

T. Cathomen et al. (eds.), *Genome Editing*, Advances in Experimental Medicine and Biology 895, DOI 10.1007/978-1-4939-3509-3_10

Introduction

Human immunodeficiency virus (HIV-1) causes a serious, life-long infection, with high rates of mortality in untreated individuals. Although the current combinations of antiretroviral drugs used to treat the infection are highly effective at suppressing HIV-1 replication, they do not ultimately cure people. This means that infected individuals have a life-long requirement for antiretroviral drug therapy, with associated high economic costs and the risk of developing drug toxicities, viral resistance, or "treatment fatigue". In fact, it is estimated that only 25 % of the HIV-infected population in the US successfully accesses therapy and achieves full viral suppression through antiretroviral drugs [1]. Therefore, alternate strategies to control and potentially cure HIV-1 infections are being considered, including those based on cell and genetic therapies.

HIV-1/AIDS has long been considered a candidate for gene therapy interventions, following early speculation that genetically modified cells could provide an 'intracellular immunization' to inhibit HIV-1 replication [2]. Some of the earliest gene therapy trials were for HIV-1, and typically used integrating retroviral vectors to allow for long-term expression of anti-HIV genes, such as the trans-dominant RevM10 protein and various RNA-based inhibitors (reviewed by Peterson et al. [3]). More recently, HIV-1 infection has proven itself uniquely suited to the first in-human use of engineered nucleases, based on CCR5 gene knockout by zinc finger nucleases (ZFNs) [4]. Beyond CCR5 disruption, future applications of engineered nucleases in anti-HIV therapies could exploit homologous recombination to insert anti-HIV genes at the disrupted CCR5 locus and thereby create a combinatorial gene therapy. Alternatively, the gene disrupting capabilities of the nucleases could be used to disable the integrated HIV-1 genomes present in infected cells, an attractive option for removing latently infected cells. The potential applications of engineered nucleases for HIV-1 therapy that will be discussed in this review are summarized in Fig. 1.

Disruption of the CCR5 Co-Receptor by ZFNs

Engineered Nucleases and DNA Repair

Engineered nucleases such as zinc finger nucleases (ZFNs), TAL effector nucleases (TALENs), homing endonucleases and the CRIPSR/Cas9 system all function in basically the same way, by creating a double-stranded break (DSB) in the DNA sequence to which they are targeted, which is then acted on by cellular DNA repair pathways. While the homing endonucleases and Cas9 contain natural endonuclease activities, ZFNs and TALENs are based on a modular design that links engineered DNA binding domains to the non-specific cleavage domain from a homodimeric type IIS nuclease, such as the *Fok*I restriction enzyme. Different cellular pathways

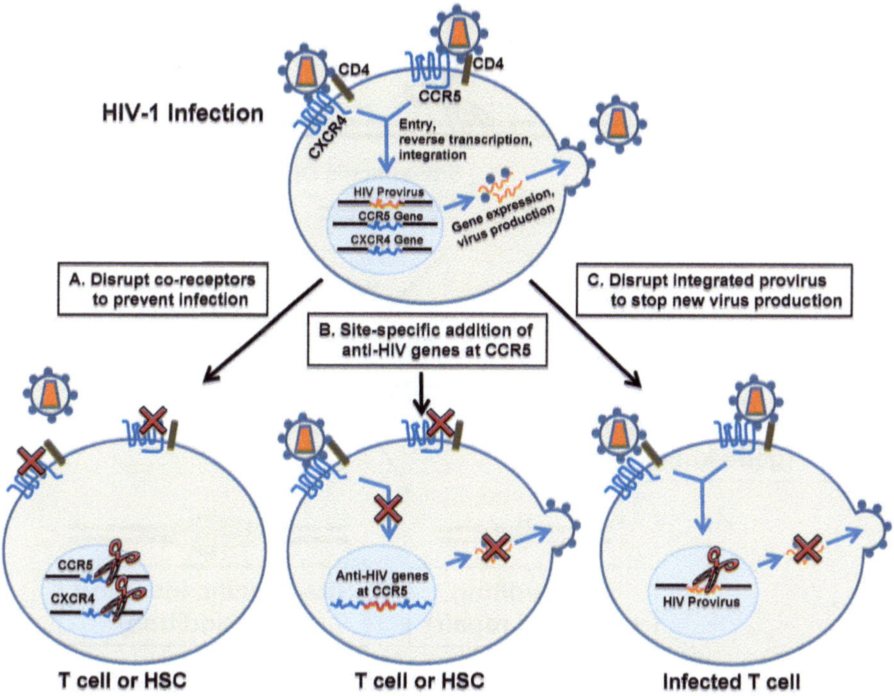

Fig. 1 Potential uses of engineered nucleases as anti-HIV therapies. The main stages of the HIV-1 life-cycle are shown. HIV-1 enters a target cell by binding to both CD4 and a co-receptor protein such as CCR5 or CXCR4, the preference for which is determined by the HIV-1 surface glycoprotein. Following entry, the viral RNA genome is reverse transcribed into a DNA form that permanently integrates into the host cell's genome. From here, the integrated provirus acts like a cellular gene, transcribing both mRNA to create HIV-1 proteins, and new viral RNA genomes, that are assembled together into particles that bud from the cell surface. (**a**) Engineered nucleases are used to disrupt the cellular co-receptor genes, CCR5 or CXCR4, in either CD4+ T cells (both genes), or their hematopoietic stem cell (HSC) precursors (CCR5 only) and thereby block HIV-1 entry. (**b**) Additional anti-HIV genes are inserted at the disrupted CCR5 locus, blocking other stages of the HIV-1 life-cycle. (**c**) Engineered nucleases are targeted to HIV-1 sequences, and thereby inactivate an integrated HIV-1 genome in an infected cell

can repair the DSBs so created, including the non-homologous end joining (NHEJ) pathway, where the frequent outcome is an insertion or deletion (indel) that can thereby lead to gene knockout (Fig. 2). Alternatively, DSBs can be more precisely repaired by recombination with a homologous sequence, such as a sister chromatid. Such homology directed repair (HDR) can also copy information from a homologous 'donor sequence', introduced into the cell at the same time as the engineered nuclease, and coding for any specific changes that are desired. The end result of this process can be a small genetic edit, for example to repair a point mutation in a defective gene, or the site-specific addition of a larger stretch of new genetic material at the site of the DSB. In this way, engineered nucleases can be used to direct three different outcomes: gene disruption, gene editing or gene addition (Fig. 2).

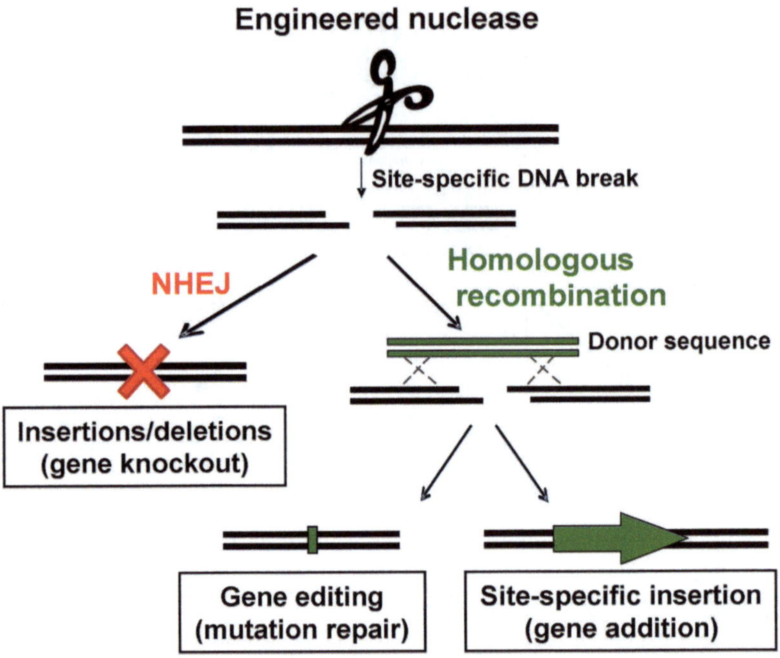

Fig. 2 Alternate outcomes following the action of engineered nucleases. Engineered nucleases create a double-stranded break (DSB) in the targeted gene. If the non-homologous end joining (NHEJ) repair pathway is used, a frequent outcome is disruption of the gene. Alternatively, DSBs can be repaired by homologous recombination, and the introduction of a homologous donor sequence into the cell can hijack this pathway to introduce a desired genetic edit, or to promote the site-specific addition of new genetic material at the site of the DSB

The repair of a DSB is committed to one of the available cellular pathways at an early time point. If exposed DNA ends are protected by the Ku70/80 complex, the NHEJ pathway is used [5], while HDR is initiated if a resection event occurs that exposes tracts of single-stranded DNA [6]. The choice of HDR or NHEJ is also influenced by the phase of the cell cycle, with HDR only occurring during and shortly after DNA replication in S and G2 [6, 7], when the required factors are available in active (phosphorylated) forms [8], and sister chromatids are in closer proximity and more able to serve as homology templates [9]. In contrast, NHEJ predominates in G1, although it is active throughout the whole cell cycle.

In hematopoietic stem cells (HSC), an important therapeutic target cell for many gene therapy applications, NHEJ is far more common than HDR [10, 11], which biases the outcome of nuclease activities. In addition, the co-introduction of DNA donor sequences along with a nuclease, as is needed to promote HDR gene editing, can result in significant toxicity to these cells. Because of these factors, NHEJ-mediated gene disruption is the most easily achieved result for engineered nucleases in HSC. However, therapeutic applications of gene disruption are likely to be limited. In this regard, the application of engineered nucleases to HIV-1 disease has found a uniquely suitable target in the CCR5 gene (Fig. 2).

Rationale for CCR5 Disruption as an Anti-HIV Therapy

CCR5 is a chemokine receptor that also functions as the major entry co-receptor used by HIV-1, in concert with the primary receptor, CD4 [12] (Fig. 1). However, its functions are not essential in humans, since a relatively high frequency of the population (~1 %) is homozygous for the defective CCR5Δ32 allele. Such individuals are correspondingly almost completely resistant to HIV-1 [13, 14] while not exhibiting any significant phenotypic defects [13, 15]. These properties have encouraged the development of a class of anti-HIV drugs targeted to CCR5, such as Maraviroc [16], which further established CCR5 as a therapeutic target.

In addition, compelling human data exists for the ability of CCR5-negative cells to confer HIV-resistance, from the well-documented case of the so-called 'Berlin patient'. This individual was an HIV-positive leukemia patient who received an HSC transplant, as part of his leukemia treatment, from a donor who was homozygous for the CCR5Δ32 mutation. Following the treatment, the Berlin patient has been HIV-1 free for 8 years, and is considered the first documented case of an HIV-1 cure [17–19]. Although the treatment included aggressive ablative regimens to destroy the leukemia, which likely also significantly reduced the burden of HIV-infected cells, the essential role of the CCR5Δ32 genotype of the donor is illustrated by the fact that other HIV patients undergoing bone marrow transplantation without a CCR5Δ32 donor have not been similarly cured [20, 21]. These findings support the idea that blocking CCR5 expression in a patient's own cells by genetic methods could also result in the control of HIV-1 replication.

Methods to block CCR5 have previously included the use of anti-CCR5 shRNAs or ribozymes, delivered using integrating retroviral or lentiviral vectors [22–27]. Such approaches can provide significant, albeit incomplete, inhibition, and will require the life-long expression of the anti-CCR5 moiety in the engineered cell. While the use of integrating vectors provides for this possibility, it is known that gene expression from such vectors can be silenced over time and that integrating viral vectors provide an unknown risk of insertional mutagenesis [28]. Because of these factors, there has been much interest in the use of engineered nucleases as an approach to disable the CCR5 gene. Since expression of the nucleases would only need to be a transient event, this approach can make use of potentially safer, non-permanent and non-integrating vector systems, yet still result in a permanent genetic change [29–35]. Both the mature CD4+ T cells that HIV-1 infects, as well as the precursor HSC that give rise to these cells, are considered suitable target cells for CCR5 engineering.

First to Clinic: CCR5 Knockout in T Cells Using ZFNs

CD4+ T cells are the primary target of HIV-1 infection, although other CD4+ cells such as macrophages are also infected by the virus. T cells are also an excellent clinical target cell for genetic manipulation, since considerable experience already

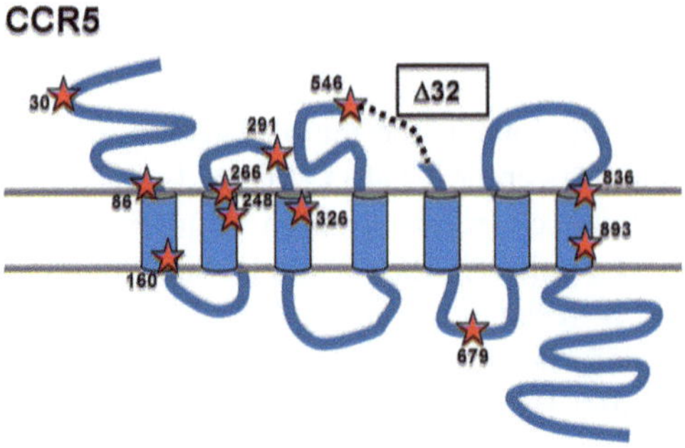

CCR5 target (~ nucleotide position)	Engineered nuclease	References
30	ZFN	*[103]
30	CRISPR/Cas9	**[49]
86	CRISPR/Cas9	[50]
160	ZFN	[28-34,47,75,101]
160	TALEN	[46, 47]
248	CRISPR/Cas9	[50]
266	ZFN	[37, *103]
291	CRISPR/Cas9	[50]
326	HE	[51]
546	TALEN	[45, 46]
836	ZFN	*[103]
893	ZFN	[37]

Fig. 3 Target sites for engineered nucleases in CCR5. The approximate location of the target sites for a series of engineered nucleases are shown superimposed on a schematic of the CCR5 protein, and described more fully in the associated table. Also indicated is the location of the 32 bp deletion that produces a prematurely truncated and defective protein from the CCR5Δ32 allele. *Kim et al. [119] tested 315 ZFN pairs against 33 sites, and the 3 shown had the highest activity with lowest off-target effects. **Cho et al. [53] tested 10 CRISPRs against CCR5, and the one shown had the highest activity and lowest off-target effects. HE, homing endonuclease

exists from their use in anti-cancer applications, and the procedures to harvest, expand, genetically modify, store, and re-infuse these cells back into patients are well established [30, 36]. In addition, and contrary to initial expectations, genetically marked T cells have been shown to persist for at least 11 years in vivo, suggesting that any such modified T cells could confer a relatively long-lasting effect [37]. These factors made ZFN engineering of T cells for CCR5 disruption as an anti-HIV therapy an obvious first clinical application of engineered nucleases.

ZFNs against CCR5 were first described by Mani et al. [38], who constructed two different ZFN pairs targeted to the second and seventh transmembrane domains of CCR5 respectively (Fig. 3). Each ZFN pair consisted of monomers containing three zinc fingers, and was therefore capable of recognizing 9-bp target sequences on either side of the DSB site. Using these ZFNs, the authors were able to disrupt CCR5 plasmid DNA in an in vitro system. Since then, multiple studies have shown that ZFNs work well in a variety of human cells. Many of these studies have used a ZFN pair that creates a DSB centered at approximately nucleotide 160 in the open-reading frame, which maps to the first transmembrane domain of the CCR5 protein, and which contains four zinc fingers in each monomer [29–35]. Fortuitously, for a gene knockout approach, the most common indel resulting from this reagent is a 5 bp insertion that occurs in ~10–30 % of edited alleles and creates two in-frame stop codons shortly after the target site [31, 33, 35].

The predicted ability of CCR5 disruption by ZFNs to impact HIV-1 replication has been demonstrated in an escalating series of studies in T cell lines [29, 31], primary T cell cultures [29–31, 39], and immune-deficient mice, engrafted with human T cells [29, 31, 39]. In the first such report, Perez et al. [31] showed that a CCR5 disruption rate of 2.4 % of alleles in the PM1 T cell line was increased to 73 % following infection of the culture with a CCR5-using (R5-tropic) strain of HIV-1. The authors further reported achieving CCR5 disruption rates between 28 and 33 %

when the ZFNs were delivered to primary T cells using adenoviral vectors based on the Ad5/F35 variant [40]. When such modified T cells were engrafted into NOG mice and challenged with HIV-1, plasma viremia was reduced 7.2-fold compared to mice infused with unmodified T cells, and rates of CCR5 disruption at the end of the HIV-1 challenges increased by threefold (8.5–27.5 %). These pre-clinical studies demonstrated both the feasibility of using engineered nucleases to disrupt the CCR5 gene in primary human T cells, as well as the anti-HIV consequences of such engineering. The studies also formed the basis for a series of T cell based human clinical trials in HIV-infected individuals, using the same combination of the Ad5/F35 vector and the site 160 ZFN pair (Table 1).

The initial clinical trials of ZFN-mediated CCR5 disruption are based on a strategy of delivering CCR5 ZFNs to a patient's own (autologous) T cells, which are then expanded ex vivo using CD3/CD28 stimulation. This protocol has been reported to allow the generation of up to 3×10^{10} T cells, with CCR5 disruption rates of 30–36 %, when measured at 10 days post transduction [30]. ZFNs were initially delivered as Ad5/35 vectors, but more recent trials are using mRNA electroporation. The first completed trial, NCT00842634, involved 12 patients who were each infused with 10^{10} T cells containing CCR5 modification rates between 10.9 and 27.7 % [4]. The procedure was well tolerated, with only one patient reporting an adverse reaction at time of infusion. CCR5-modified T cells were found to persist for several months, with a calculated half-life of 48 weeks. For 6 of the patients, a planned analytical treatment interruption (ATI) was initiated 4 weeks after infusion, involving cessation of antiretroviral therapy (ART) for 12 weeks, and this was completed by 4 of the 6 individuals. The rational for an ATI is that withdrawal of ART usually causes a rapid increase in HIV-1 viremia [41], and this may provide a selection pressure to increase the frequency of the CCR5-negative T cells. In addition, since this HIV-1 rebound follows a fairly characteristic pattern, an ATI provides an opportunity to evaluate whether the presence of CCR5-disrupted cells is influencing the ability of the patients to control HIV-1.

During the ATI, some indications of anti-HIV effects were observed. Although both CCR5-modified and unmodified CD4 T cells numbers declined in response to the rebound of HIV-1 viremia during the ATI, the rate of decline for the CCR5-modified T cells was significantly slower than the rate for the non-modified cells (−1.81 cells/mm^3/day compared to −7.25 cells/mm^3/day, $p=0.02$), although this did not reach significance when mean values were considered ($p=0.08$). In addition, in one of the patients undergoing ATI, HIV-1 RNA levels did not rebound until week 6 post ATI initiation, and then decreased to undetectable levels, even before therapy was re-started. Since this patient was heterozygous for the CCR5Δ32 allele, it is possible that this 'half-way there' genetic background facilitated the production of homozygous CCR5-negative cells by the ZFNs. This hypothesis is being further tested in clinical trial NCT01044654, in which 10 individuals who are CCR5Δ32 heterozygotes have been treated (Table 1). As a further refinement, the most recent T cell trials have also included treatment with cyclophosphamide (Cytoxan). This pre-conditioning treatment is expected to transiently reduce the patients' T cell numbers prior to infusion of the ZFN-engineered cells, and thereby facilitate greater engraftment of the modified T cells.

Table 1 Clinical trials of HIV-infected individuals using ZFN-modified autologous T cells[a]

Clinical trial	Status	Cohorts/study populations	Dose[b] of cells ($\times 10^9$)	ZFN delivery method, cell type	Additional treatments	Subjects enrolled
NCT00842634	Completed[c]	1. Multidrug resistant, virologic failure	5–10	Adenovirus, CD4 cells	ATI (cohort 2)	0
		2. On ART, aviremic, CD4 counts >450	5–10			6
		3. On ART, aviremic, CD4 counts <500	5–10			6
NCT01044654	Completed	1. On ART, aviremic, CD4 counts 200–500	5–10	Adenovirus, CD4 cells	ATI (cohort 5)	3
		2. On ART, aviremic, CD4 counts 200–500	20			3
		3. On ART, aviremic, CD4 counts 200–500	30			3
		4. Multidrug resistant, virologic failure	5–30			0
		5. CCR5Δ32 heterozygotes, on ART, aviremic, CD4 counts >500	5–30			10
NCT01252641	Completed	Not on ART, CD4 counts >500	5–30	Adenovirus, CD4 cells		N/A
NCT01543152	Recruiting	1. On ART, aviremic, CD4 counts >500	5–30	Adenovirus, CD4 cells (except cohort 3*, CD4/CD8 T cells)	200 mg Cytoxan	3
		2. On ART, aviremic, CD4 counts >500	5–30		0.5 g/m^2 Cytoxan	6
		3. On ART, aviremic, CD4 counts >500	5–30		1.0 g/m^2 Cytoxan	3
		4. On ART, aviremic, CD4 counts >500	5–30		2.0 g/m^2 Cytoxan	3
		5. On ART, aviremic, CD4 counts >500	5–30		1.5 g/m^2 Cytoxan	3
		3*. On ART, aviremic, CD4 counts >500	5–30		1.0 g/m^2 Cytoxan	Up to 8
					ATI (all cohorts)	
NCT02225665	Recruiting	1. On ART, aviremic, CD4 counts >500	Up to 40, in 2 doses	mRNA, CD4/CD8 cells	1.0 g/m^2 Cytoxan (both cohorts)	3–6
		2. On ART, aviremic, CD4 counts >500	Up to 40, in 3 doses		ATI (both cohorts)	3–6
NCT02388594	Recruiting	1. On ART, aviremic, CD4 counts >450	5–30 (estimated)	mRNA, CD4 cells	Cytoxan treatment (cohorts 2 and 3)	Up to 15
		2. On ART, aviremic, CD4 counts >450			ATI (all cohorts)	
		3. On ART, aviremic, CD4 counts >450				

[a]Information obtained from website Clinicaltrials.gov, and publically reported by sponsors
[b]All ZFN-modified T cells given as single dose infusions unless noted
[c][4]
ART anti-retroviral therapy, *ATI* analytical treatment interruption, *N/A* not available

CCR5 Disruption in HSC Using ZFNs

Engineered nucleases could also be used to disrupt the CCR5 gene in HSC, since these stem cells give rise to all lineages of the immune system, including CD4+ T cells. Although HSC are more challenging to work with than mature T cells, the longer life-span of the cells could allow for a one-shot treatment, while the fact that HSC give rise to both myeloid and lymphoid lineages means that non-T cell targets of HIV-1 such as macrophages would also be protected.

We previously reported on the ability of the site 160 ZFN pair to modify human HSC, isolated as CD34+ cells from umbilical cord blood [35]. By using plasmid DNA electroporation (Nucleofection) to introduce the ZFNs into the HSC, an average disruption rate of 17 % of the CCR5 alleles was achieved. These modified HSC were then used to engraft immune-deficient NSG mice, where they were found to differentiate comparably to unmodified HSC into various lineages of the human immune system, including CD4+ T cells. Following infection of the mice with R5-tropic HIV-1, protection of CD4+ T cells in the blood and lymphoid tissues was observed in the animals receiving ZFN-treated HSC compared to control HSC, which rapidly lost CD4+ T cells. Analysis of the blood and lymphoid tissues of the mice at 8–12 weeks post-infection also revealed a strong selection by HIV-1 for cells that were CCR5-negative. During the HIV-1 challenges, both unmodified and ZFN-engineered HSC cohorts of mice were equally infectable by HIV-1, but mice in the ZFN cohort were eventually able to suppress HIV-1, in both blood and tissues. This fits with the hypothesis that even a minority of CCR5-negative cells would be selected for during an active HIV-1 infection, and could eventually increase to a frequency where they were able to impact virus replication.

Although these observations support the use of ZFN-modified HSC as an alternative to engineering mature T cells, protocols will be needed that can deliver the nucleases to HSC while ensuring no or minimal impact on the ability of the cells to function as stem cells. In addition, although many pre-clinical studies rely on human HSC isolated from cord blood or fetal liver specimens, the clinical target cell for autologous HSC gene therapies will be the bone marrow resident HSC, which have distinct properties compared to other sources [42, 43]. These adult HSC are typically isolated by mobilization of the cells into the peripheral blood by treatment with cytokines (G-CSF), chemotherapeutic agents (cyclophosphamide), or small molecules such as the CXCR4 antagonist, AMD3100 [44]. We previously reported the modification of up to 25 % of CCR5 alleles in adult CD34+ HSC mobilized by G-CSF and treated with the same Ad5/F35 site 160 ZFN vectors as used in the current T cell trials (Table 1) [33]. However, adenoviral vector transduction of HSC proved to be more challenging than T cells, and transient low-dose treatment of the cells with PMA was necessary to achieve CCR5 disruption rates greater than 5 %. The associated toxicity cost when higher levels of gene disruption were achieved suggests that adenoviral vectors may not be optimal for this cell type.

In contrast, we have recently identified mRNA electroporation as a relatively non-toxic and efficient way to deliver ZFNs and TALENs to HSC, resulting in up to 50 % CCR5 disruption with minimal toxicity [45]. This method can also be used at large-scale, allowing a full patient dose of cells to be treated in one batch, and thereby increasing the clinical utility of this procedure.

CCR5 Disruption Using Other Engineered Nucleases

The recent progress to clinical trials of therapies based on ZFN modification of CCR5 is encouraging the development of other CCR5-specific reagents based on alternate engineered nucleases. The relative ease of design of TALENs compared to ZFNs is allowing the rapid evaluation of a range of CCR5-targeted TALENs, with variations in both the site of CCR5 that is targeted, as well as the basic design of the TALEN protein (e.g. length of the C-terminus). Miller et al. [46] first reported up to 15–20 % disruption of CCR5 in the K562 cell line using TALENs with either C28 or C63 backbone designs, respectively. For each backbone architecture, several left and right TALEN combinations were tested, allowing varying spacer lengths to be accommodated between the DNA sequences recognized by each monomer. All of the combinations evaluated were designed against sequences in the second extracellular domain of CCR5, close to the site of the natural Δ32 mutation (Fig. 3). Recently, we found that a C17 backbone TALEN also targeted close to the Δ32 site was able to achieve 60 % disruption in HSC [47]. Similarly, a C17 backbone TALEN targeting site 157, which overlaps the region recognized by the site 160 ZFNs, has also been evaluated by Mussolino et al., with the TALEN producing 17 % disruption of the CCR5 gene in 293 T cells [48]. Interestingly, when these authors looked at off-target disruption at the homologous CCR2 gene, they reported that such events were lower in cells transfected with the TALEN pair than the site 160 ZFN pair. In a follow up study, in which toxicity was more rigorously tested, the authors also noted lower overall cytotoxicity in the TALEN-treated cells [49]. Although both these TALENs and ZFNs targeted a similar region in CCR5, the non-identical nature of the target sequences recognized by the two reagents could have created such differences. In addition, while off-target and cytotoxicity analyses are a vital part of the evaluation of different candidate nucleases, especially for clinical applications, these analyses also need to be performed in the proposed clinical target cell, such as primary human T cells or HSC. Finally, it has been reported that CCR5 site 157 targeted TALENs can be introduced into cells by electroporation of mRNA, and thereby achieve disruption in both the PM1 cell line (up to 94 %) and primary T cells (between 17 and 46.8 %) [50].

It is possible that DNA breaks generated by different classes of nucleases will be qualitatively different, even when using the same FokI nuclease to introduce the break. For example, after analyzing a total of 1,456 mutant sequences at 122 target sites reported in 43 independent studies, Kim et al. found differences in the gene disruptions caused by ZFNs and TALENs. Specifically, they noted that while ZFNs cause an even distribution of insertions and deletions at the DSB site, TALENs almost exclusively cause deletions [51]. While the reason for this difference has not been determined, the authors speculated that the smaller spacer region between the two monomers in a ZFN pair may create a more defined cleavage site that is more prone to insertions, while the much larger spacers between the two regions bound by TALEN monomers could allow more heterogeneous cleavage sites, and result in more deletions. As noted earlier, the action of the widely used CCR5 site 160 ZFN pair frequently results in a 5 bp insertion that creates two in-frame stop codons, and which thereby terminates translation of the CCR5 protein

Homing endonucleases have also been adapted to target CCR5. These nucleases recognize stretches of 18–34 bp and cleave the target DNA in a manner similar to a restriction enzyme. Hundreds of homing endonucleases with different specificities have been reported, including I-Sce1, from baker's yeast, and I-Cre1 from green algae chloroplasts. Although they can be very efficient at cleaving double stranded DNA, they do not use a specific code that is easily engineered to recognize a desired target site, as is the case for ZFNs or TALENs. This makes it more challenging to retarget them to specific sites such as CCR5. To date, one homing endonuclease has been reported that disrupts CCR5, based on I-CreI and targeting the third transmembrane domain of the protein (Fig. 3) [52]. The authors used combinatorial assembly of archived I-Cre1 derivatives to generate a final nuclease that matched the original I-Cre1 binding site at only 4 out of 24 bases. Interestingly, by co-expressing a DNA end-processing enzyme, the TREX2 exonuclease, the authors were also able to increase CCR5 disruption rates from 5 to 37 % in HSC, when both proteins were co-expressed from a lentiviral vector. This TREX2 co-expression also improved gene disruption rates for CCR5-targeted ZFNs and TALENs by almost threefold, suggesting a general approach to increasing gene disruption [52]. While the difficulty in engineering homing endonucleases is currently a bottleneck to their more widespread use, the small size of these proteins facilitates their delivery via viral vectors, which could make them attractive tools for certain gene therapy applications.

Finally, for the CRISPR/Cas9 system, since target site specificity is provided by a complementary RNA, these reagents are the simplest class of engineered nucleases to construct and evaluate, and lend themselves to high throughput analyses of different sites within a targeted gene. For example, Cho et al. [53] designed CRISPRs against 10 different sites in CCR5 and were able to disrupt up to 60 % of alleles in K562 cells using a CRISPR directed against the first extracellular domain of CCR5 (Fig. 3). Similarly, Cradick et al. described CRISPRs that could disrupt CCR5 at high levels (21–77 %) in 293T cells, targeting the N-terminal side of the second transmembrane domain and the second extracellular domain [54]. CRISPRs have also been shown to work in HSC. In a recent study, HSC were transfected with a Cas9 plasmid also expressing GFP. Cas9-expressing cells were then specifically FACS sorted based on GFP expression and electroporated with a CCR5 guide RNA plasmid. Subsequently, 25.8–30 % of colonies derived from these HSCs were found to have CCR5 disrupted at both alleles [55]. CRISPR/Cas9 components targeted to CCR5 have also been delivered via Ad5/F35 or lentiviral vectors: Li et al. were able to disrupt CCR5 at an average of 30.3 % of alleles in primary T cells using Ad5/F35 vectors [56], while Wang et al. achieved up to 43 % CCR5 negative cells when using lentiviral vector delivery to the CEMss-CCR5 T cell line [57].

CXCR4 is an Additional Target for Gene Knockout

Therapies based on engineered nucleases to disable the CCR5 co-receptor will always run into two potential limitations; (1) both copies of the gene need to be disabled to produce a CCR5-negative cell, and (2) certain strains of HIV-1 can use alternate co-receptors, such as CXCR4, to enter cells. While such X4-tropic viruses are not as common as R5-tropic strains, they emerge in ~50 % of HIV patients towards the later stages of the disease, and there is concern that therapies or drugs targeted towards CCR5 could speed this process or selection. This issue could be addressed by using CXCR4-targeted nucleases in addition to CCR5 reagents, although the important role that CXCR4 plays in HSC homing to the bone marrow stem cell niche means that CXCR4 disruption could only be feasible for T cells, and not HSC.

Targeting CXCR4 in T cells has been described by two groups. Wilen et al. showed that T cells modified by CXCR4 ZFNs were protected from X4-tropic HIV-1 infection in vitro, and that this provided partial protection in vivo in NSG mice transplanted with CXCR4 ZFN-modified T cells [58]. However, this protection was lost over time due to the emergence of R5-tropic strains of HIV-1. The authors also demonstrated that they could modify T cells obtained from homozygous CCR5Δ32 individuals, and that these cells were protected against both R5 and X4-tropic HIV-1. Yuan et al. further confirmed the finding that modification of CXCR4 with ZFNs protected cells in vitro and in vivo and additionally reported that this approach was more efficient at providing protection against X4-tropic HIV-1 than adenovirus vector delivery of shRNA against CXCR4 [59].

Taking this approach another step forward, Didigu et al. were able to co-transduce both a T cell line and primary T cells with two sets of ZFNs targeted against both CCR5 and CXCR4 [29]. These dual-disrupted T cells were rapidly enriched after infection with both R5 and X4-tropic viruses in vitro, confirming their protection. Consistent with the previous studies, this protection was also observed in vivo, when mice were transplanted with the T cells, previously infected with both R5 and X4-tropic viruses. Here, mice receiving the dual-ZFN treated T cells were found to have up to 200 times higher CD4 counts in the blood at day 55, compared to mice receiving unmodified cells, or those disrupted for CCR5 only [29]. This approach could therefore represent an improvement over disruption of CCR5 alone in T cells, and further suggests that CXCR4 disruption in T cells could be combined with CCR5 disruption in HSC. It should be noted, however, that while no toxic effects were observed in these model systems, the impact of disrupting CXCR4 in mature T cells is unknown, and further testing of this method will be required before use in humans.

Beyond Gene Knockout: Site-Specific Gene Addition and Combination Anti-HIV Therapies

As noted above, the ability of HIV-1 to adapt to use co-receptors other than CCR5 may limit the effectiveness of therapies relying solely on CCR5 disruption. Here, similar to the NIH guidelines for treatment with CCR5 inhibitor drugs [60], patients would first need to be screened to confirm that they do not already harbor viruses with X4 tropism. Beyond this limitation, CCR5 knockout is expected to only be effective in cells with homozygous disruptions, although a single allele knockout may mimic CCR5Δ32 heterozygotes, who do have a better prognosis than homozygous wild-type individuals [61–64]. Because of this limitation, the use of engineered nucleases to promote homologous recombination in order to insert additional anti-HIV genes into the CCR5 locus could provide a more effective strategy than CCR5 disruption alone (Fig. 1).

There are a number of anti-HIV molecules that are already being considered for more conventional HIV-1 gene therapies based on integrating retroviral and lentiviral vectors, and which could also be adapted for a gene knock-in approach. These include trans-dominant versions of HIV-1 proteins such as the RevM10 mutant [65, 66] and modified, HIV-resistant versions of cellular restriction factors such as TRIM5α and APOBEC3G. For example, although HIV-1 is naturally resistant to the human form of TRIM5α, it is inhibited by the Rhesus macaque ortholog [67], and engineered human-rhesus TRIM5α derivatives restore this anti-viral activity [68, 69]. Similarly, although the HIV-1 Vif protein normally degrades APOBEC3G, a cytosine deaminase that causes G→A mutations in the viral genome during reverse transcription, the D128K point mutant is resistant to Vif and thereby inhibits HIV-1 replication [70, 71]. Alternative candidates for insertion at the CCR5 locus include RNA based therapeutics such as ribozymes or small interfering RNAs (siRNA) directed against different HIV-1 targets such as Rev [72], Vif [73], Nef [74], Gag [75], Env [76, 77], or Tat [72, 73, 78], as well as the C46 peptide inhibitor that potently blocks HIV-1 fusion [79]. The feasibility of inserting three different anti-HIV genes (human-rhesus TRIM5α, APOBEC3G D128K, and Rev M10) at the CCR5 locus has been demonstrated in a cell line model [80].

For HIV-1 therapies and beyond, the use of site-specific gene addition in the primary human cells that would be used clinically will require significant increases in the efficiency of the process, which has proven to be challenging. In a pioneering study, Lombardo et al. [81] reported up to 50 % integration of a GFP cassette at the CCR5 site 160 in immortalized cell lines when using three non-integrating lentiviral vectors (IDLVs) to deliver each ZFN monomer and a homologous donor sequences. However, the difficulty of achieving such an outcome with human HSC was shown by the same vectors achieving only 0.11 % targeted GFP addition. Subsequently, the same group reported using adenovirus 5 vectors to deliver the CCR5 ZFNs, combined with IDLVs for the donor sequence, and were able to achieve up to 5 % targeted addition in primary human T cells at the site 160 locus [82].

We have previously reported on the ability of site 160 ZFNs, delivered to human HSC as plasmid DNA, to achieve mean gene disruption rates of 17 % of CCR5 alleles [35]. When a CCR5 homology donor plasmid containing an internal PGK-GFP expression cassette was co-introduced into the cells with the ZFNs, although we observed an increase in DNA-mediated toxicity, the viable human HSC in the population were able to engraft NSG mice, and give rise to mature human cells stably expressing GFP at frequencies up to 5 % in multiple lineages (Fig. 4a). When humanized mice are challenged with an R5-tropic strain of HV-1, the HIV-mediated destruction of CD4 T cells typically causes selection of phenotypically

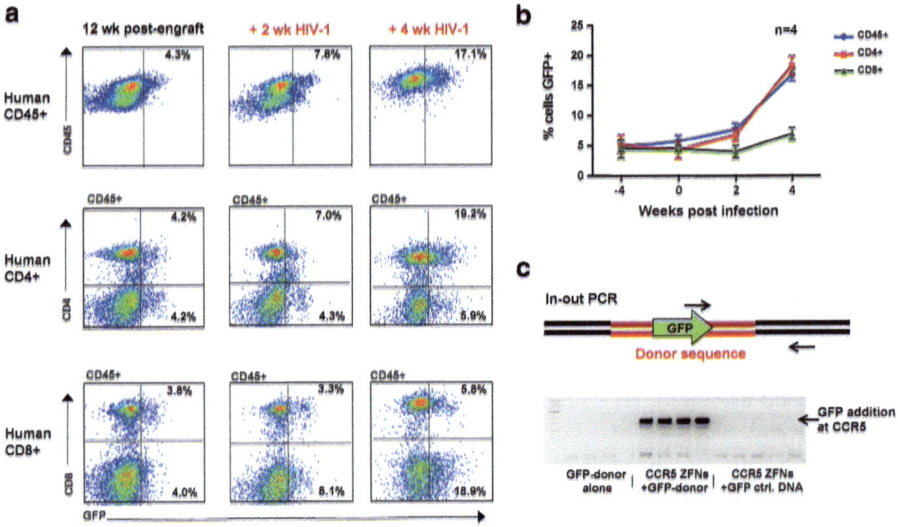

Fig. 4 Targeted gene addition at the CCR5 locus. Human cord blood HSC were electroporated with plasmid DNAs expressing CCR5 site 160 ZFNs and a donor sequence containing a PGK-GFP cassette flanked by CCR5 homologous sequences. The cells were engrafted into neonatal NSG mice, as described previously [35]. At 12 weeks of age, the mice were infected with R5-tropic HIV-1. (**a**) At the indicated time points, blood was analyzed by FACS for GFP expression in total human leukocytes (CD45 cells), and in the CD4 and CD8 subsets. (**b**) Following HIV-1 infection, the frequency of GFP-expressing cells increased in the CD4 fraction, which represent the target cells of HIV-1 infection, as well as the total CD45 group, but not in the CD8 fraction. (**c**) The site-specific addition of the GFP cassette at the CCR5 locus was confirmed by a specific in-out PCR analysis of clones derived from the input HSC, where one of the PCR primers is specific for the GFP expression cassette, and one is specific for CCR5, but beyond the region contained in the donor sequence

CCR5-negative cells, created by the action of the nucleases [35]. In this experimental situation where GFP was inserted at the CCR5 locus, HIV-1 infection also co-selected for GFP-expressing cells in the CD4 cell fraction only (Fig. 4a, b). This observation is in keeping with the GFP cassette being specifically integrated at the CCR5 locus, which was also confirmed by PCR analysis (Fig. 4c).

Despite these promising developments, and the expected ability of HIV-1 itself to act as a selection agent to increase the frequency of cells with CCR5-specific gene addition, the currently achievable levels of targeted addition in HSC cannot match the efficiency of more standard integrating viral vectors for the delivery of genes, and higher rates of targeted integration, especially into primary HSC, may be required. Recent promising progress has been made by introducing homologous donors using non-integrating lentiviral [83, 84] and AAV vectors [85] and strategies that can promote HDR over NHEJ in human cells are also being explored [86, 87].

Disrupting CCR5 in iPSC

HSC are challenging to culture and expand in vitro without losing their ability to differentiate, which limits the potential to pre-select genome modified cells prior to infusion into a patient. In the future, it may be possible to derive fully functional HSC from patient-specific induced pluripotent stem cells (iPSC), which would thereby also allow for the generation of a more homogenous population of modified cells. Towards this goal, several groups have recently reported modifying iPSC: Ye et al. used CRISPRs and TALENs to generate iPSC that contained the exact CCR5Δ32 mutation while maintaining pluripotency [88], while Yao et al. disrupted CCR5 in human embryonic stem cells and iPSC and were able to differentiate them into CD34+ cells and more differentiated hematopoietic lineages [89].

Disrupting HIV-1 Genomes with Engineered Nucleases

Progress to Developing HIV-Specific Nucleases

An additional application of engineered nucleases to combat HIV-1 would be to engineer the reagents to target the HIV-1 genome itself (Fig. 1). Although antiretroviral drugs are highly effective and capable of suppressing HIV-1 to undetectable levels, they do not completely eliminate the virus, which then usually rebounds if therapy is stopped. The source of this rebounding virus are the so-called latent HIV reservoirs—cells in the body where the virus lies dormant, unperturbed by antiretroviral drugs, but from which the virus can be activated at a later date. Among the most prominent reservoirs are memory CD4 T cells [90–92]. Methods to remove or disable the virus in these cells could allow drug-free suppression of HIV-1, or even a cure.

There have been several reports in which different types of engineered nucleases have been developed that recognize the HIV-1 genome [93–96]. In most of these studies, the targeted sequence has been the HIV-1 long terminal repeat (LTR), which is present at both ends of the integrated DNA version of the virus. The LTRs play multiple roles, with the 5′ sequence driving transcription of the viral genome and the 3′ sequence regulating transcription termination and polyadenylation. Intact LTRs are also necessary for correct RNA processing to allow successful reverse transcription and integration during the viral life-cycle. In addition, the presence of two copies of the LTR in the HIV-1 genome provides anuclease with double the target sequences, as well as the possibility of a double cut and re-ligation event, leading to the permanent excision of the intervening HIV-1 genome.

One of the first reports describing disruption of an integrated HIV-1 genome through engineering recognition of specific HIV-1 sequences used a modified cre-recombinase [93]. This enzyme usually recognizes 34 bp loxP sites, but was evolved to recognize a similar sequence that is present in the LTRs of the clade A

HIV-1 strain, TZB0003 [97]. The evolved recombinase, named Tre recombinase, was able to disrupt integrated HIV-1 genomes in HeLa cells, and reduce the production of new HIV-1 particles [93]. The Tre recombinase was subsequently evaluated in an in vivo model [95], where primary T cells or HSC were transduced with lentiviral vectors expressing the recombinase under control of an HIV-inducible promoter. This design used a Tre-resistant HIV-1 LTR promoter containing tandem copies of the HIV-1 Tar element, which responds to the viral transactivator, Tat, produced upon HIV-1 infection. Consequently, the recombinase would only be expressed in cells infected with HIV-1, so reducing the potential for toxicity. In this proof of concept experiment, cells transduced with the vectors at between 30 and 60 % were engrafted into $Rag2^{-/-}\gamma c^{-/-}$mice, which were then challenged with a replication-competent R5-tropic HIV-1 reporter virus, modified to contain the Tre recognition sites in its LTRs. The animals receiving the Tre recombinase-containing cells had increasing or stable levels of human CD45 and CD4 cells, and virus levels that decreased approximately 1-log between weeks 2 and 12 post infection. This was in contrast to mice receiving non-transduced cells, which showed decreasing levels of human CD45 and CD4 cells and increasing or stable HIV-1 levels over the same time period. While this is an intriguing approach, the broader application of the technology will require the development of recombinases that recognize sequences that are more divergent from the natural loxP sequence and, preferably, are highly conserved across multiple strains of HIV-1. Such engineering challenges may be facilitated by the availability of new design tools [98]. In this regard, another Cre recombinase derivative (uTre) has recently been described that recognizes an HIV LTR sequence that is conserved in 94 % of the major HIV subtypes A, B, and C, which would make it much more therapeutically relevant [99].

For the classes of engineered nucleases with less restrictions on target site selection (ZFNs, TALENs and CRISPR/Cas9), several different regions of the HIV-1 genome could provide target sites in addition to the LTRs. Figure 5 displays an entropy analysis of sequence variation in the HIV-1 genome, highlighting regions of conservation that could provide a source of such target sites. ZFNs have already been described targeting the LTRs, using a site that is highly conserved in viral isolates from patient samples [94]. These reagents were reported to disrupt up to 60 % of GFP-expressing HIV-1 genomes present in a Jurkat T cell line following DNA electroporation, and to reduce viral production from infected primary human PBLs (29 % reduction) or CD4+ T cells (31 % reduction). Similarly, TALENs have also been used to target HIV DNA. Ebina et al. were able to significantly disrupt HIV sequences in a Jurkat cell line model of an integrated latent HIV genome, where LTR-driven GFP expression occurs after stimulation with TNFα. By electroporating in mRNA for a TALEN pair targeting the LTR, they saw reduction in the levels of GFP-positive cells following stimulation from 63 % down to 4.3 %, and reported complete removal of the HIV genome in 53 % of the cells [100].

Finally, CRISPRs are also being used to target integrated HIV-1 DNA. Here, the potential for these reagents to be multiplexed [101] and thereby target more than one HIV-1 sequence could prove to be a distinct advantage against a virus known for its ability to rapidly evolve resistance to other therapies. Ebina et al. designed two CRISPR gRNAs against the TAR and NFκB regions of the LTR, which were tested for their ability to block expression from an integrated LTR-GFP reporter sequence that also contained the Tat gene (JLAT cells) [96]. The utility of the reagents to disrupt LTR function was demonstrated following three rounds of electroporation with the CRISPR/Cas9 reagents, where GFP expression was reduced in activated cells by 25–32 % [96]. Similarly, Zhu et al. tested 10 different gRNAs against the HIV LTR and Pol regions in JLAT cells and were able to reduce the levels of GFP+ cells by more than 24-fold when used in combination [102]. Liao et al. also found that introducing multiple CRISPR gRNAs was better able to disrupt integrated HIV sequences when compared to single gRNAs, and they further demonstrated that cells expressing anti-HIV CRISPRs were protected from HIV infection [103]. Finally, Hu et al. reported an effect in CHME-5 cells, a microglial cell line of HIV latency, where GFP expression after activation was reduced from 76 to 17.1 % or 3.9 % of cells for two different LTR-directed gRNAs [104].

Challenges for the Use of Anti-HIV Nucleases

While the studies described above represent important first steps toward using engineered nucleases to disrupt HIV-1, much work will be needed to deliver these reagents to the cells in patients that harbor latent HIV-1 genomes, including those cells where latency may involve chromatin condensation. In addition, even when highly conserved HIV-1 sequences are targeted (Fig. 5), the potential for HIV-1 to mutate and evolve resistance against other therapies is well established, and it should be assumed that this will also be the case for HIV-specific engineered nucleases. In this regard, targeting multiple sequences simultaneously may be advantageous, including all known variants known to be tolerated at a specific site, although this would greatly increase the complexity of the therapy. In addition, in common with all therapeutic applications using nucleases, the potential for off-target effects will need to be characterized. Finally, unless a lie-in-wait protection approach is to be used, as described above for the Tat-inducible Tre-recombinase system, methods to deliver the reagents to HIV-infected cells in vivo will need to be developed. This requirement will be especially challenging for latently infected cells, since the lack of expression of HIV-1 genes means that there are no obvious signs to indicate that the cells harbor an HIV-1 genome. Thus, bulk delivery to all T cells, or select delivery through the use of surrogate markers of reservoir cells [90], may be needed.

Considerations for the Clinical Use of Engineered Nucleases

Strategies to Deliver Engineered Nucleases to Human Cells

The delivery of engineered nucleases to cells presents certain challenges, including the requirement for the co-delivery of more than one component for the ZFN, TALEN and CRISPR/Cas9 nucleases, and the additional inclusion of a donor

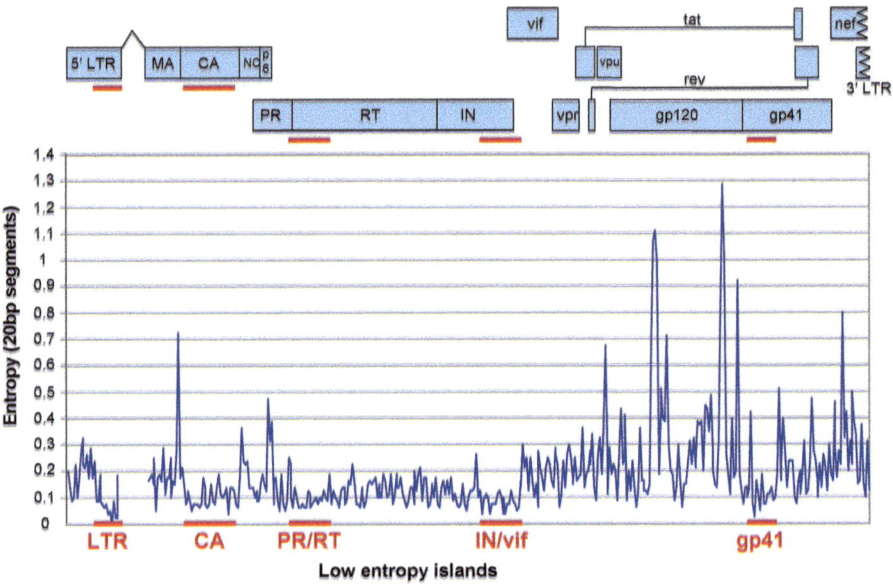

Fig. 5 Identification of low entropy islands in the HIV-1 genome. 102 LTR and 442 coding region sequences from the Los Alamos HIV-1 sequence database were queried using the Entropy-one tool (www.hiv.lanl.gov). Entropy measures the amount of sequence variation at each position, with lower entropy scores indicating higher conservation. The average entropy score across sequential 20 bp segments was calculated and plotted, to highlight islands (red lines) of consistent low entropy (at or below 0.1)

sequence if HDR-mediated engineering is desired. Beyond this, the extensive regions of repeat sequences present in the DNA binding domains of ZFNs and TALENs can lead to instability. However, since the permanent genetic changes that result from the action of engineered nucleases only require their transient expression in target cells, a wide variety of delivery systems could potentially be used.

Reflecting their earlier development and adoption for clinical applications, ZFNs are the class of engineered nucleases that have been evaluated in the widest array of delivery methods. These include packaging in standard viral vectors such as adenovirus, adeno-associated virus (AAV) and IDLVs [31, 33, 81, 82, 105–107] as well as less frequently used systems such as baculovirus vectors [32]. In a simpler delivery approach, nucleases have also been introduced into cells following electroporation of both DNA and mRNA [32, 35, 45, 47, 81, 82]. Interestingly, ZFN proteins are also capable of direct uptake by cells, with one study reporting CCR5 disruption rates up to 27 % in cell lines, and 8 % in primary human CD4 T cells, using this approach [34]. The authors also reported lower rates of off-target disruption at 9 sites with homology to CCR5 when compared to the rates occurring following transfection of ZFN DNA into 293T cells although, as noted, this may reflect lower concentrations of the ZFNs when delivered by this route. In a similar approach, Chen et al. developed a system whereby ZFN proteins are attached to the transferrin molecule through a cleavable disulfide bond [108]. Here, the ZFN proteins are taken up into cells by transferrin receptor-mediated endocytosis, and the system appears to function in a variety of cell types, reflecting the broad distribution of this receptor. Such protein-based delivery systems may be simpler to use than viral based systems and, by reducing the time of exposure of the cells to ZFN proteins, may indeed reduce off-target effects, which are a function of both the concentration of the nucleases, and the time of exposure.

Although TALENs and ZFNs share many common design features, TALEN delivery is likely to be more complicated than ZFNs. First, TALEN constructs with the commonly used C63 backbone are roughly 1 kb larger (per monomer) than ZFNs, which complicates their use in size constrained vectors such as AAV. In addition, the 1 base pair recognition code used by the TALE repeat component is in contrast to the 3 bp recognition sequence of each zinc finger module, so that TALENs directed to the same target sequence contain three times as many repeating units as ZFNs. Such designs can cause instability in certain vector types, especially lentiviral vectors [107], where the high level of homology between the TALEN subunits can lead to rearrangements and deletions as a result of strand switching between the subunits of the two packaged vector RNAs during reverse transcription. A possible solution to this problem will be to make numerous silent mutations in each of the repeat units of the TALE to decrease homology. Alternatively, it was found in the same study that Ad5 vectors worked well for delivering TALENs, producing 47–55 % gene disruption of the AAVS1 locus in HeLa cells, immortalized myoblasts and bone marrow-derived mesenchymal stem cells [107]. Recently another option was described in which the lentiviral vector delivering the TALEN was mutated to be unable to perform reverse transcriptase. The resulting RNA genome was able to express an encoded TALEN following addition of an IRES and poly A sequence, to mimic a mRNA [109].

For all classes of engineered nucleases, current technologies are limited almost exclusively to methods of ex vivo delivery, but future broader utilization of the reagents will benefit from in vivo methods of delivery, as are being developed for siRNAs [110], and may be possible for certain viral vectors. For example, different subtypes of AAV display different in vivo tropisms, and AAV vectors have been shown to be capable of co-delivering both ZFN monomers and a donor sequence to mouse hepatocytes following intraperitoneal injection of hepatotropic AAV vectors [105]. In addition, recent advances in retargeting lentiviral vectors, based on including scFvs or ligands to cell surface proteins, could be exploited to allow the delivery of engineered nucleases to specific cell types in the future [111–117].

Off-Target Effects and Toxicity

A major concern for any clinical application of engineered nucleases is the potential for adverse events, including cellular toxicity [118], or nuclease modification at off-target sites [31, 33, 119–121]). Toxicity may result from expression of the nucleases themselves, be a consequence of the cellular response to DNA damage, or result from a combination of factors that stress the cells, including the delivery method used. Off-target gene modifications can occur at sites predicted by bioinformatics to be highly homologous to the desired target sequence, as well as at sites that are revealed by in vitro screens [122], or by DSB site capture in cells following NHEJ-mediated integration of IDLVs [123]. Since such off-target events and toxicities may be cell-type specific, cell line studies may not predict the outcome for primary human cells. Thus pre-clinical toxicology studies should include both the evaluation of activity at predicted off-target sites, as well as more general analyses to evaluate the tumorigenic potential of patient-sized doses of ZFN-treated T cells or HSC, using both in vitro cellular assays and following engraftment into immune-deficient mice [30, 42].

Recently, several new methods to detect off-target events have been described. For example, high-throughput genome-wide translocation sequencing (HTGTS) was used to identify translocation junctions in cells treated with I-sce meganuclease [124] and this method has been adapted and enhanced by Frock et al. to include linear-amplification-mediated PCR (LAM-PCR) to identify off-target sites created by CRISPRs and TALENs [125]. This study confirmed previous findings that CRISPR-mediated off-target activity can be reduced by using CRISPR nickases [53], and also found that TALEN-mediated off-target effects were mostly due to homodimers [125].

Other genome-wide off-target detection methods include GUIDE-seq (genome-wide, unbiased identification of DSBs enabled by sequencing), which is based on detection of double-stranded oligodeoxynucleotide tags inserted into the DSBs [126], and the BLESS assay (breaks labeling, enrichments on streptavidin and next-generation sequencing), which labels DSBs with biotinylated oligonucleotides [127, 128]. Yet another method is digenome-seq (digested genome sequencing), which works by nuclease treatment of genomic DNA in vitro, followed by whole genome sequencing of the resulting fragments [129]. The value of these assays is that they don't rely on bioinformatics predictions and so can potentially detect cryptic off-target sites in a more unbiased way. However, the assays are still imperfect since the HTGTS, GUIDE-seq and digenome-seq all produced different sets of off-target sites for a VEGF-A CRISPR target site (Reviewed in [120]).

For the commonly used site 160 CCR5 ZFNs, off-target disruption is most commonly observed at the highly homologous CCR2 gene, and can reach 9–10 % of the rates at the CCR5 locus [31, 33, 49, 119, 122, 123], and increasing as the expression of ZFNs is increased. In this way, it may be useful to consider nuclease activity as having a practical plateau, with a maximum amount of on-target disruption and acceptable amount of off-target activity. Meanwhile, methods to reduce off-target activity of nucleases are also being developed. For example, both ZFNs and TALENs are made more specific through the use of engineered obligate heterodimers of the Fok1 endonuclease, which limits the possible pair combinations of the individual component monomers to the desired heterodimer [130]. In a related approach, CRISPR/Cas9 specificity can be improved by requiring two guide RNA binding events and using a catalytically inactive Cas9 protein fused to Fok1 moieties [131]. Finally, altering the nature of the DNA-binding RVDs in TALENs can also result in reduced off-target effects by increasing on-target specificity [47].

Summary

The life-cycle of HIV-1 provides several opportunities for interventions by therapies based on engineered nucleases. The requirement of the virus for a cellular co-receptor, CCR5, that is non-essential to its human host, provides an obvious application for the gene disrupting capabilities of nucleases, and the use of ZFNs in this regard is currently the most clinically advanced application of this new class of genome modifying tools. Beyond that, gene disruption could be combined with HDR to insert additional anti-HIV genes at the CCR5 locus. In addition, the integrated HIV-1 genomes that persist in patients' cells despite antiretroviral therapy can also be considered as a genetic target for disruption, although the challenges of delivering nucleases to the cells that harbor such latent genomes will be formidable. Finally, as in all gene therapy approaches that create HIV-resistant cells, it is anticipated that HIV-1 could be harnessed to assist in its own demise, by enabling selection for the engineered, HIV-resistant cells.

Acknowledgements We gratefully acknowledge the expertise of our collaborators at Sangamo BioSciences, including Michael Holmes, Jianbin Wang and Philip Gregory, and thank Liz Wolffe for her help compiling Table 1. This work was supported by the James B. Pendleton Charitable Trust, NIH grants HL073104, AI110149 and HL129902, and the California HIV/AIDS Research Program grant ID12-USC-245.

References

1. http://aids.gov/federal-resources/policies/care-continuum, 2013.
2. Baltimore D. Gene therapy. Intracellular immunization. Nature. 1988;335(6189):395–6.
3. Peterson CW, et al. Combinatorial anti-HIV gene therapy: using a multipronged approach to reach beyond HAART. Gene Ther. 2013;20(7):695–702.
4. Tebas P, Stein D, Tang WW, Frank I, Wang SQ, Lee G, et al. Gene editing of CCR5 in autologous CD4 T cells of persons infected with HIV. N Engl J Med. 2014;370:10.
5. Sun J, et al. Human Ku70/80 protein blocks exonuclease 1-mediated DNA resection in the presence of human Mre11 or Mre11/Rad50 protein complex. J Biol Chem. 2012;287(7):4936–45.
6. Symington LS, Gautier J. Double-strand break end resection and repair pathway choice. Annu Rev Genet. 2011;45:247–71.
7. Chapman JR, Taylor MR, Boulton SJ. Playing the end game: DNA double-strand break repair pathway choice. Mol Cell. 2012;47(4):497–510.
8. Escribano-Diaz C, et al. A cell cycle-dependent regulatory circuit composed of 53BP1-RIF1 and BRCA1-CtIP controls DNA repair pathway choice. Mol Cell. 2013;49(5):872–83.
9. Heyer WD, Ehmsen KT, Liu J. Regulation of homologous recombination in eukaryotes. Annu Rev Genet. 2010;44:113–39.
10. Mohrin M, et al. Hematopoietic stem cell quiescence promotes error-prone DNA repair and mutagenesis. Cell Stem Cell. 2010;7(2):174–85.
11. Rossi DJ, et al. Hematopoietic stem cell quiescence attenuates DNA damage response and permits DNA damage accumulation during aging. Cell Cycle. 2007;6(19):2371–6.
12. Deng H, et al. Identification of a major co-receptor for primary isolates of HIV-1. Nature. 1996;381(6584):661–6.

13. Samson M, et al. Resistance to HIV-1 infection in Caucasian individuals bearing mutant alleles of the CCR-5 chemokine receptor gene. Nature. 1996;382(6593):722–5.
14. Novembre J, Galvani AP, Slatkin M. The geographic spread of the CCR5 Delta32 HIV-resistance allele. PLoS Biol. 2005;3(11), e339.
15. Dean M, et al. Genetic restriction of HIV-1 infection and progression to AIDS by a deletion allele of the CKR5 structural gene. Hemophilia Growth and Development Study, Multicenter AIDS Cohort Study, Multicenter Hemophilia Cohort Study, San Francisco City Cohort, ALIVE Study. Science. 1996;273(5283):1856–62.
16. Wood A, Armour D. The discovery of the CCR5 receptor antagonist, UK-427,857, a new agent for the treatment of HIV infection and AIDS. Prog Med Chem. 2005;43: 239–71.
17. Allers K, et al. Evidence for the cure of HIV infection by CCR5Delta32/Delta32 stem cell transplantation. Blood. 2011;117(10):2791–9.
18. Yukl SA, et al. Challenges in detecting HIV persistence during potentially curative interventions: a study of the Berlin patient. PLoS Pathog. 2013;9(5), e1003347.
19. Hutter G, et al. Long-term control of HIV by CCR5 Delta32/Delta32 stem-cell transplantation. N Engl J Med. 2009;360(7):692–8.
20. Hutter G, Zaia JA. Allogeneic haematopoietic stem cell transplantation in patients with human immunodeficiency virus: the experiences of more than 25 years. Clin Exp Immunol. 2011;163(3):284–95.
21. Hayden EC. Hopes of HIV cure in 'Boston patients' dashed. Nature News 2013.
22. Ringpis GE, et al. Engineering HIV-1-resistant T-cells from short-hairpin RNA-expressing hematopoietic stem/progenitor cells in humanized BLT mice. PLoS One. 2012;7(12), e53492.
23. Lee MT, et al. Inhibition of human immunodeficiency virus type 1 replication in primary macrophages by using Tat- or CCR5-specific small interfering RNAs expressed from a lentivirus vector. J Virol. 2003;77(22):11964–72.
24. ter Brake O, et al. Lentiviral vector design for multiple shRNA expression and durable HIV-1 inhibition. Mol Ther. 2008;16(3):557–64.
25. Qin XF, et al. Inhibiting HIV-1 infection in human T cells by lentiviral-mediated delivery of small interfering RNA against CCR5. Proc Natl Acad Sci U S A. 2003;100(1):183–8.
26. An DS, et al. Optimization and functional effects of stable short hairpin RNA expression in primary human lymphocytes via lentiviral vectors. Mol Ther. 2006;14(4):494–504.
27. Li MJ, et al. Inhibition of HIV-1 infection by lentiviral vectors expressing Pol III-promoted anti-HIV RNAs. Mol Ther. 2003;8(2):196–206.
28. Wu C, Dunbar CE. Stem cell gene therapy: the risks of insertional mutagenesis and approaches to minimize genotoxicity. Front Med. 2011;5(4):356–71.
29. Didigu CA, et al. Simultaneous zinc-finger nuclease editing of the HIV coreceptors ccr5 and cxcr4 protects CD4+ T cells from HIV-1 infection. Blood. 2014;123(1):61–9.
30. Maier DA, et al. Efficient clinical scale gene modification via zinc finger nuclease-targeted disruption of the HIV co-receptor CCR5. Hum Gene Ther. 2013;24(3):245–58.
31. Perez EE, et al. Establishment of HIV-1 resistance in CD4+ T cells by genome editing using zinc-finger nucleases. Nat Biotechnol. 2008;26(7):808–16.
32. Lei Y, et al. Gene editing of human embryonic stem cells via an engineered baculoviral vector carrying zinc-finger nucleases. Mol Ther. 2011;19(5):942–50.
33. Li L, et al. Genomic editing of the HIV-1 coreceptor CCR5 in adult hematopoietic stem and progenitor cells using zinc finger nucleases. Mol Ther. 2013;21(6):1259–69.
34. Gaj T, et al. Targeted gene knockout by direct delivery of zinc-finger nuclease proteins. Nat Methods. 2012;9(8):805–7.
35. Holt N, et al. Human hematopoietic stem/progenitor cells modified by zinc-finger nucleases targeted to CCR5 control HIV-1 in vivo. Nat Biotechnol. 2010;28(8):839–47.
36. Kalos M, June CH. Adoptive T cell transfer for cancer immunotherapy in the era of synthetic biology. Immunity. 2013;39(1):49–60.
37. Scholler J, et al. Decade-long safety and function of retroviral-modified chimeric antigen receptor T cells. Sci Transl Med. 2012;4(132):132ra53.

38. Mani M, et al. Design, engineering, and characterization of zinc finger nucleases. Biochem Biophys Res Commun. 2005;335(2):447–57.
39. Yi G, et al. CCR5 gene editing of resting CD4(+) T cells by transient ZFN expression from HIV envelope pseudotyped nonintegrating lentivirus confers HIV-1 resistance in humanized mice. Mol Ther Nucleic Acids. 2014;3, e198.
40. Shayakhmetov DM, et al. Efficient gene transfer into human CD34(+) cells by a retargeted adenovirus vector. J Virol. 2000;74(6):2567–83.
41. Chun TW, et al. Relationship between pre-existing viral reservoirs and the re-emergence of plasma viremia after discontinuation of highly active anti-retroviral therapy. Nat Med. 2000;6(7):757–61.
42. Hofer U, et al. Pre-clinical modeling of CCR5 knockout in human hematopoietic stem cells by zinc finger nucleases using humanized mice. J Infect Dis. 2013;208 Suppl 2:S160–4.
43. Lepus CM, et al. Comparison of human fetal liver, umbilical cord blood, and adult blood hematopoietic stem cell engraftment in NOD-scid/gammac–/–, Balb/c-Rag1–/–gammac–/–, and C.B-17-scid/bg immunodeficient mice. Hum Immunol. 2009;70(10):790–802.
44. Lapid K, et al. Egress and mobilization of hematopoietic stem and progenitor cells: a dynamic multi-facet process. Cambridge: Harvard Stem Cell Institute; 2008. StemBook [Internet].
45. Cannon PM, et al. Electroporation of ZFN mRNA enables efficient CCR5 gene disruption in mobilized blood hematopoietic stem cells at clinical scale. Mol Ther. 2013;21:S71–2.
46. Miller JC, et al. A TALE nuclease architecture for efficient genome editing. Nat Biotechnol. 2011;29(2):143–8.
47. Llewellyn N, et al. Next generation TALENs mediate efficient disruption of the CCR5 gene in human HSCs. Mol Ther. 2013;21:S72.
48. Mussolino C, et al. A novel TALE nuclease scaffold enables high genome editing activity in combination with low toxicity. Nucleic Acids Res. 2011;39(21):9283–93.
49. Mussolino C, et al. TALENs facilitate targeted genome editing in human cells with high specificity and low cytotoxicity. Nucleic Acids Res. 2014;42(10):6762–73.
50. Mock U, et al. mRNA transfection of a novel TAL effector nuclease (TALEN) facilitates efficient knockout of HIV co-receptor CCR5. Nucleic Acids Res. 2015;43(11):5560–71.
51. Kim Y, Kweon J, Kim JS. TALENs and ZFNs are associated with different mutation signatures. Nat Methods. 2013;10(3):185.
52. Certo MT, et al. Coupling endonucleases with DNA end-processing enzymes to drive gene disruption. Nat Methods. 2012;9(10):973–5.
53. Cho SW, et al. Analysis of off-target effects of CRISPR/Cas-derived RNA-guided endonucleases and nickases. Genome Res. 2014;24(1):132–41.
54. Cradick TJ, et al. CRISPR/Cas9 systems targeting beta-globin and CCR5 genes have substantial off-target activity. Nucleic Acids Res. 2013;41(20):9584–92.
55. Mandal PK, et al. Efficient ablation of genes in human hematopoietic stem and effector cells using CRISPR/Cas9. Cell Stem Cell. 2014;15(5):643–52.
56. Li C, et al. Inhibition of HIV-1 infection of primary CD4+ T cells by gene editing of CCR5 using adenovirus-delivered CRISPR/Cas9. J Gen Virol. 2015;96(8):2381–93.
57. Wang W, et al. CCR5 gene disruption via lentiviral vectors expressing Cas9 and single guided RNA renders cells resistant to HIV-1 infection. PLoS One. 2014;9(12), e115987.
58. Wilen CB, et al. Engineering HIV-resistant human CD4+ T cells with CXCR4-specific zinc-finger nucleases. PLoS Pathog. 2011;7(4), e1002020.
59. Yuan J, et al. Zinc-finger nuclease editing of human cxcr4 promotes HIV-1 CD4(+) T cell resistance and enrichment. Mol Ther. 2012;20(4):849–59.
60. http://aidsinfo.nih.gov/guidelines/html/1/adult-and-adolescent-arv-guidelines/8/co-receptor-tropism-assays. Guidelines for the use of antiretroviral agents in HIV-1-infected adults and adolescents. 2013.
61. Meyer L, et al. Early protective effect of CCR-5 delta 32 heterozygosity on HIV-1 disease progression: relationship with viral load. The SEROCO Study Group. AIDS. 1997; 11(11):F73–8.

62. de Roda Husman AM, et al. Association between CCR5 genotype and the clinical course of HIV-1 infection. Ann Intern Med. 1997;127(10):882–90.
63. Meyer L, et al. CCR5 delta32 deletion and reduced risk of toxoplasmosis in persons infected with human immunodeficiency virus type 1. The SEROCO-HEMOCO-SEROGEST Study Groups. J Infect Dis. 1999;180(3):920–4.
64. Ioannidis JP, et al. Effects of CCR5-Delta, 32, CCR2–64I, and SDF-1 3′A alleles on HIV-1 disease progression: an international meta-analysis of individual-patient data. Ann Intern Med. 2001;135(9):782–95.
65. Bevec D, et al. Inhibition of human immunodeficiency virus type 1 replication in human T cells by retroviral-mediated gene transfer of a dominant-negative Rev trans-activator. Proc Natl Acad Sci U S A. 1992;89(20):9870–4.
66. Bonyhadi ML, et al. RevM10-expressing T cells derived in vivo from transduced human hematopoietic stem-progenitor cells inhibit human immunodeficiency virus replication. J Virol. 1997;71(6):4707–16.
67. Stremlau M, et al. The cytoplasmic body component TRIM5alpha restricts HIV-1 infection in Old World monkeys. Nature. 2004;427(6977):848–53.
68. Sawyer SL, et al. Positive selection of primate TRIM5alpha identifies a critical species-specific retroviral restriction domain. Proc Natl Acad Sci U S A. 2005;102(8):2832–7.
69. Anderson J, Akkina R. Human immunodeficiency virus type 1 restriction by human-rhesus chimeric tripartite motif 5alpha (TRIM 5alpha) in CD34(+) cell-derived macrophages in vitro and in T cells in vivo in severe combined immunodeficient (SCID-hu) mice transplanted with human fetal tissue. Hum Gene Ther. 2008;19(3):217–28.
70. Schrofelbauer B, Chen D, Landau NR. A single amino acid of APOBEC3G controls its species-specific interaction with virion infectivity factor (Vif). Proc Natl Acad Sci U S A. 2004;101(11):3927–32.
71. Xu H, et al. A single amino acid substitution in human APOBEC3G antiretroviral enzyme confers resistance to HIV-1 virion infectivity factor-induced depletion. Proc Natl Acad Sci U S A. 2004;101(15):5652–7.
72. Anderson J, et al. Safety and efficacy of a lentiviral vector containing three anti-HIV genes--CCR5 ribozyme, tat-rev siRNA, and TAR decoy--in SCID-hu mouse-derived T cells. Mol Ther. 2007;15(6):1182–8.
73. Kumar P, et al. T cell-specific siRNA delivery suppresses HIV-1 infection in humanized mice. Cell. 2008;134(4):577–86.
74. ter Brake O, et al. Evaluation of safety and efficacy of RNAi against HIV-1 in the human immune system (Rag-2(−/−)gammac(−/−)) mouse model. Gene Ther. 2009;16(1):148–53.
75. Novina CD, et al. siRNA-directed inhibition of HIV-1 infection. Nat Med. 2002;8(7):681–6.
76. Kohn DB, et al. A clinical trial of retroviral-mediated transfer of a rev-responsive element decoy gene into CD34(+) cells from the bone marrow of human immunodeficiency virus-1-infected children. Blood. 1999;94(1):368–71.
77. Humeau LM, et al. Efficient lentiviral vector-mediated control of HIV-1 replication in CD4 lymphocytes from diverse HIV+ infected patients grouped according to CD4 count and viral load. Mol Ther. 2004;9(6):902–13.
78. Michienzi A, et al. A nucleolar TAR decoy inhibitor of HIV-1 replication. Proc Natl Acad Sci U S A. 2002;99(22):14047–52.
79. Zahn RC, et al. Efficient entry inhibition of human and nonhuman primate immunodeficiency virus by cell surface-expressed gp41-derived peptides. Gene Ther. 2008;15(17):1210–22.
80. Voit RA, et al. Generation of an HIV resistant T-cell line by targeted "stacking" of restriction factors. Mol Ther. 2013;21(4):786–95.
81. Lombardo A, et al. Gene editing in human stem cells using zinc finger nucleases and integrase-defective lentiviral vector delivery. Nat Biotechnol. 2007;25(11):1298–306.
82. Lombardo A, et al. Site-specific integration and tailoring of cassette design for sustainable gene transfer. Nat Methods. 2011;8(10):861–9.
83. Genovese P, et al. Targeted genome editing in human repopulating haematopoietic stem cells. Nature. 2014;510(7504):235–40.

84. Hoban MD, et al. Correction of the sickle cell disease mutation in human hematopoietic stem/progenitor cells. Blood. 2015;125(17):2597–604.
85. Wang J, Exline CM, et al. Highly efficient homology-driven genome editing in CD34+ hematopoietic stem/progenitor cells by combining zinc finger nuclease mRNA and AAV donor delivery. Nat Biotechnol. 2015;33(12):1256–1263.
86. Chu VT, et al. Increasing the efficiency of homology-directed repair for CRISPR-Cas9-induced precise gene editing in mammalian cells. Nat Biotechnol. 2015;33(5):543–8.
87. Maruyama T, et al. Increasing the efficiency of precise genome editing with CRISPR-Cas9 by inhibition of nonhomologous end joining. Nat Biotechnol. 2015;33(5):538–42.
88. Ye L, et al. Seamless modification of wild-type induced pluripotent stem cells to the natural CCR5Delta32 mutation confers resistance to HIV infection. Proc Natl Acad Sci U S A. 2014;111(26):9591–6.
89. Yao Y, et al. Generation of CD34+ cells from CCR5-disrupted human embryonic and induced pluripotent stem cells. Hum Gene Ther. 2012;23(2):238–42.
90. Chomont N, et al. HIV reservoir size and persistence are driven by T cell survival and homeostatic proliferation. Nat Med. 2009;15(8):893–900.
91. Chun TW, et al. Quantification of latent tissue reservoirs and total body viral load in HIV-1 infection. Nature. 1997;387(6629):183–8.
92. Finzi D, et al. Latent infection of CD4+ T cells provides a mechanism for lifelong persistence of HIV-1, even in patients on effective combination therapy. Nat Med. 1999;5(5):512–7.
93. Sarkar I, et al. HIV-1 proviral DNA excision using an evolved recombinase. Science. 2007;316(5833):1912–5.
94. Qu X, et al. Zinc-finger-nucleases mediate specific and efficient excision of HIV-1 proviral DNA from infected and latently infected human T cells. Nucleic Acids Res. 2013;41(16):7771–82.
95. Hauber I, et al. Highly significant antiviral activity of HIV-1 LTR-specific tre-recombinase in humanized mice. PLoS Pathog. 2013;9(9), e1003587.
96. Ebina H, et al. Harnessing the CRISPR/Cas9 system to disrupt latent HIV-1 provirus. Sci Rep. 2013;3:2510.
97. Blackard JT, et al. Transmission of human immunodeficiency type 1 viruses with intersubtype recombinant long terminal repeat sequences. Virology. 1999;254(2):220–5.
98. Surendranath V, et al. SeLOX—a locus of recombination site search tool for the detection and directed evolution of site-specific recombination systems. Nucleic Acids Res. 2010;38(Web Server issue):W293–8.
99. Karpinski J, et al. Universal Tre (uTre) recombinase specifically targets the majority of HIV-1 isolates. J Int AIDS Soc. 2014;17(4 Suppl 3):19706.
100. Ebina H, et al. A high excision potential of TALENs for integrated DNA of HIV-based lentiviral vector. PLoS One. 2015;10(3), e0120047.
101. Cong L, et al. Multiplex genome engineering using CRISPR/Cas systems. Science. 2013;339(6121):819–23.
102. Zhu W, et al. The CRISPR/Cas9 system inactivates latent HIV-1 proviral DNA. Retrovirology. 2015;12:22.
103. Liao HK, et al. Use of the CRISPR/Cas9 system as an intracellular defense against HIV-1 infection in human cells. Nat Commun. 2015;6.
104. Hu W, et al. RNA-directed gene editing specifically eradicates latent and prevents new HIV-1 infection. Proc Natl Acad Sci U S A. 2014;111(31):11461–6.
105. Li H, et al. In vivo genome editing restores haemostasis in a mouse model of haemophilia. Nature. 2011;475(7355):217–21.
106. Joglekar AV, et al. Integrase-defective lentiviral vectors as a delivery platform for targeted modification of adenosine deaminase locus. Mol Ther. 2013;21(9):1705–17.
107. Holkers M, et al. Differential integrity of TALE nuclease genes following adenoviral and lentiviral vector gene transfer into human cells. Nucleic Acids Res. 2013;41(5), e63.
108. Chen Z, et al. Receptor-mediated delivery of engineered nucleases for genome modification. Nucleic Acids Res. 2013;41(19), e182.

109. Mock U, et al. Novel lentiviral vectors with mutated reverse transcriptase for mRNA delivery of TALE nucleases. Sci Rep. 2014;4:6409.
110. Khatri N, et al. In vivo delivery aspects of miRNA, shRNA and siRNA. Crit Rev Ther Drug Carrier Syst. 2012;29(6):487–527.
111. Morizono K, et al. Lentiviral vector retargeting to P-glycoprotein on metastatic melanoma through intravenous injection. Nat Med. 2005;11(3):346–52.
112. Lin AH, et al. Receptor-specific targeting mediated by the coexpression of a targeted murine leukemia virus envelope protein and a binding-defective influenza hemagglutinin protein. Hum Gene Ther. 2001;12(4):323–32.
113. Frecha C, et al. A novel lentiviral vector targets gene transfer into human hematopoietic stem cells in marrow from patients with bone marrow failure syndrome and in vivo in humanized mice. Blood. 2012;119(5):1139–50.
114. Anliker B, et al. Specific gene transfer to neurons, endothelial cells and hematopoietic progenitors with lentiviral vectors. Nat Methods. 2010;7(11):929–35.
115. Paraskevakou G, et al. Epidermal growth factor receptor (EGFR)-retargeted measles virus strains effectively target EGFR- or EGFRvIII expressing gliomas. Mol Ther. 2007;15(4):677–86.
116. Kneissl S, et al. Measles virus glycoprotein-based lentiviral targeting vectors that avoid neutralizing antibodies. PLoS One. 2012;7(10), e46667.
117. Kneissl S, et al. CD19 and CD20 targeted vectors induce minimal activation of resting B lymphocytes. PLoS One. 2013;8(11), e79047.
118. Alwin S, et al. Custom zinc-finger nucleases for use in human cells. Mol Ther. 2005;12(4):610–7.
119. Kim HJ, et al. Targeted genome editing in human cells with zinc finger nucleases constructed via modular assembly. Genome Res. 2009;19(7):1279–88.
120. Koo T, Lee J, Kim JS. Measuring and reducing off-target activities of programmable nucleases including CRISPR-Cas9. Mol Cells. 2015;38(6):475–81.
121. Hendel A, et al. Quantifying on- and off-target genome editing. Trends Biotechnol. 2015;33(2):132–40.
122. Pattanayak V, et al. Revealing off-target cleavage specificities of zinc-finger nucleases by in vitro selection. Nat Methods. 2011;8(9):765–70.
123. Gabriel R, et al. An unbiased genome-wide analysis of zinc-finger nuclease specificity. Nat Biotechnol. 2011;29(9):816–23.
124. Chiarle R, et al. Genome-wide translocation sequencing reveals mechanisms of chromosome breaks and rearrangements in B cells. Cell. 2011;147(1):107–19.
125. Frock RL, et al. Genome-wide detection of DNA double-stranded breaks induced by engineered nucleases. Nat Biotechnol. 2015;33(2):179–86.
126. Tsai SQ, et al. GUIDE-seq enables genome-wide profiling of off-target cleavage by CRISPR-Cas nucleases. Nat Biotechnol. 2015;33(2):187–97.
127. Crosetto N, et al. Nucleotide-resolution DNA double-strand break mapping by next-generation sequencing. Nat Methods. 2013;10(4):361–5.
128. Ran FA, et al. In vivo genome editing using Staphylococcus aureus Cas9. Nature. 2015;520(7546):186–98.
129. Kim D, et al. Digenome-seq: genome-wide profiling of CRISPR-Cas9 off-target effects in human cells. Nat Methods. 2015;12(3):237–43, 1 p following 243.
130. Doyon Y, et al. Enhancing zinc-finger-nuclease activity with improved obligate heterodimeric architectures. Nat Methods. 2011;8(1):74–9.
131. Tsai SQ, et al. Dimeric CRISPR RNA-guided FokI nucleases for highly specific genome editing. Nat Biotechnol. 2014;32(6):569–76.

Strategies to Determine Off-Target Effects of Engineered Nucleases

Eli J. Fine, Thomas James Cradick, and Gang Bao

Abstract Genome editing is greatly facilitated by using engineered nucleases to specifically cleave a pre-selected DNA sequence. Cellular repair of the nuclease-induced DNA breaks by either non-homologous end joining (NHEJ) or homology-directed repair (HDR) allows genome editing in a wide range of organisms and cell lines. However, if a nuclease cleaves at genomic locations other than the intended target, known as "off-target sites", it can lead to mutations, chromosomal loss or rearrangements, causing gain/loss of function and cytotoxicity. Although zinc finger nucleases (ZFNs), TAL effector nuclease (TALENs), and CRISPR/Cas9 systems have been used successfully to create specific DNA breaks in cells, they lack perfect specificity and may result in off-target cleavage. Methods have been developed to predict and to quantify the off-target cleavage events, which are very important for optimizing nuclease design and determining if the gene editing approaches are highly specific. These methods have the potential to significantly facilitate the design of engineered nucleases for genome editing applications.

Keywords Gene editing • Nucleases • Off-target • Specificity • TAL effector nuclease (TALEN) • Zinc finger nuclease (ZFN) • CRISPR/Cas9

Abbreviations

bp	Base pairs (of nucleic acid)
Cas9	CRISPR associated protein 9, an endonuclease from *Streptococcus pyogenes*
CCR5	The chemokine (C-C motif) receptor 5 gene
CRISPR	Clustered regularly interspaced short palindromic repeats

E.J. Fine • T.J. Cradick • G. Bao, Ph.D. (✉)
Department of Biomedical Engineering, Georgia Institute of Technology and Emory University, 313 Ferst Dr., Atlanta, GA 30332, USA
e-mail: ejfine@gmail.com; tj@alum.mit.edu; gang.bao@bme.gatech.edu

T. Cathomen et al. (eds.), *Genome Editing*, Advances in Experimental Medicine and Biology 895, DOI 10.1007/978-1-4939-3509-3_11

CRISPR/Cas9	The combination of guide RNA and associated protein required for DNA cleavage
DSB	Double strand break
FokI	An endonuclease derived from *Flavobacterium okeanokoites*
gRNA	Guide RNA, used in the CRISPR/Cas9 system
HBB	The hemoglobin beta gene, also known as beta-globin
HDR	The homology directed DNA repair pathway
IDLV	Integrase deficient lentiviral vector
indel	A short insertion, deletion, or combined insertion and deletion of DNA
NGS	Next generation sequencing, platforms such as Illumina fall in this category
NHEJ	The non-homologous end-joining DNA repair pathway
RVD	Repeat variable di-residue, the two amino acids in TAL repeats that specify the DNA base
SMRT	Single molecule real-time, a third generation sequencing platform
TALEN	Transcription activator-like effector nuclease
ZFN	Zinc finger nuclease

Introduction

Engineered nucleases cleave their target site DNA to facilitate genome editing, which occurs using cellular repair pathways. DNA cleavage dramatically increases the frequency of gene editing through homologous recombination (HR) with a supplied donor DNA [1, 2], and efficient gene knockout can be achieved by triggering the non-homologous end-joining (NHEJ) repair pathway that results in small insertions and deletions (indels) at the site of the break. Of course, ensuring only one cleavage spot in a large genome presents challenges. Engineered nucleases need to minimize off-target cleavage, which can lead to mutations, chromosomal loss or rearrangements, changes in gene regulation and cytotoxicity. There have been a number of nucleases used to create these specific breaks, including meganucleases [3, 4], zinc finger nucleases (ZFNs) [5–7], TAL effector nuclease (TALENs) [8–11] and most recently CRISPR/Cas9 [12–14]. Each of these types of nuclease lacks perfect specificity and has been found to have off-target cleavage, even naturally evolved meganucleases, also known as homing endonucleases, such as I-SceI [15].

Cleavage by each of these nuclease families generally results in DNA double-strand breaks at their target site and at off-target genomic sites. Different families of nucleases are prone to cleave different types and numbers of off-target sites. Within a family, variations in targeting also lead to dramatic changes in off-target site cleavage. Therefore, it is important to choose the variant optimal for a given application in order to best balance activity and specificity. It is also important to understand the

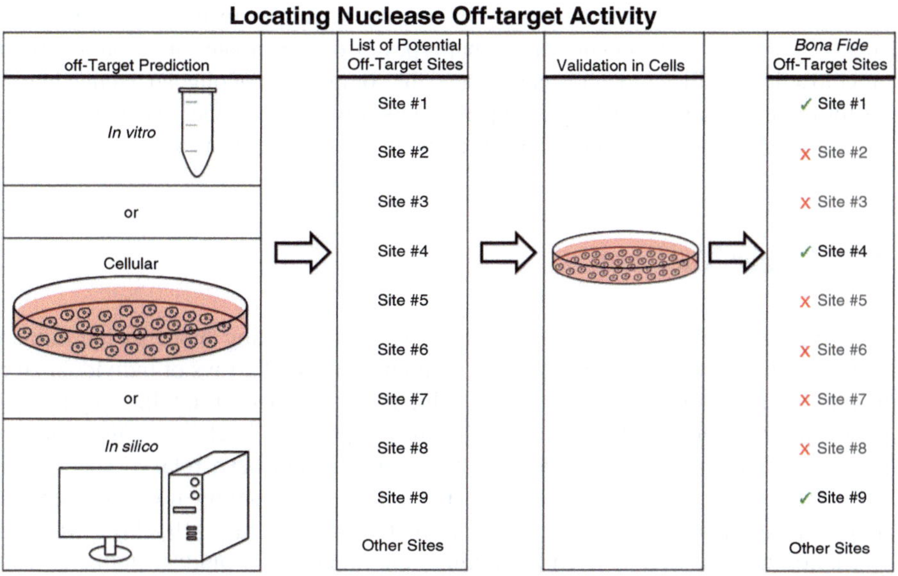

Fig. 1 Locating nuclease off-target activity. Off-target sites are predicted through either in vitro, cellular, or in silico techniques, which generate a list of potential off-target sites. Genomic DNA is harvested from cells treated with nucleases, the sites are amplified, tested and a subset validated as bona fide off-target sites

aspects of each variant because it will impact the search parameters for off-target sites throughout the genome.

There are various strategies to determine off-target effects of nucleases. The simplest are bulk measures of overall off-target activity through observing effects on the overall cell population, such as an increase in DNA repair foci, cell cycle dysregulation, or an increase in cell death. More intricate strategies focus on finding the specific sites of off-target cleavage and measuring the nuclease-induced mutation rates at those sites. Most methods to locate site-specific off-target activity follow a common framework (Fig. 1). First, a list of potential off-target sites are predicted in the genome based on experimental characterization of the nucleases (in cells or in vitro) or through in silico modeling. Then, these loci are analyzed in cells treated with nucleases in order to identify which sites are *bona fide* locations of nuclease off-target activity and which sites are false positive predictions.

Factors Influencing Off-Target Activity

When choosing engineered nucleases systems, it is important to assess the acceptable risk and (Fig. 1) level of off-target effects that is tolerable for each application. Each nuclease family, and variations of those nucleases, offers tradeoffs between activity and specificity. Furthermore, when using strategies to determine off-target effects, it is important to understand what types of off-target sites are more likely in order to adjust in silico modeling parameters, the composition of oligonucleotide libraries for in vitro analysis, or the search parameters used to locate the potential cleavage site within the genomic region identified by an in vivo technique accordingly.

Zinc Finger Nucleases

C_2H_2-type zinc fingers are the most abundant DNA-binding motif in the human genome [16]. Zinc finger DNA binding domains consist of arrays of individual zinc finger modules each comprising ~30 aa, which form a ββα-fold stabilized by hydrophobic interactions and the chelation of a zinc ion [17, 18]. Each zinc finger domain generally recognizes 3–4 nucleotides, but some fingers have relaxed specificity. Zinc fingers are modular and can be assembled for targeting novel sequences, though positional effects and context limit the success rate [19, 20]. Artificial restriction enzymes are created by linking zinc finger DNA binding domains with the catalytic domain from the bacterial endonuclease FokI, creating zinc finger nucleases (ZFNs) [21, 22]. This work developed from the observation that the catalytic domain of FokI was sequence non-specific, separable and could be replaced with other DNA binding domains [21]. Specificity is greatly increased by the requirement for correct binding of two ZFNs to half-sites with correct orientation and spacing. This allows the two FokI domains to dimerize and cleave the intervening sequence [23]. The individual zinc fingers can be changed, or the framework can remain constant and the contact residues in the zinc finger changed to direct ZFNs to novel sequences [6, 24]. ZFNs have been shown to have off-target cleavage and cytotoxicity [25–30]. Research continues to improve specificity through modifications and further refinement of each aspect of the ZFNs, as described below.

Protein-DNA Interactions

A number of methods were used in hopes of decreasing ZFN off-target cleavage and toxicity. Much of this research was directed at improving specificity through better understanding of protein-DNA binding and use of more specific nucleases [31]. Selection experiments were used to identify tighter binding zinc finger domains using phage display [32–35] and bacterial systems, which may also take into account specificity through competition with bacterial chromosomes [36–38]. Several selection-based platforms have allowed selection of many zinc fingers domains, but remain difficult methods [39, 40]. Targeting non guanine-rich sequences remains more difficult for ZFNs. Overall though, methods to improve ZFN specificity have failed to make dramatic improvements.

Poly-Zinc Finger Arrays

Increasing the number of zinc fingers in artificial transcription factors or ZFNs can lead to high affinity DNA binding. Four and six finger zinc finger domains were often constructed linking series of two finger units [41]. Several groups have increased the number of zinc finger domains to increase DNA binding domains' affinity and/or specificity [42, 43], though there has been little comparative work to validate the approach by comparing arrays targeting the same location with different number of fingers. A six-finger zinc finger linked to a repression domain was shown to have high specificity in down regulating only its targeted gene [44]. As this was a repression domain, off-target binding at silent genes could not be observed, nor was this compared to similarly targeted repressors containing fewer fingers. A recent comparison of on- and off-target cleavage directly compared a three-finger and a four-finger ZFN pair targeting overlapping sites. The four-finger ZFN pair had higher on-target cleavage, and lower off-target cleavage [45]. However, additional finger domains do not always ensure high specificity; a substantial amount of off-target activity was observed in cultured rat cells for a pair of five-finger CompoZr ZFNs designed by Sigma-Aldrich [46]. Sensitive sequencing methods will allow further work comparing ZFNs with different number of fingers.

Protein Linker Domains

Research was also conducted to optimize the use of linkers between individual zinc finger domains within a ZFP or ZFN [41, 47], but no general rules were derived applicable to all situations. An alternative approach was to increase specificity by decreasing non-specific DNA interactions between the linker and DNA [48]. When the linker between the zinc finger domains and the FokI domain was varied, it changed the half-site spacing requirement, impacting locating on- and off-target cleavage sites [49].

Transcription Activator-Like Effector Nucleases

Transcription activator-like effectors (TALEs) are a family of DNA binding proteins, discovered in the plant pathogen *Xanthomonas* [10, 11, 50, 51]. Artificial restriction enzymes were created by linking the TALE DNA binding domains with the same sequence non-specific catalytic domain used in ZFNs from the bacterial endonuclease FokI, thus creating TALENs [52]. As with ZFNs, the key component of their specificity is the requirement for correct binding of two TALENs to two half-sites with correct orientation and spacing [23].

Protein-DNA Interactions

Each TALE DNA-binding domain contains repeats of 33–35 amino-acids that differ primarily in the 12th and 13th position of each repeat—termed the repeat-variable di-residues (RVDs)[50]. TALEN specificity is conferred through the RVDs using a DNA binding code that was derived from naturally occurring TALE target sites [53]. TALE or TALEN DNA binding domains can be easily designed because a simple one-to-one relationship exists between each RVD and the preferred nucleotide [10, 11, 54]. Although binding to adenine, cytosine and thymine is straightforward, there are several RVDs that bind to guanine; the most common RVD, "NN" (Asparagine-Asparagine), binds to both guanine and adenine with nearly equal affinity. TALENs containing the "NN" RVD tend to have better activity, but show more off-target activity than those TALENs containing the "NK" (Asparagine-Lysine) RVD [45].

Protein Linker Domains

Non-canonical linkers within the DNA binding domain of TALENs have not yet been investigated, but there are several alternative linkers that have been tested to connect the TAL repeats to the FokI domain. These different linkers consist of different lengths of the natural *AvrBs4* TALE backbone that are preserved C-terminally from the end of the TAL repeats; for example, the commonly used +63 linker preserves the first 63 amino acids from the *AvsBs4* backbone before truncating the backbone and fusing it to the FokI domain. Although some publications found that certain linkers can provide higher TALEN activity at the intended target site, results have been mixed [54, 55], except for strong evidence that the full *AvsBs4* backbone (termed the +231 linker) yields much lower activity than shorter truncations [52, 56]. However, it is clear that different linkers have a dramatic effect on the range of spacing distances between the two TALEN half-sites over which effective cleavage can occur.

A design tradeoff exists between choosing a linker that allows a wider half-site spacer range and one that allows a more limited range. A wider range allows more flexibility in choosing an intended TALEN target site. However, this creates a greater potential for off-target activity, as mismatched half-sites throughout the genome separated by a larger number of spacer distances could be susceptible to TALEN cleavage. The +63 linker is the most commonly used length, and generally allows cleavage with spacers ranging from 10 to 30 bp, although detectable activity has been observed with a spacer as small as 6 bp [55] and as large as 39 bp [54]. A +47 linker was shown to provide a somewhat stricter range than the +63 linker, but was still quite tolerant of spacers from 12 to 21 bp [55]. Using a +18 linker was reported to only allow cleavage with spacers from 13 to 17 bp (and greatly reduced, but detectable, cleavage at 24 bp) [54]. A +17 linker was reported to provide a similar limitation in spacers, allowing 9–18 bp spacers as well as 21 bp [55]. If limiting off-target activity is a major concern for the intended application, then a shorter (+17 or +18) linker is recommended.

One group attempted to re-engineer the C-terminal linker to provide greater specificity [57]. They replaced either three or seven cationic amino acids with glutamine in order to limit non-specific interaction of the C-terminal domain with DNA. While the results in vitro were very promising (showing little reduction in on-target activity coupled with a substantial decrease in off-target cleavage), the on-target activity in cultured cells was severely compromised. Nevertheless, further work may show that this approach is beneficial in situations with an extremely low tolerance for off-target cleavage.

Modifications to the FokI Domain of ZFNs and TALENs

Both ZFNs and TALENs employ the catalytic domain from the FokI endonuclease to cleave DNA. Although attempts have been made to attach ZFNs or TALENs to alternate cleavage domains, such as PvuII [58] or I-TevI [59], FokI remains the most widely used. Several improvements and alterations have been made to the FokI domain that should be considered for use in different genome engineering applications.

Obligate Heterodimers

ZFNs were originally designed to function as heterodimeric pairs made up of a "left" ZFN and a "right" ZFN with identical FokI cleavage domains. If an off-target site consisted of two "left" or two "right" right half-sites, it was therefore possible for the FokI domains from two ZFNs to homodimerize and create a DSB, doubling the number of potential off-target binding configurations. In practice, homodimers often account for a disproportionately large fraction of off-target sites [28, 45], because one ZFN is much less specific than the other. To address this issue, the FokI dimerization interface was re-engineered to inhibit homodimerization [60, 61]. While effective at reducing or eliminating homodimeric off-target effects [60, 61], some earlier versions were found to reduce on-target activity as well [29]. Heterobligate constructs greatly reduced off-target events at homodimeric sites, but some thorough studies found that ZFNs that were thought to be obligate heterodimers had rare off-target cleavage and mutagenesis at homodimeric sites [27, 28]. The most prevalent obligate heterodimeric FokI pair in current use is the "ELD/KKR" set, containing triple mutations in each FokI domain, which has been found to often have comparable on-target activity to the wild-type FokI domain both for ZFNs [62] and TALENs [63].

Sharkey Enhancement

Directed evolution was applied to the FokI domain to enhance ZFN cleavage activity [64]. Several mutants were found that increased activity up to ~10-fold in mammalian cells and some were also compatible with the obligate heterodimer FokI architectures. The most efficient of these mutants (S418P, K441E) was called "Sharkey". Although on-target activity is increased, Sharkey has been found to increase bulk measures of ZFN off-target activity in cells, even when used in conjunction with obligate heterodimer architecture [25], and so its use is cautioned when off-target effects are a major concern. A thorough comparison of the use of Sharkey vs. wild-type TALENs has not yet been reported.

Nick Only

If the desired DNA repair pathway for an application is gene editing through homology directed repair (HDR), as opposed to mutagenic non-homologous end-joining (mutNHEJ), an option to reduce off-target activity is the use of "nick-only" enzymes. The D450A mutation in FokI inactivates the catalytic domain of that ZFN and renders it unable to cleave the DNA strand. Therefore, when one active and one inactive ZFN bind and the two FokI domains dimerize, they do not create a double strand break, but break the backbone of only one of the strands of DNA, creating a "nick". Resolution of a DNA nick is much less likely to use the NHEJ pathway and therefore much less likely to result in mutations [65]. ZFNickases have been shown in human cells to be able to induce HDR at the target site while only creating a very low level of mutNHEJ events compared to standard ZFNucleases [65]. The absolute rate of HDR caused by the nickases however was substantially reduced compared to nucleases. Although not yet demonstrated, the combination of ZFNickases with obligate heterodimer architecture should theoretically drastically reduce or even eliminate any off-target activity. TALENickases have not yet been reported, although unpublished data indicate that using a TALEN pair with one inactivated (D450A) FokI domain may still result in mutNHEJ at the on- and off-target sites in some instances.

CRISPR/Cas9 Systems

Another type of engineered nuclease was developed that does not rely on protein-DNA interactions to direct specific cleavage, but instead was based on the discovery of a bacterial defense system that uses RNA-guided DNA cleaving enzymes and clustered, regularly interspaced, short palindromic repeats (CRISPR) [66–70]. CRISPR provides an exciting alternative to ZFNs and TALENs, as the CRISPR-associated (Cas) protein remains the same for different gene targets; only the short sequence of a single guide RNA (sgRNA) needs be changed to redirect the site-specific cleavage [12, 14]. Each construct is made by simply adding a pair of annealed oligonucleotides. This is in stark contrast to ZFNs and TALENs, which require redesign of the protein—and reconstruction of the plasmids—in order to target novel sequences [71].

gRNA-DNA and PAM Interactions

Early studies on CRISPR/Cas9 systems suggested high off-target activity could occur with nonspecific binding of the guide strands to DNA sequences with base pair mismatches at positions distal from the protospacer adjacent motif (PAM) region [12, 13, 72, 73]. CRISPR systems have also been found to cleave sequences with mismatches closer to the PAM site as well [55, 65, 74–76]. A clear understanding of "rules" governing CRISPR off-target cleavage has not been achieved and there appear to be strong influences of the specific sequence of the guide RNA (gRNA). However, a general trend has been observed that mismatches in the "seed" region closer to the PAM are less tolerated than distant mismatches. The PAM appears to be the most critical region for determining specificity, with a single mismatch being able to completely abolish cleavage at that location in most cases [77, 78]. The preferred PAM sequence for the most commonly used Cas9 variant (derived from *Streptococcus pyogenes*, also called SpCas9) is any nucleotide followed by a pair of guanines (the motif NGG), although off-target sites with NAG PAMs also seem to be well tolerated [79–81]. Different CRISPR systems target different PAMs, varying in length and sequence preference, and further work on Cas molecules with longer PAM requirements may allow for increased specificity [82]. Recent work has also suggested that it may be possible to lower off-target activity by deploying separate transactivating CRISPR RNA (tracrRNA) and CRISPR RNA (crRNA) rather than a single gRNA (sgRNA) [83].

Modifications to Cas9

Although studies are rapidly revealing more about the structure [84] and mechanism [85] of Cas9 cleavage, thus far Cas9 has not yet been re-engineered in the variety of ways as FokI, however there have been several modifications developed that can mitigate off-target activity. Similar to FokI, "nick only" mutants of Cas9 have been generated by inactivating one of the cleavage domains of Cas9 through either the D10A mutation or the H840A mutation [86, 87]. These CRISPR nickases still trigger on-target HDR, while greatly reducing off-target levels of NHEJ. Pairs of CRISPR nickases have been used to generate offset nicks that are repaired by NHEJ similar to DSBs [86, 87]. As single nicks at off-target sites cause mutations at a much lower frequency than DSBs caused by nucleases (50–1500 fold lower [86]), using nickase pairs decreases standard Cas9 off-target activity by requiring that two off-target sites exist within ~100 bp of each other in order for a double strand break to occur. Fusion proteins consisting of FokI and a catalytically inactivated Cas9 have also been generated which require dimerization of the two FokI domains for cleavage in a similar manner to ZFNs and TALENs [88, 89]. While single nicking by Cas9 at an off-target site can still induce low rates of NHEJ, these RNA-guided FokI Nucleases (RFNs) have been shown to have even greater specificity and no detectable NHEJ was observed by deep sequencing at sites previously identified as having low off-target activity by Cas9 nicking (using the same gRNA).

Bulk Assays of Off-Target Activity

If nuclease activity can be observed at the intended target site and gross cytotoxicity is not readily visible in treated cells, bulk assays can be used as a first step in determining the level of off-target activity. These assays provide information on whether the nucleases are causing DNA breaks in sufficient quantity throughout the genome to adversely affect the properties of the bulk cell population. A common negative control for these assays is the meganuclease I-SceI, available in a mammalian expression plasmid from AddGene (#21299). The high level of specificity seen with I-SceI does not typically induce signals of bulk off-target activity.

γH2AX Foci

Double strand breaks in the genome cause the rapid appearance of phosphorylated histone H2AX (γH2AX) at the site of the DNA break [64]. Substantial nuclease off-target activity will result in the appearance of many of these foci that can be visualized using microscopy or flow cytometry. A commercial kit for flow cytometry analysis is available from Millipore (Catalog #17-344) and its use for detecting nuclease off-target activity was described by Guo et al. [64].

Cell Cycle Disregulation

DSBs occurring throughout the genome due to nuclease off-target activity can prevent transition between different phases of the cell cycle, until the breaks are repaired. HeLa FUCCI cells (originally described by Sakaue-Sawano et al. [90] and available from Amalgaam) are an easily transfectable system that offer fluorescent readouts to determine which of three phases of the replication cycle a cell is in: G_1, G_1/S, or S/G_2/M. The use of this assay for detecting nuclease off-target activity was described by Mussolino et al. [30].

Apoptosis and Cell Viability

If cells are unable to repair the DSBs caused by nuclease off-target activity, the cells may commence apoptosis or show other signs of reduced viability. An increase in AnnexinV-positive cells is observed with increased apoptosis [30]. Alternatively, cells that uptake the dye 7-aminoactinomycin D (7-AAD) are no longer viable. Although 7-AAD cannot distinguish between apoptosis and necrosis, it is a very cheap and easy method that has the advantage over the propidium iodide (PI) viability dye because the fluorescent spectrum of 7-AAD does not overlap with many fluorescent proteins commonly used as transfection controls [91].

Loss of Fluorescence

If nucleases are being delivered as DNA into cells, then a common and straightforward assay involves co-delivering a fluorescent protein such as eGFP [74]. The percentage of cells that are fluorescent is then measured at two different time points (commonly 2 and 5 days after transfection). Presumably, if cells were fluorescent at an earlier timepoint, then they also received the nucleases. Therefore, the percent loss of fluorescence (i.e. 40 % at Day 5 divided by 80 % at Day 2 equals 50 % loss of fluorescence) can be used as an indication of how toxic the nucleases. This approach measures the effect of cells within a population which received nuclease exhibit reduced viability compared to cells in the same population that did not receive the nuclease.

Experimental-Based Off-Target Site Prediction Methods

Most previous studies of nuclease off-target activity have used experimental characterization of the specific nuclease in order to predict potential off-target sites. Although experimental-based prediction methods are generally quite effective (nearly all publications employing them have located at least one *bona fide* nuclease off-target site), they are very technically challenging, costly, and time intensive. Because of the difficulty of implementing these techniques, most have never been replicated outside of the original laboratories in which they were developed.

SELEX

Systematic Evolution of Ligands by Exponential Enrichment (SELEX) is an established technique for determining nucleic acid sequences that have high affinity for target molecules. This approach has been used to ascertain nuclease specificity in works from the laboratories of Sangamo Biosciences. SELEX (Fig. 2) has been used to determine the binding preference of ZFNs [92, 93] and TALENs [52, 75, 76] and subsequently to guide bioinformatics searches for potential genomic off-target sites. The general approach is to (1) genetically tag the nuclease with an affinity molecule, such as hemagglutinin (HA), (2) express the nuclease in vitro, (3) incubate it with a semi-randomized library of oligonucleotides (usually biased towards the expected binding site of the nuclease), (4) capture the nuclease protein using antibodies, and (5) PCR the bound DNA fragments to amplify them. Steps (3)–(5) are then repeated for multiple rounds of enrichment with the PCR products from step (5) replacing the initial semi-randomized library in step (3). After the desired number of selection rounds, the PCR amplicons are sequenced to determine the identity of the selected DNA. SELEX typically yields 20–50 unique sequences that were bound by the nucleases or DNA binding domains [52]; if too few or too many unique sequences are found, amplicons from prior rounds can be sequenced or additional rounds of selection can be carried out. These sequences can be compiled to form position weight matrices (PWMs) indicating the binding preferences of the nuclease at each position (Fig. 2).

Once PWMs for each nuclease half-site have been established, the genome can be searched bioinformatically and scores calculated for each position. Each potential bi-partite nuclease off-target site—one half-site, separated by an appropriate length spacer sequence, and the other half-site—can be given a score by calculating the product of the values of the PWM for each nucleotide comprising the potential off-target site (Note: the authors from Sangamo did not publish their formula for generating a score from the PWMs, but calculating the product of all positions provides a close approximation [94]). All sites in the genome can then be ranked and a subset can be chosen for further investigation.

This technique has faced several criticisms, but has proven remarkably robust at finding off-target sites for both ZFNs and TALENs. Drawbacks of this technique

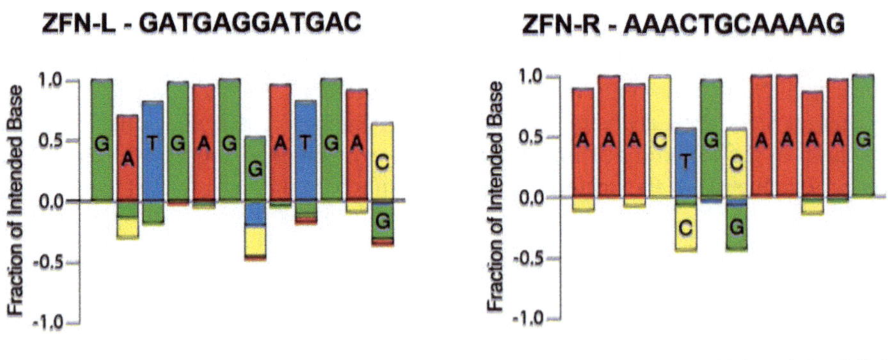

Fig. 2 ZFN targeting specificity. A position weight matrix defining the specificity of the CCR5 ZFNs based on SELEX analysis. From Perez, EE, et al. (2008) Establishment of HIV-1 resistance in CD4+ T cells by genome editing using zinc-finger nucleases. Nat Biotechnol, 26, 808-816. With permission from Nature Publishing Group

include the fact that it only provides information about the binding preferences of each nuclease half-site, therefore ignoring interactions between the two half-sites required for nuclease cleavage. Another limitation is that it is performed completely in vitro, therefore ignoring changes that may occur to the protein in the cellular environment as well as ignoring any factors that may affect the genomic DNA at the potential off-target sites in the cells, such as chromatin structure, accessibility, and methylation status. Finally, because the starting oligonucleotide library is only semi-randomized, this method is biased towards finding sites with relatively high homology to the intended nuclease target. Nevertheless, this has been the most published experimental characterization technique for successfully finding nuclease off-target sites and is one of only two experimental-based prediction technique thus far published that has found *bona fide* TALEN off-target sites [75, 76].

Bacterial One-Hybrid

The bacterial one-hybrid (B1H) approach is similar to SELEX in that it analyzes the binding preferences of nuclease monomers. To begin, a library of reporter plasmids is generated with a semi-randomized (biased towards the intended nuclease target) region upstream of the reported gene. This library is co-transformed into bacteria along with a plasmid encoding the nuclease DNA binding domain fused to a transcriptional activator [29]. *E. coli* colonies expressing the reporter gene, due to the nuclease DNA binding domain having sufficient affinity for the sequence in the plasmid to be able to activate the gene, are selected and the plasmid is sequenced to determine the sequence of the semi-randomized binding site. All sequences recovered from the *E. coli* colonies are compiled to create a PWM that can be used to screen the genome for potential off-target sites in the same way as the SELEX method. Using B1H for nuclease off-target prediction was developed by Scot Wolfe's laboratory which has been the only group to employ this method so far, and only for predicting off-target activity of ZFNs [29]. This approach faces many of the same criticisms as SELEX relating to the analysis of single monomers, but it has the advantage of being performed in a cellular (albeit bacterial and not eukaryotic) environment which may better model protein-DNA interactions than a completely in vitro analysis.

In Vitro Cleavage

Unlike the previous two prediction methods that separately characterize the DNA binding abilities of each monomer, in vitro cleavage assays explicitly investigate which DNA sequences in a random pool can be cut by a nuclease [27]. This approach has been applied to ZFNs [27], TALENs [57], and CRISPRs [95], but has only been utilized in studies published by David Liu's laboratory. In this approach, a semi-randomized oligonucleotide library is synthesized that consists of the full nuclease recognition site. For paired nucleases (such as ZFNs, TALENs, or paired CRISPR nickases), the half-sites of both monomers are included, separated by appropriate length spacer sequences. The nuclease is then expressed in vitro, and incubated with the oligonucleotide library. Several enzymatic and gel isolation steps allow separation of sequences cleaved by the nuclease. Libraries are deep sequenced before and after nuclease incubation to identity the sequences cleaved by the nuclease. A bioinformatics search is then performed to determine if any of the sites that were cleaved in vitro also exist in the genome of interest. These sites can then be assayed for off-target activity.

There are several advantages and limitations of this technique. By examining nuclease cleavage instead of merely binding, insights were gained in the original study [27] that led to the hypothesis of an "energy compensation" model of dimeric ZFN interactions where larger numbers of mismatches in one half-site can be compensated by few or no mismatches in the other half-site. However, as this technique is performed entirely in vitro, effects of the cellular environment on the nuclease and genomic DNA are not accounted for. Furthermore, since the oligonucleotide library is semi-randomized, the analysis is biased towards finding sites with higher levels of homology to the intended nuclease target site.

An extension to this approach was recently developed to make better use of the large amount of data generated [26]. The original applications of this method searched through genomes to find exact matches to sequences that had been cleaved [27, 95], but those sequences that matched the genome were only a small fraction of the total sequences that the nuclease was shown to be able to cleave. By applying a Bayesian machine learning algorithm to the full list of sequences that the *CCR5* and *VEGF* ZFNs were confirmed to cleave in the original study, classifiers were developed for each nuclease that could generate a score for any given sequence predicting the likelihood of cleavage. The full genome was then screened bioinformatically—in a similar manner to the PWM screening in the SELEX and B1H methods—for sites that scored highly by the classifier. The analysis of off-target activity at the sites predicted by this method demonstrated that it could locate *bona fide* off-target sites with relatively low sequence homology and sites that have low activity. Impressively, this method also appears to have a fairly low false discovery rate that resulted in the analysis locating a large number of new *bona fide* off-target sites for both the *CCR5* and *VEGF* ZFNs [26] and two TALENs targeting *CCR5* and *ATM* [57]. Unfortunately, the incredibly difficult and time consuming nature of this approach that must be performed for each nuclease to be studied—conducting the in vitro cleavage experiments and then subsequently building a Bayesian classifier using machine learning—will likely severely limit the number of nucleases that are studied using this method.

IDLV LAM-PCR

Integrase-Deficient Lentiviral Vector Linear Amplification Mediated Polymerase Chain Reaction (IDLV LAM-PCR) is one of the two off-target prediction methods that is performed in the full intracellular environment. This approach was developed by Christof von Kalle's laboratory [28]. In this approach, the cells are transduced by an IDLV encoding a selectable marker, such as green fluorescent protein (GFP). Because the virus is integrase deficient, its ability to integrate into the genome is severely limited. Therefore, cultured dividing cells would rapidly dilute the IDLV gene sequence after several weeks, as it is not replicated during cell division. If nucleases are added, the resulting DSB can lead to a much higher efficiency of IDLV integration into the cellular genome. In this case then, after culturing dividing cells for several weeks, a larger fraction of nuclease treated cells express the selectable marker compared to control cells. These cells are then selected and viral integration site analysis is performed. Briefly, their approach was to use LAM-PCR on the genomic DNA using primers that bind to the long terminal repeat (LTR) regions of the IDLV. The amplicons resulting from LAM-PCR include a portion of the genomic sequence flanking the LTR, and therefore the location of the integration site of the IDLV can be deduced by high-throughput sequencing of the amplicons. Clustered integration site (CLIS) analysis is performed to filter out much of the random integration by imposing a criteria that two independent integration sites must be observed within 500 bp of each other in order for that locus to be considered a potential site of nuclease activity (although fragile sites in the genome are also prone to clustered integration sites). The next step is to search the sequence space surrounding the CLIS locus for a region with homology to the intended nuclease target that might be a location of *bona fide* nuclease cleavage activity; random sequence space has an expected level of ~45 % homology to the nuclease target [28], so sequences with >60 % homology to the nuclease target are likely candidates. The predicted off-target sites can then be interrogated in cells treated with nucleases without the IDLV.

The major limitation of this approach is its lack of sensitivity. This drawback is an inherent part of the method since the underlying process relies on the relatively rare event of the IDLV being captured during the repair of a DSB. Consequently, many off-target sites, especially those with lower activity, are overlooked; the IDLV LAM-PCR analysis of the hetero-obligate *CCR5* ZFNs [28] predicted only four of the 38 known off-target sites [26, 45]. This method has been successfully used to locate ZFN [28], TALEN [96], and CRISPR [96] off-target sites and provides an unbiased survey of any highly active off-target sites in the full intracellular environment with the cell's genome in its native structure. Because it is not biased by oligonucleotide selection libraries, this method was able to locate *bona fide* off-target sites with extremely low (66 %) homology to the intended nuclease target [28]. As this method lacks sensitivity, it may not be optimal for testing nucleases for potential use as human therapeutics—or other applications where even rare off-target cleavage could cause adverse events—but it remains a highly useful research tool because its unbiased nature allows it to uncover sites that might not fit standard models of nuclease specificity used to guide the generation of oligonucleotide libraries or in silico searches.

ChIP-Seq

Chromatin immunoprecipitation followed by deep sequencing (ChIP-Seq) is a well-established method for determining what sequences in a genome a certain protein binds. ChIP-Seq involves genetically tagging the nuclease with an affinity epitope (commonly hemagglutinin) and catalytically inactivating the nuclease (so that it binds but does not cleave DNA), expressing the modified nuclease in cells, cross-linking the protein and DNA together, shearing the genomic DNA into smaller fragments, purifying the nuclease (and the DNA fragments to which it is cross-linked) using antibodies (immunoprecipitation), sequencing the DNA fragments bound to the nuclease, and then mapping those sequences to the genome. In early 2013 it was noted how well the idea of ChIP-Seq seemed to be suited to facilitating an unbiased genome-wide survey of nuclease off-target activity in living cells [71], but the results of recent studies thus far have not been as promising as initially hoped.

Dimeric nucleases—such as ZFNs, TALENs, RFNs, and paired Cas9 nickases—present special challenges for ChIP-Seq. As noted in an unsuccessful attempt in late 2013 to use ChIP-Seq to identify off-target sites of the *CCR5* ZFNs: "thousands of high affinity monomeric target sites may exist in the genome, however a monomer is not sufficient to generate a lesion. Alternatively, dimeric ZFN sites that are bound weakly by both monomers may be sufficient to cleave DNA at a low frequency but may not bind stably enough to be detected reliably via ChIP" [26].

Because Cas9 can act as a monomeric nuclease, three groups attempted to use ChIP-Seq to locate CRISPR/Cas9 off-target sites in early 2014. While Cas9 binding was observed at many (up to thousands, depending on the gRNA used) sites throughout the genome other than the intended target site, off-target nuclease activity (NHEJ) was only found at a tiny fraction of the sites interrogated by two of the groups [97, 98], indicating that this method has a very high false positive rate as a method of discovery of off-target nuclease activity. The other group (Kuscu et al.) claimed to have found *bona fide* nuclease activity at 53 % of the sites predicted by their ChIP-Seq experiments [99]. However, large discrepancies in the results of Kuscu et al. when applying different window sizes to look for nuclease-induced mutations (reported at ranging from 12 to 53 % of predictions being *bona fide* off-target sites) and their unorthodox method of detecting indels (using the CIGAR output from Bowtie2 rather than analyzing the pairwise alignment between sequencing read and template sequence) raise questions about the validity of their findings which were in marked contrast to the findings of the other two research groups.

In summary, ChIP-Seq has not yet become a reliable method to predict off-target nuclease activity. Although it identifies the on-target site with reasonably accuracy (although sometimes off-target sites with no detectable nuclease activity are given higher scores via ChIP-Seq binding [99], ChIP-Seq may be fundamentally ill-suited to identifying off-target sites of nuclease activity. Emerging evidence into the mechanism of Cas9 shows that it uses a multi-step approach to DNA cleavage [97]: binding to many points along a chromosome as it pauses in the search for a target matching the gRNA but only cleaving when a good match is found. ChIP-Seq detects all binding events which leads to very high false positive prediction rates of nuclease activity. Without a method to discriminate between binding and cleavage events, ChIP-Seq may be incapable of detecting sites of low frequency off-target cleavage because the ChIP-Seq signal at those sites may not rise above the background noise.

Additional Genome-Wide Tools for CRISPR Off-Target Site Identification

Following the failure of ChIP-Seq, several other methods were recently developed to attempt to locate genome-wide off-target sites in an "unbiased" (not guided by homology to the intended target) manner. The digestion of genomic DNA was also used to predict off-target cleavage sites [100]. An alternate method identified sites using high-throughput, genome-wide translocation sequencing (HTGTS), which identified potential off-target cleavage sites through the translocations they make possible, using linear-amplification PCR and NGS [101]. However, it is important to note that the potential off-target sites predicted by HTGTS in the manuscript were never assayed for evidence of NHEJ necessary to confirm them as bona fide off-target sites. Although orthogonal evidence was presented to show that at least some of the predicted sites were likely to be bona fide, more thorough studies are needed to determine aspects of the methods performance such as its false positive discovery rate.

The final method in this category that has been published is called Genome-wide Unbiased Identifications of DSBs Evaluated by Sequencing (GUIDE-Seq) [102]. This method is similar to the IDLV approach but uses short double stranded oligonucleotides instead of a full-sized virus. Although a seemingly minor alteration, this had a dramatic impact on the sensitivity of the approach (the major drawback of the IDLV method), allowing identification of off-target sites with NHEJ frequencies on the order of tenths of a percent. GUIDE-Seq retained the low false positive rate that was an advantage of the IDLV system and additionally is the only method thus far that yields a substantial correlation between the predictive ranking of the off-target sites and the NHEJ frequencies experimentally observed at those sites. On the other hand, a comparison of the sites in the exhaustive output by the COSMID bioinformatics tool to those identified by the first gRNA in the GUIDE-Seq manuscript (VEGF#1) and found that 92 % of the total off-target activity (defined by the number of GUIDE-Seq reads) occurred at locations within the top 25 ranked sites predicted by COSMID (Fig. 3). With the continual improvement of in silico predictive

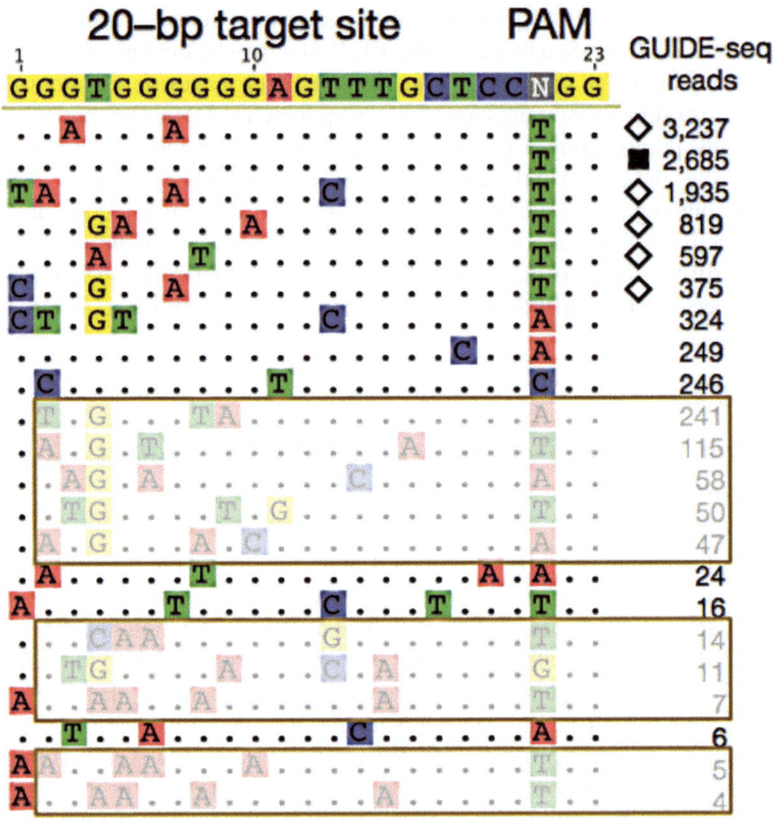

Fig. 3 Comparison of Off-target Site Identified by COSMID and GUIDE-seq. The top 25 ranked sites in the COSMID output for VEGF gRNA#1 account for 92 % of the off-target sequencing reads observed using GUIDE-seq. The sites given by COSMID are uncovered, while those sites that were located using GUIDE-seq, but not the output of COSMID are *boxed* and *faded*. These sites were not in the output of COSMID as they had four, or in one case five, mismatches between the gRNA and the genomic sequence. The seventh locus in the list was output by COSMID with a gRNA deletion and two mismatches. From Tsai SQ, et al. (2014) GUIDE-seq enables genome-wide profiling of off-target cleavage by CRISPR-Cas nucleases. Nat Biotechnol, 33, 187-197. With permission from Nature Publishing Group

methods for CRISPR off-target activity, it remains to be seen which genome engineering applications will require a thorough enough analysis of off-target activity to justify the added expense and time of using an experimental-based predictive method (such as GUIDE-Seq) as opposed to an in silico predictive approach (such as COSMID). Comparison of these two methods on the same sample would prove most informative. It remains to be seen if any additional sites predicted by COSMID will be validated as bona fide sites, using COSMID or any of these or newer methods. The individual methods may locate additional bona fide off-target cleavage sites as was shown with the *CCR5* directed ZFNs [45, 101].

In Silico Off-Target Site Prediction Methods

While experimental-based off-target prediction methods are reasonably accurate and have provided the bulk of total nuclease off-target sites discovered so far, the shifting landscape of nuclease design and testing increasingly magnifies the disadvantages of those approaches. Accurate in silico prediction methods also provide the possibility of a dramatic increase in the total number of off-target analyses that are performed, due to the substantial cost and implementation advantages of in silico over experimental-based methods.

All but two of the experimental-based prediction methods (IDLV LAM-PCR and ChIP-Seq) rely on a semi-randomized library of oligonucleotides. However, the degree to which a library can be randomized is directly dependent on the length of the sequence; shorter sequences can have a higher degree of randomization than longer sequences given a finite total number of molecules allowed in the library. Most of the experimental-based off-target analyses were performed with three-finger or four-finger ZFNs, which specify 9–10 or 12–13 bp, respectively. The trend over the last several years has been towards longer target sequence lengths; 6-finger ZFNs with each monomer targeting 18 bp are now the standard for the CompoZr nucleases created by Sigma-Aldrich, TALEN monomers are commonly designed to target 16–20 bp sequences, and CRISPR/Cas9 systems target 22 bp stretches (including the protospacer and PAM). With oligonucleotide libraries that are more homologous to the intended target, the benefit of these experimental-based approaches is greatly reduced. Indeed, of the three TALEN off-target sites that were located using SELEX experiments, two of the three were readily predicted by relatively simple search algorithms based on sequence homology [45, 75, 76]. Furthermore, the in vitro cleavage analysis of CRISPR/Cas9 systems [95] located fewer bona fide off-target sites than prediction using naïve homology based searches of the genome [78, 103].

Another major trend that (Fig. 4) disadvantages the IDLV LAM-PCR and ChIP-Seq prediction methods is the shift towards higher overall nuclease specificity. While the advantage of those methods is their unbiased approach that does not rely on oligonucleotide libraries, the disadvantage is their relative lack of sensitivity to sites of low frequency off-target activity. While many ZFN off-target sites with >1 % activity have been discovered [26–30, 45], TALENs and paired CRISPR nickases appear to be much more specific. To date, only one TALEN off-target site—lacking high (<87 %) sequence homology to the intended target–has been discovered with >1 % activity [57]. Paired CRISPR nickases have been developed and shown to lower off-target activity at selected sites 50–1500-fold compared to single CRISPR nucleases yielding activities substantially lower than 1 % at all sites discovered thus far [86]. RNA-guided FokI nucleases (RFNs) are another recent development which offers further improvements in the specificity of CRISPR systems [88, 89].

The final major trend that disadvantages experimental-based prediction methods is a shift in the bottleneck of the nuclease development process. It is a major challenge for most laboratories to identify ZFN binding helices that efficiently cleave their target sites (Fig. 4a) [20]. If a considerable amount of time and effort is required

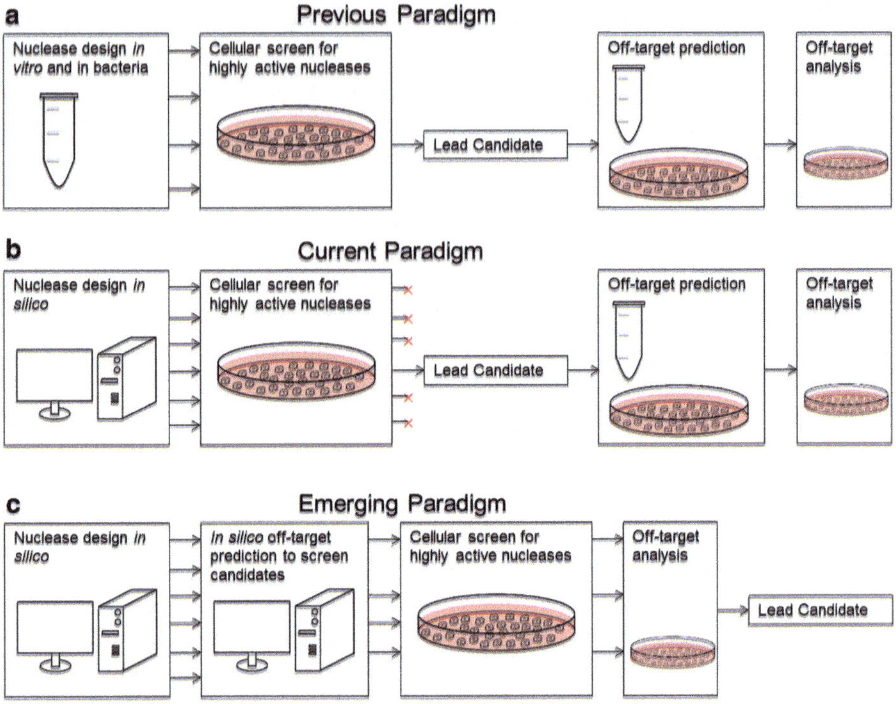

Fig. 4 Paradigm shifts in the nuclease development process. (**a**) Prior to recent developments in TALEN and CRISPR design, creating nucleases was a difficult process and typically only a small fraction of the nucleases were highly active when tested in cells. Experimental-based off-target prediction methods could then be used on that small subset of candidates in order to identify a lead candidate that showed the least lowest off-target activity. (**b**) Simple in silico design rules now result in large numbers of highly active nucleases, but experimental-based off-target prediction methods are prohibitively challenging and resource intensive to conduct, so many active nucleases are arbitrarily removed (*red* 'x') from further analysis. (**c**) Development of improved in silico off-target prediction methods allows pre-screening for designs with lower numbers of likely off-target sites and off-target analysis of all highly active nucleases designed in order to make a more informed choice of the nuclease with the fewest off-target effects

to develop a single active nuclease, then it is a judicious use of resources to apply an experimental-based off-target prediction method to that nuclease to determine the off-target profile. However, with current techniques, highly active TALENs and CRISPRs can be easily designed to cleave almost any sequence of interest [12, 104]. When dozens of highly active nucleases to address the gene of interest can be generated within a few weeks by a single researcher, it becomes impractical to use experimental-based prediction methods to analyze each nuclease in order to determine which has the lowest amount of off-target cleavage. These constraints result in an arbitrary elimination of most of the active nuclease candidates without consideration of the results of an off-target analysis (Fig. 4b). However, with potential off-target sites predicted rapidly in silico, many more nucleases can be assayed for their level of off-target cleavage, resulting in a more informed choice of a final lead candidate nuclease for the genome engineering application (Fig. 4c).

In Silico Tools for ZFN Off-Target Prediction

PROGNOS

The Predicted Report Of Genomewide Nuclease Off-target Sites (PROGNOS) is an online tool that models potential off-target activity of both ZFNs and TALENs (Fig. 5a) [45]. For ZFNs, it weighs several factors including sequence homology, the distance of mismatches from the FokI domain, the prevalence of guanosine residues, the energy compensation model of dimeric ZFN cleavage, and the binding

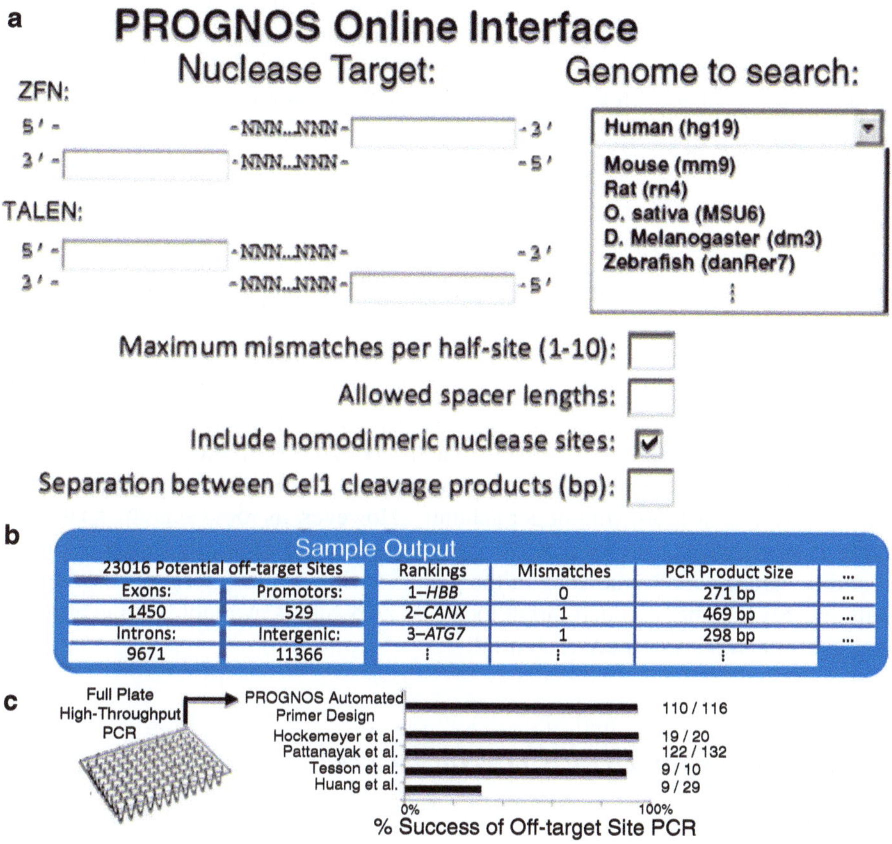

Fig. 5 Off-target analysis using PROGNOS. (**a**) Parameters such as the nuclease target site, the genome of interest, and allowed spacer distances are entered into the online interface. (**b**) PROGNOS provides a list of all sites in the genome matching the search parameters as well as a rank-ordered list of the most likely off-target locations according to its algorithms. (**c**) PCR primer sequences are automatically designed, according to input specifications, which can be used to amplify potential off-target sites in a high-throughput manner. Primers designed by PROGNOS have comparable PCR success rates to manually designed primers in other publications. From Fine EJ, et al. (2014) An online bioinformatics tool predicts zinc finger and TALE nuclease off-target cleavage. Nucleic Acids Res, 42, e42. With permission from Oxford University Press

energy of individual zinc finger subunits in order to provide a ranked list of potential off-target sites in a genome (Fig. 5b). PROGNOS listed more than half of all previously discovered ZFN off-target sites in the top subset of its rankings, located a novel off-target site for the well-studied *CCR5* ZFNs, and located several off-target sites for newly designed 3-finger ZFNs [45], 4-finger ZFNs [30, 45], and 5-finger CompoZr ZFNs designed by Sigma-Aldrich [105]. To aid in off-target analysis, PROGNOS also generates PCR primer sequences for different off-target detection methods which can be used in under a single set of thermocycler conditions in a high-throughput manner with a high success rate (Fig. 5c). The major disadvantage of PROGNOS compared to other in silico prediction tools is that it has longer run times on the online server, making pre-screening of large numbers of potential nucleases less feasible. For researchers who are interested and have some programming knowledge, PROGNOS can also be downloaded and run locally from the command line for faster performance. PROGNOS is available online at http://baolab.bme.gatech.edu/Research/BioinformaticTools/prognos.html or http://bit.ly/PROGNOS.

ZFN-Site

ZFN-Site is a web interface that searches multiple genomes for nuclease off-target sites based on the target sequence or known nuclease specificity [94]. ZFN-site uses the FetchGWI search engine to provide a quick, exhaustive search that does not miss potential off-target sites. The output of located sites includes links to genome browsers, facilitating off-target cleavage site screening. A major limitation of ZFN-site is that it only allows up to two mismatches per ZFN half-site in its search algorithm, though this is less than are found in many of the *bona fide* off-target sites of ZFNs with four or more zinc finger subunits. However, its rapid search capabilities make it an excellent tool for quickly screening potential ZFN target sites to ensure that there are no additional sites in the genome with two or fewer mismatches per half-site besides the intended target location. ZFN-site is available online at http://ccg.vital-it.ch/tagger/targetsearch.html.

In Silico Tools for TALEN Off-Target Prediction

PROGNOS

As described in the ZFN section above, PROGNOS is an online tool that models potential off-target activity of both ZFNs and TALENs [45]. For TALENs, it weighs several factors including sequence homology, interactions between the TALEN dimers, the distance of mismatches from the N-terminus, compensation effects from "strong" RVDs (i.e. NN and HD) that flank mismatches, and RVD-nucleotide binding preferences derived from SELEX analysis of engineered TAL domains in order to provide a ranked list of potential off-target sites in a genome (Fig. 5b). Of the three *bona fide* TALEN off-target sites found using the SELEX prediction method [75, 76], PROGNOS listed two of the three in the top subset of its rankings. PROGNOS was used to locate eight additional *bona fide* off-target sites for seven newly designed TALENs [30, 45]. To aid in off-target analysis, PROGNOS also generates PCR primer sequences for different off-target detection methods which can be used in under a single set of thermocycler conditions in a high-throughput manner with a high success rate (Fig. 5c).

TALENoffer

TALENoffer is another web tool for predicting TALEN off-target sites [106]. It uses a model based on the RVD-nucleotide binding preferences of natural TAL effectors to predict potential off-target sites. Although it substantially outperforms the older TALE-NT web tool [107] in terms of accurately locating bona fide off-target sites [106], it was shown to be less accurate than PROGNOS [45], and has yet to be used to locate any novel *bona fide* off-target sites. However, TALENoffer performs genomewide searches much faster than PROGNOS (and faster than TALE-NT), making it an excellent tool to use for quickly screening potential TALEN binding sites to ensure that there are no other sites in the genome that score as a high potential for off-target activity. TALENoffer is available online at http://galaxy2.informatik.uni-halle.de:8976/.

In Silico Tools for CRISPR Off-Target Prediction

Although CRISPR/Cas systems have proven able to effectively cleave their target sites in a wide range of organisms, off-target cleavage remains a major concern. Determining the possible off-target sites is critical for choosing guide sequences and for thorough testing after use. Possible off-target genomic cleavage sites are identified by comparing the guide strand to each site in the genome adjacent to the specified PAMs. Generally the user starts by entering a number of search criteria. Off-target programs output the genomic sites most similar to the input guide sequence within the specified constraints if found adjacent to an allowed PAM. Currently available programs differ markedly in the search method, features and ranking of the output. Ranking generally conforms to the concept that mismatches are better tolerated further from the PAM, whereas cleavage is less likely with mismatches between the guide and target sequences that are closer to the PAM. These tools will continue to improve, particularly in terms of rankings, as discussed below.

Ranking

As the number of sites that can be screened for off-target activity is limited, it is worthwhile to prioritize or rank the genomic sites matching the user-supplied criteria. Ranking is important due to the high number of bona fide off-target events that have been observed even with greater than three mismatches between the guide RNA and genomic sequence [77, 78, 103]. To order the output based on each putative site's probability of being cleaved, many of the programs take into consideration the number and location of mismatches between the complementary genomic sequence and the guide strand and specified PAMs. In addition to ranking the output for a given gRNA, some programs can scan a genomic region and rank the identified guide strands based on the total number of off-target sites found or using a scoring system. Data from continued use of these tools to guide off-target analysis will permit better ranking of off-target sites and choice of guide strands.

An early program, the Optimized CRISPR design tool, weighs the effects of mismatches at different positions based on experimental testing of several sgRNAs in human cells [78]. Other tools provide less complex scoring systems that correlate with the observation that mismatches between the gRNA and genomic sequence are less well tolerated closer to the PAM than on the 5' end. Therefore the ranking systems used in a number of tools are based on the sum of weight factors specifying the number and locations of the mismatches between the gRNA and the genomic loci. Using the same guide RNA with different tools will result in similar, but divergent ranked output, as shown in Fig. 6 [108].

At this point, all rankings are based solely on the sequence and do not take into account any factors related to the genomic context. Therefore identical sequences receive the same ranking. We and others have seen dramatic differences in the level

Loci ID	COSMID (Rank)	Cas Online Designer (Rank)	ZiFit	CRISPR Design Tool	Cas Offinder (Sorted)	Mutation Rate (%)	Gene
R01_OT2	2-6	2-7	15	-	18-139	43.6	None
R01	1	1	on	1	on	35.2	HBB
R01_OT10	7	2-7	3	-	3-17	23.4	None
R01_OT1	2-6	2-7	16	-	-	21.8	None
R01_OT5	2-6	2-7	5	-	3-17	15.9	None
R01_OT7	143-145	73-76	24	-	18-139	12.9	SECISBP2
R01_OT4	2-6	2-7	7	-	3-17	10.8	None
R01_OT8	355-357	238-241	34	-	18-139	6.6	VTI1A
R01_OT6	143-145	73-76	25	-	18-139	2.7	FSTL5

Fig. 6 Comparison of online search tools' output. The observed mutation rates at on- and off-target sites for gRNA R-01 that contain two mismatches are listed by decreasing T7EI activity. Sites with matching sequences (outside first base) have their names in *bold* with matching colors. Annotated genes corresponding to the sites are listed to right. Off-target analysis was performed with different online search tools. If the specified sites were predicted by a given tool, listed on top (such as Cas OFFinder), the sites' rankings in the output of the tool (if sortable) are shown. Sites not predicted by that tool are indicated by a *dash* in a *grey* box

of cleavage of a small number of identical sequences at different sites [108]. In one instance, a high-ranking off-target site (OT2) has higher cleavage than the on-target site, while another location is below detection (OT3) (Fig. 6). Scoring may therefore represent the opportunity for cleavage or the highest level of cleavage that might be seen under different conditions or in different cell types. This type of data highlights the role of genomic context and the difficulty in predicting the level of cleavage at a site based solely on the similarity of the gRNA and genomic sequences. To model this type of blocking, developmental or cell-specific data would have to be incorporated, such as the possible locations of accessible chromatin, chromatin methylation and transcription factor binding sites; the growth of the ENCODE database to contain much of this information offers hope that future predictive tools will be able to incorporate these factors.

Exhaustive Searches to Find All Sites

A recent comparison with other tools and *bona fide* off-target cleavage sites found that some tools failed to exhaustively output all sites in the genome matching the user input [108]. Some off-target search tools vary in their inclusion of non-coding regions including CG or other repeat regions that may occur at very high numbers. Therefore care must be taken to determine the parameters used by programs so that thousands of off-target sites are not excluded because of their location.

We recently published that genomic DNA sequences can be cleaved that are longer ('DNA bulge') or shorter ('RNA bulge') than the RNA guide strand at levels higher than the matching site that lack mismatches [109]. Due to the high levels of off-target activity at the 'bulge' sites observed in this study, the COSMID tool was developed to enable including these sites in genomic searches and Cas-OFFinder will be adding this option (personal communication). Including the option for bulges can greatly increases the number of sites output, so it remains to be seen how extensive the off-target events are across these sites.

Search Features

There is a steady stream of new CRISPR tools and/or changes and improvements to the existing tools. Current off-target search tools (Table 1) vary in their ability to exhaustively output all genomic sites matching the user-supplied criteria, as described above, and vary in their range of features. Some tools scan input sequences or genomic regions and then output the located guide strands ranked by estimated off-target sites [110], while others focus on exhaustive searches for possible off-target sites for an entered guide strand. While a number of features are found on several tools, some are not, such as the ability of E-CRISPR to search for off-target sites in a user provided sequence, such as would be used to determine if a knocked-in gene could be cleaved [111]. Jack Lin's CRISPR/Cas9 gRNA finder has links to a pair of RNA secondary structure web tools, though it only links to non-exhaustive

Table 1 CRISPR off-target analysis tools

Downloadable programs	
CasOT [113]	eendb.zfgenetics.org/casot/index.php
CRISPRseek [120]	bioconductor.org/packages/release/bioc/html/CRISPRseek.html
sgRNAcas9 [110]	http://www.biootools.com/col.jsp?id=140
Web-based tools	
Cas-OFFinder [121]	rgenome.net/cas-offinder/
Cas9 Design [115]	cas9.cbi.pku.edu.cn/index.jsp
COD (Cas9 and Off-target Designer)	cas9.wicp.net
COSMID [108]	crispr.bme.gatech.edu
CHOPCHOP [116]	chopchop.rc.fas.harvard.edu
CRISPR Design Tool	www.broadinstitute.org/mpg/crispr_design
CRISPR Optimal Target Finder	tools.flycrispr.molbio.wisc.edu/targetFinder/
Drosophila CRISPR gRNA design search	flyrnai.org/crispr2/
E-CRISP [111]	www.e-crisp.org/E-CRISP
Optimized CRISPR Design [78]	crispr.mit.edu
Jack Lin's CRISPR/Cas9 gRNA finder [110]	spot.colorado.edu/~slin/cas9.html
ZiFiT [115]	zifit.partners.org/ZiFiT/ChoiceMenu.aspx

Software programs that search for possible off-target sites, including three programs that can be downloaded and 12 programs that have web interfaces

Users provide individual guide strands or genomic target regions that are searched for guide strands. Putative genomic off-target sites are then output enabling choice of guide strand or testing after CRISPR/Cas use

BLAST and BLAT searches. If a tool lacks a desired genome, many sites will add them upon request. A number of tools can locate off-target sites for two correctly spaced binding sites that might allow cleavage by CRISPR nickase [86, 112] pairs or CRISPR FokI [88, 89] pairs [113–115], which is easier than combining the output from two individual runs, such as the excel output of COSMID [108].

Off-target analysis is included in some CRISPR design tools that search the specified genomic region, ranking the potential guide sequences in that region based on their predicted number of off-target sites in the selected genome. Although it is easy to rule out gRNA that have very similar off-target sites in the genome, ranking and comparing sites requires making a number of assumptions to total the combined effects of multiple putative off-target cleavage sites. These CRISPR design tools speed target site selection by providing a ranked list of the sites with the least potential for off-target events [78, 110, 114–116]. These tools markedly differ in how they screen for off-target sites and in the top sites they output, so re-scanning using an exhaustive search tool is required to ensure the optimal target site is selected.

Measuring the scale of off-target activity by testing a given number of sites is hampered by several factors. As described above, off-target tools varied on how exhaustively they located all the sites matching the specified criteria. Experimental results have indicated that many *bona fide* off-target cleavage sites are not located by these methods [102]. In addition, next generation sequencing is often needed to provide a sufficient number of reads for precise measurement, as off-target events can be missed using mutation detection assays, which generally cannot detect activity below 1–2 % modification rates [55, 117]. Although there are these limitations, comparisons of this type can provide a comparative readout of the specificity of individual guide strands. One zinc finger nuclease (ZFN) pair has been studied using a number of bioinformatics and experimentally driven off-target analysis techniques, each of which returned a portion of the total sequence validated off-target sites found combining the methods [45], so it is likely that a number of methods may need to be used to locate all the sites of CRISPR off-target cleavage, but there will continue to be improvement in bioinformatics and experimentally derived methods.

Off-Target Activity Detection Methods

All the methods listed previously are predictive only, and any potential off-target sites must be validated using separate assays to test for the presence of *bona fide* nuclease activity. Although some of the predictive methods have substantially lower false discovery rates than others, all of them have predicted sites at which no off-target activity was observed during subsequent validation assays. If no off-target activity is detected at a site with a particular assay, a more sensitive assay can be employed if warranted. Several methods to assay for off-target activity are outlined below in order of increasing sensitivity and difficulty. The first step in each of these detection methods is amplifying the region of interest surrounding the potential off-target site using PCR. Different detection methods have different requirements for the PCR amplicons, particularly in terms of length, so care must be taken when designing PCR primers for each different method. In order to limit the occurrence of mutations during the PCR reaction, high fidelity polymerases should always be used. With all of these detection methods, it is critical to perform analyses in parallel on samples from mock treated cells in order to assess any level of background signal at a particular off-target site for that assay. A possible exception to this requirement, because of its extremely low error rate, is TOPO sequencing. In the unlikely event of very high observed rates of identical mutated sequences, TOPO sequencing should also be conducted on mock treated cells to rule out genetic polymorphisms at that site.

Mismatch Detection Enzymes

These enzymatic assays are commonly used techniques to provide a quick answer about off-target activity at a specific location. However, their sensitivity is fairly low, typically having a limit of detection on the order of ~1 %. In this method, amplicons are denatured and allowed to re-anneal such that heteroduplexes form, with one strand from a wild-type amplicon and one strand from an amplicon containing a nuclease-induced mutation. Heteroduplexes also form from two different mutations. Adding an enzyme that selectively cleaves these heteroduplexes, such as Surveyor Nuclease (also known as the Cel1 enzyme and available from Transgenomic) or T7 Endonuclease I (also known as T7EI and available from New England BioLabs) allows direct quantification of band intensity by agarose or polyacrylamide gel electrophoresis. The percentage of alleles in the sample that contained nuclease-induced mutations can be calculated from the intensity of the bands on the gel [117, 118]. This assay generally works best with amplicons ranging in size from 300 to 500 bp. Because of the need for quantification of three different bands on the gel—the parental band and the two smaller products resulting from the mismatch detection enzyme cleaving at the point containing nuclease-induced mutations—amplicon sizes must be carefully chosen so that all three bands can be fully separated during electrophoresis; general guidelines are to ensure that the point of nuclease cleavage within the amplicon is >45 bp away from the midpoint of the amplicon and >90 bp from either end of the amplicon. A step-by-step protocol for using the T7EI enzyme to detect nuclease off-target activity is available [119].

TOPO Sequencing

TOPO sequencing provides an advantage over mismatch detection enzymes in that the actual sequences are read and compiled. The overall process is to use the TOPO TA kit (available from Invitrogen) to clone the amplicons into a plasmid that is transformed into *E. coli*, then to prepare plasmids from several of the individual *E. coli* colonies, then to perform Sanger sequencing on the plasmids, and finally to analyze the sequencing reads for evidence of nuclease-induced mutations. Because Sanger sequencing allows relatively long read lengths (typically >750 bp), if amplicons of ~450–550 bp are designed with the nuclease cleavage site near the midpoint of the amplicon, larger indels that might be missed with NGS methods that use short read lengths can be detected. Because Sanger sequencing is extremely accurate, the TOPO sequencing process also avoids the tendency of NGS to have errors with certain types of sequences (such as GC-rich stretches or homopolymer sequences) and background noise that can complicate locating (Fig. 6) small indels, such as those generated by CRISPR/Cas9 [77]. Because of the low-throughput nature of TOPO sequencing, the sensitivity is usually low, but can be raised by prepping and sequencing of many colonies.

However, for clonal populations of cells or analyzing organisms grown from embryos injected with nucleases, TOPO sequencing can very useful because a low detection limit is not needed and the sequences of all alleles can be determined. To analyze clonal populations of bi-allelic cells, amplification products are cloned into plasmids and transformed into *E. coli*. At least seven colonies should be sequenced in order to have sufficient confidence ($p<0.01$) that at least one sequencing read is obtained from each allele. After microinjection into embryos, nucleases can still cause mutations in the two-cell stage of the embryo thereby allowing for the possibility of four different allelic variants in the whole organism. Therefore, in samples of genomic DNA obtained from whole organisms, at least 17 *E. coli* colonies should be sequenced in order to have confidence that all four allelic variants are represented ($p<0.01$).

SMRT Sequencing

Single molecule real-time (SMRT) sequencing is a recently developed "third-generation" sequencing platform that offers greatly improved sensitivity over TOPO sequencing and mismatch detection enzymes. Approximately 25,000 high quality sequencing reads can be obtained during a single sequencing run for a consumable reagents cost of ~$250 (although sequencing cores may charge various additional machine usage fees). This allows for the analysis of 24 potential off-target sites for a nuclease with a sensitivity of ~0.1 %, making this platform an attractive option for laboratories conducting analyses of limited numbers of nucleases. Another advantage of the SMRT platform for laboratories making their first foray into high-throughput sequencing is the simplicity of the process. There are minimal constraints on the design of amplicons compared with platforms like Illumina; amplicons should be between 270 and 400 bp and the nuclease target site should be >40 bp from each end. As long as those two requirements are met, amplicons from multiple off-target sites can be pooled together (in roughly equimolar quantities) for a single sequencing run. There are no special requirements for the primer sequences and all steps can be performed as for a standard PCR reaction. If possible, better results can be achieved if the amplicons are near the shorter end of the range (sequencing reads will have higher quality on average) and the nuclease site is positioned near the midpoint of the amplicon (larger deletion events can be observed). A description of using SMRT sequencing to analyze nuclease off-target activity is available in Fine et al. [45].

Illumina Sequencing

For cases where sensitivity is critical, Illumina sequencing provides the highest level of sequencing throughput for the lowest cost per read. Whereas SMRT provides high quality reads in chunks of 25,000 per run, Illumina starts at ~10 million high quality reads per run and increases from there. If samples from many different experiments are multiplexed together, substantial cost savings relative to SMRT sequencing can be realized while obtaining similar sensitivity. However, great care must be taken when preparing amplicons for Illumina sequencing. There are special adapter sequences that must be present on the ends of both primer sequences and a second round of PCR with a separate set of primers must be performed. Recommended amplicon lengths vary substantially depending on the type of Illumina chemistry used (single-end vs paired-end sequencing, and 100 vs. 250 vs. 300 bp read lengths). When attempting to perform Illumina sequencing for the first time, it is highly recommended to consult with sequencing core facilities or other experienced individuals to fully understand the Illumina process and all the steps required as there are many common errors that can result in the sequencing run providing no useful information.

Conclusion

Engineered nucleases are valuable research tools, and becoming significantly more promising for therapeutic use. The genome-editing field has garnered increasing attention and excitement with each newly developed class of engineered nuclease; however, each of these nucleases lacks perfect specificity. The potentially disastrous consequences of off-target cleavage require new methods and assays to accurately predict and quantify off-target events in hopes of choosing the most specific and safe nucleases to use, and to ensure that nuclease-treated cells have minimal off-target effects. The methods developed to detect and quantify off-target cleavage activity will also help improve the specificity of future nucleases.

Acknowledgements This work was supported by the National Institutes of Health as an NIH Nanomedicine Development Center Award [PN2EY018244 to GB]. E.J.F. was additionally supported by the National Science Foundation Graduate Research Fellowship [DGE-1148903].

References

1. Smithies O, Gregg RG, Boggs SS, Koralewski MA, Kucherlapati RS. Insertion of DNA-sequences into the human chromosomal beta-globin locus by homologous recombination. Nature. 1985;317:230–4.
2. Capecchi MR. Altering the genome by homologous recombination. Science. 1989;244:1288–92.
3. Marcaida MJ, Muñoz IG, Blanco FJ, Prieto J, Montoya G. Homing endonucleases: from basics to therapeutic applications. Cell Mol Life Sci. 2010;67:727–48.
4. Silva G, Poirot L, Galetto R, Smith J, Montoya G, Duchateau P, Pâques F. Meganucleases and other tools for targeted genome engineering: perspectives and challenges for gene therapy. Curr Gene Ther. 2011;11:11–27.
5. Porteus MH, Carroll D. Gene targeting using zinc finger nucleases. Nat Biotechnol. 2005;23:967–73.
6. Urnov FD, Miller JC, Lee YL, Beausejour CM, Rock JM, Augustus S, Jamieson AC, Porteus MH, Gregory PD, Holmes MC. Highly efficient endogenous human gene correction using designed zinc-finger nucleases. Nature. 2005;435:646–51.
7. Chandrasegaran S, Smith J. Chimeric restriction enzymes: what is next? Biol Chem. 1999;380:841–8.

8. Li T, Huang S, Jiang WZ, Wright D, Spalding MH, Weeks DP, Yang B. TAL nucleases (TALNs): hybrid proteins composed of TAL effectors and FokI DNA-cleavage domain. Nucleic Acids Res. 2011;39:359–72.
9. Christian M, Cermak T, Doyle EL, Schmidt C, Zhang F, Hummel A, Bogdanove AJ, Voytas DF. Targeting DNA double-strand breaks with TAL effector nucleases. Genetics. 2010;186:757–61.
10. Boch J, Scholze H, Schornack S, Landgraf A, Hahn S, Kay S, Lahaye T, Nickstadt A, Bonas U. Breaking the code of DNA binding specificity of TAL-type III effectors. Science. 2009;326:1509–12.
11. Moscou MJ, Bogdanove AJ. A simple cipher governs DNA recognition by TAL effectors. Science. 2009;326:1501.
12. Cong L, Ran FA, Cox D, Lin S, Barretto R, Habib N, Hsu PD, Wu X, Jiang W, Marraffini LA, et al. Multiplex genome engineering using CRISPR/Cas systems. Science. 2013;339:819–23.
13. Jiang W, Bikard D, Cox D, Zhang F, Marraffini LA. RNA-guided editing of bacterial genomes using CRISPR-Cas systems. Nat Biotechnol. 2013;31:233–9.
14. Jinek M, Chylinski K, Fonfara I, Hauer M, Doudna JA, Charpentier E. A programmable dual-RNA-guided DNA endonuclease in adaptive bacterial immunity. Science. 2012;337:816–21.
15. Petek LM, Russell DW, Miller DG. Frequent endonuclease cleavage at off-target locations in vivo. Mol Ther. 2010;18:983–6.
16. Tadepally HD, Burger G, Aubry M. Evolution of C2H2-zinc finger genes and subfamilies in mammals: species-specific duplication and loss of clusters, genes and effector domains. BMC Evol Biol. 2008;8:176.
17. Miller J, McLachlan AD, Klug A. Repetitive zinc-binding domains in the protein transcription factor IIIA from Xenopus oocytes. EMBO J. 1985;4:1609–14.
18. Frankel AD, Berg JM, Pabo CO. Metal-dependent folding of a single zinc finger from transcription factor IIIA. Proc Natl Acad Sci U S A. 1987;84:4841–5.
19. Wolfe SA, Grant RA, Elrod-Erickson M, Pabo CO. Beyond the "recognition code": structures of two Cys2His2 zinc finger/TATA box complexes. Structure. 2001;9:717–23.
20. Ramirez CL, Foley JE, Wright DA, Muller-Lerch F, Rahman SH, Cornu TI, Winfrey RJ, Sander JD, Fu F, Townsend JA, et al. Unexpected failure rates for modular assembly of engineered zinc fingers. Nat Methods. 2008;5:374–5.
21. Li L, Wu LP, Chandrasegaran S. Functional domains in Fok I restriction endonuclease. Proc Natl Acad Sci U S A. 1992;89:4275–9.
22. Kim YG, Cha J, Chandrasegaran S. Hybrid restriction enzymes: zinc finger fusions to Fok I cleavage domain. Proc Natl Acad Sci U S A. 1996;93:1156–60.
23. Bitinaite J, Wah DA, Aggarwal AK, Schildkraut I. FokI dimerization is required for DNA cleavage. Proc Natl Acad Sci U S A. 1998;95:10570–5.
24. Porteus MH, Baltimore D. Chimeric nucleases stimulate gene targeting in human cells. Science. 2003;300:763.
25. Gaj T, Guo J, Kato Y, Sirk SJ, Iii CFB. Targeted gene knockout by direct delivery of zinc-finger nuclease proteins. Nat Methods. 2012;9:805–7.
26. Sander JD, Ramirez CL, Linder SJ, Pattanayak V, Shoresh N, Ku M, Foden JA, Reyon D, Bernstein BE, Liu DR, et al. In silico abstraction of zinc finger nuclease cleavage profiles reveals an expanded landscape of off-target sites. Nucleic Acids Res. 2013;41, e181.
27. Pattanayak V, Ramirez CL, Joung JK, Liu DR. Revealing off-target cleavage specificities of zinc-finger nucleases by in vitro selection. Nat Methods. 2011;8:765–70.
28. Gabriel R, Lombardo A, Arens A, Miller JC, Genovese P, Kaeppel C, Nowrouzi A, Bartholomae CC, Wang J, Friedman G, et al. An unbiased genome-wide analysis of zinc-finger nuclease specificity. Nat Biotechnol. 2011;29:816–23.
29. Gupta A, Meng X, Zhu LJ, Lawson ND, Wolfe SA. Zinc finger protein-dependent and -independent contributions to the in vivo off-target activity of zinc finger nucleases. Nucleic Acids Res. 2011;39:381–92.

30. Mussolino C, Alzubi J, Fine EJ, Morbitzer R, Cradick TJ, Lahaye T, Bao G, Cathomen T. TALENs facilitate targeted genome editing in human cells with high specificity and low cytotoxicity. Nucleic Acids Res. 2014;42:6762–73.
31. Elrod-Erickson M, Rould MA, Nekludova L, Pabo CO. Zif268 protein-DNA complex refined at 1.6 A: a model system for understanding zinc finger-DNA interactions. Structure. 1996;4:1171–80.
32. Choo Y, Klug A. Selection of DNA binding sites for zinc fingers using rationally randomized DNA reveals coded interactions. Proc Natl Acad Sci U S A. 1994;91:11168–72.
33. Jamieson AC, Kim SH, Wells JA. In vitro selection of zinc fingers with altered DNA-binding specificity. Biochemistry. 1994;33:5689–95.
34. Rebar EJ, Pabo CO. Zinc finger phage: affinity selection of fingers with new DNA-binding specificities. Science. 1994;263:671–3.
35. Choo Y, Klug A. Toward a code for the interactions of zinc fingers with DNA: selection of randomized fingers displayed on phage. Proc Natl Acad Sci U S A. 1994;91:11163–7.
36. Joung JK, Ramm EI, Pabo CO. A bacterial two-hybrid selection system for studying protein-DNA and protein-protein interactions. Proc Natl Acad Sci U S A. 2000;97:7382–7.
37. Joung JK. Identifying and modifying protein-DNA and protein-protein interactions using a bacterial two-hybrid selection system. J Cell Biochem Suppl. 2001;37:53–7.
38. Durai S, Bosley A, Abulencia AB, Chandrasegaran S, Ostermeier M. A bacterial one-hybrid selection system for interrogating zinc finger-DNA interactions. Comb Chem High Throughput Screen. 2006;9:301–11.
39. Maeder ML, Thibodeau-Beganny S, Osiak A, Wright DA, Anthony RM, Eichtinger M, Jiang T, Foley JE, Winfrey RJ, Townsend JA, et al. Rapid "open-source" engineering of customized zinc-finger nucleases for highly efficient gene modification. Mol Cell. 2008;31:294–301.
40. Sander JD, Dahlborg EJ, Goodwin MJ, Cade L, Zhang F, Cifuentes D, Curtin SJ, Blackburn JS, Thibodeau-Beganny S, Qi Y, et al. Selection-free zinc-finger-nuclease engineering by context-dependent assembly (CoDA). Nat Methods. 2011;8:67–9.
41. Moore M, Choo Y, Klug A. Design of polyzinc finger peptides with structured linkers. Proc Natl Acad Sci U S A. 2001;98:1432–6.
42. Wolfe SA, Ramm EI, Pabo CO. Combining structure-based design with phage display to create new Cys(2)His(2) zinc finger dimers. Structure. 2000;8:739–50.
43. Imanishi M, Hori Y, Nagaoka M, Sugiura Y. DNA-bending finger: artificial design of 6-zinc finger peptides with polyglycine linker and induction of DNA bending. Biochemistry. 2000;39:4383–90.
44. Zhang HS, Liu D, Huang Y, Schmidt S, Hickey R, Guschin D, Su H, Jovin IS, Kunis M, Hinkley S, et al. A designed zinc-finger transcriptional repressor of phospholamban improves function of the failing heart. Mol Ther. 2012;20:1508–15.
45. Fine EJ, Cradick TJ, Zhao CL, Lin Y, Bao G. An online bioinformatics tool predicts zinc finger and TALE nuclease off-target cleavage. Nucleic Acids Res. 2014;42, e42.
46. Abarrategui-Pontes C, Creneguy A, Thinard R, Fine EJ, Thepenier V, Fournier LRL, Cradick TJ, Bao G, Tesson L, Podevin G, et al. Codon swapping of zinc finger nucleases confers expression in primary cells and in vivo from a single lentiviral vector. Curr Gene Ther. 2014;14:12.
47. Choo Y, Klug A. A role in DNA binding for the linker sequences of the first three zinc fingers of TFIIIA. Nucleic Acids Res. 1993;21:3341–6.
48. Jantz D, Berg JM. Reduction in DNA-binding affinity of Cys2His2 zinc finger proteins by linker phosphorylation. Proc Natl Acad Sci U S A. 2004;101:7589–93.
49. Handel EM, Alwin S, Cathomen T. Expanding or restricting the target site repertoire of zinc-finger nucleases: the inter-domain linker as a major determinant of target site selectivity. Mol Ther. 2009;17:104–11.
50. Schornack S, Meyer A, Römer P, Jordan T, Lahaye T. Gene-for-gene-mediated recognition of nuclear-targeted AvrBs3-like bacterial effector proteins. J Plant Physiol. 2006;163:256–72.
51. Boch J, Bonas U. Xanthomonas AvrBs3 family-type III effectors: discovery and function. Annu Rev Phytopathol. 2009;48:419–36.

52. Miller JC, Tan S, Qiao G, Barlow KA, Wang J, Xia DF, Meng X, Paschon DE, Leung E, Hinkley SJ, et al. A TALE nuclease architecture for efficient genome editing. Nat Biotechnol. 2011;29:143–8.
53. Cermak T, Doyle EL, Christian M, Wang L, Zhang Y, Schmidt C, Baller JA, Somia NV, Bogdanove AJ, Voytas DF. Efficient design and assembly of custom TALEN and other TAL effector-based constructs for DNA targeting. Nucleic Acids Res. 2011;39, e82.
54. Christian ML, Demorest ZL, Starker CG, Osborn MJ, Nyquist MD, Zhang Y, Carlson DF, Bradley P, Bogdanove AJ, Voytas DF. Targeting G with TAL effectors: a comparison of activities of TALENs constructed with NN and NK repeat variable Di-residues. PLoS One. 2012;7, e45383.
55. Mussolino C, Morbitzer R, Lutge F, Dannemann N, Lahaye T, Cathomen T. A novel TALE nuclease scaffold enables high genome editing activity in combination with low toxicity. Nucleic Acids Res. 2011;39:9283–93.
56. Bedell VM, Wang Y, Campbell JM, Poshusta TL, Starker CG, Krug Ii RG, Tan W, Penheiter SG, Ma AC, Leung AYH, et al. In vivo genome editing using a high-efficiency TALEN system. Nature. 2012;491:114–8.
57. Guilinger JP, Pattanayak V, Reyon D, Tsai SQ, Sander JD, Joung JK, Liu DR. Broad specificity profiling of TALENs results in engineered nucleases with improved DNA-cleavage specificity. Nat Methods. 2014;11:429–35.
58. Schierling B, Dannemann N, Gabsalilow L, Wende W, Cathomen T, Pingoud A. A novel zinc-finger nuclease platform with a sequence-specific cleavage module. Nucleic Acids Res. 2012;40:2623–38.
59. Beurdeley M, Bietz F, Li J, Thomas S, Stoddard T, Juillerat A, Zhang F, Voytas DF, Duchateau P, Silva GH. Compact designer TALENs for efficient genome engineering. Nat Commun. 2013;4:1762.
60. Szczepek M, Brondani V, Buchel J, Serrano L, Segal DJ, Cathomen T. Structure-based redesign of the dimerization interface reduces the toxicity of zinc-finger nucleases. Nat Biotechnol. 2007;25:786–93.
61. Miller JC, Holmes MC, Wang J, Guschin DY, Lee Y-L, Rupniewski I, Beausejour CM, Waite AJ, Wang NS, Kim KA, et al. An improved zinc-finger nuclease architecture for highly specific genome editing. Nat Biotechnol. 2007;25:778–85.
62. Doyon Y, Vo TD, Mendel MC, Greenberg SG, Wang J, Xia DF, Miller JC, Urnov FD, Gregory PD, Holmes MC. Enhancing zinc-finger-nuclease activity with improved obligate heterodimeric architectures. Nat Methods. 2011;8:74–9.
63. Cade L, Reyon D, Hwang WY, Tsai SQ, Patel S, Khayter C, Joung JK, Sander JD, Peterson RT, Yeh J-RJ. Highly efficient generation of heritable zebrafish gene mutations using homo- and heterodimeric TALENs. Nucleic Acids Res. 2012;40:8001–10.
64. Guo J, Gaj T, Barbas 3rd CF. Directed evolution of an enhanced and highly efficient FokI cleavage domain for zinc finger nucleases. J Mol Biol. 2010;400:96–107.
65. Ramirez CL, Certo MT, Mussolino C, Goodwin MJ, Cradick TJ, McCaffrey AP, Cathomen T, Scharenberg AM, Joung JK. Engineered zinc finger nickases induce homology-directed repair with reduced mutagenic effects. Nucleic Acids Res. 2012;40:5560–8.
66. Bolotin A, Quinquis B, Sorokin A, Ehrlich SD. Clustered regularly interspaced short palindrome repeats (CRISPRs) have spacers of extrachromosomal origin. Microbiology. 2005;151:2551–61.
67. Horvath P, Barrangou R. CRISPR/Cas, the immune system of bacteria and archaea. Science. 2010;327:167–70.
68. Marraffini LA, Sontheimer EJ. CRISPR interference: RNA-directed adaptive immunity in bacteria and archaea. Nat Rev Genet. 2010;11:181–90.
69. Garneau JE, Dupuis M, Villion M, Romero DA, Barrangou R, Boyaval P, Fremaux C, Horvath P, Magadán AH, Moineau S. The CRISPR/Cas bacterial immune system cleaves bacteriophage and plasmid DNA. Nature. 2010;468:67–71.
70. Hale CR, Zhao P, Olson S, Duff MO, Graveley BR, Wells L, Terns RM, Terns MP. RNA-guided RNA cleavage by a CRISPR RNA-Cas protein complex. Cell. 2009;139:945–56.

71. Segal DJ, Meckler JF. Genome engineering at the dawn of the golden age. Annu Rev Genomics Hum Genet. 2013;14:135–58.
72. Gasiunas G, Barrangou R, Horvath P, Siksnys V. Cas9-crRNA ribonucleoprotein complex mediates specific DNA cleavage for adaptive immunity in bacteria. Proc Natl Acad Sci U S A. 2012;109:E2579–86.
73. Jinek M, East A, Cheng A, Lin S, Ma E, Doudna J. RNA-programmed genome editing in human cells. Elife. 2013;2, e00471.
74. Cornu TI, Cathomen T. Quantification of zinc finger nuclease-associated toxicity. Methods Mol Biol. 2010;649:237–45.
75. Tesson L, Usal C, Ménoret S, Leung E, Niles BJ, Remy S, Santiago Y, Vincent AI, Meng X, Zhang L, et al. Knockout rats generated by embryo microinjection of TALENs. Nat Biotechnol. 2011;29:695–6.
76. Hockemeyer D, Wang H, Kiani S, Lai CS, Gao Q, Cassady JP, Cost GJ, Zhang L, Santiago Y, Miller JC, et al. Genetic engineering of human pluripotent cells using TALE nucleases. Nat Biotechnol. 2011;29:731–4.
77. Cradick TJ, Fine EJ, Antico CJ, Bao G. CRISPR/Cas9 systems targeting β-globin and CCR5 genes have substantial off-target activity. Nucleic Acids Res. 2013;41:9584–92.
78. Hsu PD, Scott DA, Weinstein JA, Ran FA, Konermann S, Agarwala V, Li Y, Fine EJ, Wu X, Shalem O, et al. DNA targeting specificity of RNA-guided Cas9 nucleases. Nat Biotechnol. 2013;31:827–32.
79. Mojica FJ, Díez-Villaseñor C, García-Martínez J, Almendros C. Short motif sequences determine the targets of the prokaryotic CRISPR defence system. Microbiology. 2009;155:733–40.
80. Shah SA, Erdmann S, Mojica FJ, Garrett RA. Protospacer recognition motifs: mixed identities and functional diversity. RNA Biol. 2013;10:891–9.
81. Horvath P, Romero DA, Coûté-Monvoisin AC, Richards M, Deveau H, Moineau S, Boyaval P, Fremaux C, Barrangou R. Diversity, activity, and evolution of CRISPR loci in Streptococcus thermophilus. J Bacteriol. 2008;190:1401–12.
82. Fischer S, Maier LK, Stoll B, Brendel J, Fischer E, Pfeiffer F, Dyall-Smith M, Marchfelder A. An archaeal immune system can detect multiple protospacer adjacent motifs (PAMs) to target invader DNA. J Biol Chem. 2012;287:33351–63.
83. Cho SW, Kim S, Kim Y, Kweon J, Kim HS, Bae S, Kim J-S. Analysis of off-target effects of CRISPR/Cas-derived RNA-guided endonucleases and nickases. Genome Res. 2013;24:132–41.
84. Nishimasu H, Ran FA, Hsu PD, Konermann S, Shehata SI, Dohmae N, Ishitani R, Zhang F, Nureki O. Crystal structure of Cas9 in complex with guide RNA and target DNA. Cell. 2014;156:935–49.
85. Sternberg SH, Redding S, Jinek M, Greene EC, Doudna JA. DNA interrogation by the CRISPR RNA-guided endonuclease Cas9. Nature. 2014;507:62–7.
86. Ran FA, Hsu PD, Lin C-Y, Gootenberg JS, Konermann S, Trevino AE, Scott DA, Inoue A, Matoba S, Zhang Y, et al. Double nicking by RNA-guided CRISPR Cas9 for enhanced genome editing specificity. Cell. 2013;154:1380–9.
87. Mali P, Aach J, Stranges PB, Esvelt KM, Moosburner M, Kosuri S, Yang L, Church GM. CAS9 transcriptional activators for target specificity screening and paired nickases for cooperative genome engineering. Nat Biotechnol. 2013;31:833–8.
88. Guilinger JP, Thompson DB, Liu DR. Fusion of catalytically inactive Cas9 to FokI nuclease improves the specificity of genome modification. Nat Biotechnol. 2014;32:577–82.
89. Tsai SQ, Wyvekens N, Khayter C, Foden JA, Thapar V, Reyon D, Goodwin MJ, Aryee MJ, Joung JK. Dimeric CRISPR RNA-guided FokI nucleases for highly specific genome editing. Nat Biotechnol. 2014;32:569–76.
90. Sakaue-Sawano A, Kurokawa H, Morimura T, Hanyu A, Hama H, Osawa H, Kashiwagi S, Fukami K, Miyata T, Miyoshi H, et al. Visualizing spatiotemporal dynamics of multicellular cell-cycle progression. Cell. 2008;132:487–98.

91. Zembruski NC, Stache V, Haefeli WE, Weiss J. 7-Aminoactinomycin D for apoptosis staining in flow cytometry. Anal Biochem. 2012;429:79–81.
92. Hockemeyer D, Soldner F, Beard C, Gao Q, Mitalipova M, DeKelver RC, Katibah GE, Amora R, Boydston EA, Zeitler B, et al. Efficient targeting of expressed and silent genes in human ESCs and iPSCs using zinc-finger nucleases. Nat Biotechnol. 2009;27:851–7.
93. Perez EE, Wang J, Miller JC, Jouvenot Y, Kim KA, Liu O, Wang N, Lee G, Bartsevich VV, Lee Y-L, et al. Establishment of HIV-1 resistance in CD4+ T cells by genome editing using zinc-finger nucleases. Nat Biotechnol. 2008;26:808–16.
94. Cradick TJ, Ambrosini G, Iseli C, Bucher P, McCaffrey AP. ZFN-site searches genomes for zinc finger nuclease target sites and off-target sites. BMC Bioinformatics. 2011;12:152.
95. Pattanayak V, Lin S, Guilinger JP, Ma E, Doudna JA, Liu DR. High-throughput profiling of off-target DNA cleavage reveals RNA-programmed Cas9 nuclease specificity. Nat Biotechnol. 2013;31:839–43.
96. Wang X, Wang Y, Wu X, Wang J, Wang Y, Qiu Z, Chang T, Huang H, Lin R-J, Yee J-K. Unbiased detection of off-target cleavage by CRISPR-Cas9 and TALENs using integrase-defective lentiviral vectors. Nat Biotechnol. 2015;33:175–8.
97. Wu X, Scott DA, Kriz AJ, Chiu AC, Hsu PD, Dadon DB, Cheng AW, Trevino AE, Konermann S, Chen S, et al. Genome-wide binding of the CRISPR endonuclease Cas9 in mammalian cells. Nat Biotechnol. 2014;32:670–6.
98. O'Geen H, Henry IM, Bhakta MS, Meckler JF, Segal DJ. A genome-wide analysis of Cas9 binding specificity using ChIP-seq and targeted sequence capture. Nucleic Acids Res. 2014;43:3389–404.
99. Kuscu C, Arslan S, Singh R, Thorpe J, Adli M. Genome-wide analysis reveals characteristics of off-target sites bound by the Cas9 endonuclease. Nat Biotechnol. 2014;32:677–83.
100. Kim D, Bae S, Park J, Kim E, Kim S, Yu HR, Hwang J, Kim JI, Kim JS. Digenome-seq: genome-wide profiling of CRISPR-Cas9 off-target effects in human cells. Nat Methods. 2015;12:237–43.
101. Frock RL, Hu J, Meyers RM, Ho YJ, Kii E, Alt FW. Genome-wide detection of DNA double-stranded breaks induced by engineered nucleases. Nat Biotechnol. 2015;33:179–86.
102. Tsai SQ, Zheng Z, Nguyen NT, Liebers M, Topkar VV, Thapar V, Wyvekens N, Khayter C, Iafrate AJ, Le LP, et al. GUIDE-seq enables genome-wide profiling of off-target cleavage by CRISPR-Cas nucleases. Nat Biotechnol. 2014;33:187–97.
103. Fu Y, Foden JA, Khayter C, Maeder ML, Reyon D, Joung JK, Sander JD. High-frequency off-target mutagenesis induced by CRISPR-Cas nucleases in human cells. Nat Biotechnol. 2013;31:822–6.
104. Reyon D, Tsai SQ, Khayter C, Foden JA, Sander JD, Joung JK. FLASH assembly of TALENs for high-throughput genome editing. Nat Biotechnol. 2012;30:460–5.
105. Abarrategui-Pontes C, Creneguy A, Thinard R, Fine EJ, Thepenier V, Fournier LRL, Cradick TJ, Bao G, Tesson L, Podevin G, et al. Codon swapping of zinc finger nucleases confers expression in primary cells and in vivo from a single lentiviral vector. Curr Gene Ther. 2014;14:365–76.
106. Grau J, Boch J, Posch S. TALENoffer: genome-wide TALEN off-target prediction. Bioinformatics. 2013;29:2931–2.
107. Doyle EL, Booher NJ, Standage DS, Voytas DF, Brendel VP, Vandyk JK, Bogdanove AJ. TAL effector-nucleotide targeter (TALE-NT) 2.0: tools for TAL effector design and target prediction. Nucleic Acids Res. 2012;40:W117–22.
108. Cradick TJ, Qui P, Lee CM, Fine EJ, Bao G. COSMID: a Web-based tool for identifying and validating CRISPR/Cas Off-target sites. Mol Ther Nucleic Acids. 2014;3:e214.
109. Lin Y, Cradick TJ, Brown MT, Deshmukh H, Ranjan P, Sarode N, Wile BM, Vertino PM, Stewart FJ, Bao G. CRISPR/Cas9 systems have off-target activity with insertions or deletions between target DNA and guide RNA sequences. Nucleic Acids Res. 2014;42:7473–85.
110. Xie S, Shen B, Zhang C, Huang X, Zhang Y. sgRNAcas9: A software package for designing CRISPR sgRNA and evaluating potential off-target cleavage sites. PLoS One. 2014;9:e100448.

111. Babu M, Beloglazova N, Flick R, Graham C, Skarina T, Nocek B, Gagarinova A, Pogoutse O, Brown G, Binkowski A, et al. A dual function of the CRISPR-Cas system in bacterial antivirus immunity and DNA repair. Mol Microbiol. 2011;79:484–502.
112. Shen B, Zhang W, Zhang J, Zhou J, Wang J, Chen L, Wang L, Hodgkins A, Iyer V, Huang X, et al. Efficient genome modification by CRISPR-Cas9 nickase with minimal off-target effects. Nat Methods. 2014;11:399–402.
113. Xiao A, Cheng Z, Kong L, Zhu Z, Lin S, Gao G, Zhang B. CasOT: a genome-wide Cas9/gRNA off-target searching tool. Bioinformatics. 2014;21:2014.
114. Sander JD, Maeder ML, Reyon D, Voytas DF, Joung JK, Dobbs D. ZiFiT (Zinc Finger Targeter): an updated zinc finger engineering tool. Nucleic Acids Res. 2010;38(Suppl):W462–8.
115. Ma M, Ye AY, Zheng W, Kong L. A guide RNA sequence design platform for the CRISPR/Cas9 system for model organism genomes. Biomed Res Int. 2013;2013:270805.
116. Montague TG, Cruz JM, Gagnon JA, Church GM, Valen E. CHOPCHOP: a CRISPR/Cas9 and TALEN web tool for genome editing. Nucleic Acids Res. 2014;42:W401–7.
117. Guschin DY, Waite AJ, Katibah GE, Miller JC, Holmes MC, Rebar EJ. A rapid and general assay for monitoring endogenous gene modification. Methods Mol Biol. 2010;649:247–56.
118. Ran FA, Hsu PD, Wright J, Agarwala V, Scott DA, Zhang F. Genome engineering using the CRISPR-Cas9 system. Nat Protocols. 2013;8:2281–308.
119. Fine EJ, Cradick TJ, Bao G. Identification of Off-Target Cleavage Sites of Zinc Finger Nucleases and TAL Effector Nucleases Using Predictive Models. In: Storici F, editor, Gene correction. Totowa: Humana Press; 2014. Vol. 1114, pp. 371–83.
120. Zhu LJ, Holmes BR, Aronin N, Brodsky MH. CRISPRseek: a bioconductor package to identify target-specific guide RNAs for CRISPR-CAS9 genome-editing systems. PLoS One. 2014;9, e108424.
121. Bae S, Park J, Kim JS. Cas-OFFinder: a fast and versatile algorithm that searches for potential off-target sites of Cas9 RNA-guided endonucleases. Bioinformatics. 2014;30:1473–5.

Cellular Engineering and Disease Modeling with Gene-Editing Nucleases

Mark J. Osborn and Jakub Tolar

Abstract Two rapidly evolving technologies are set to intersect at the crossroads of the future of medicine: the knowledge of how to induce and maintain cellular pluripotency, and the ability to precisely manipulate the genome with engineered nucleases. Together, these two advances have significant potential in the development of the next generation of cell and gene therapies. This review will discuss human and animal models of stem cells and the application of engineered nucleases for precision gene targeting and control. For animal studies and models, nucleases have allowed for greater flexibility and expandability. Previously untargetable regions of the murine genome are now accessible via engineered nucleases. Prior to the availability of gene editing proteins, the entire rat genome was largely refractory to gene targeting and manipulation. The ability to engineer larger animals may reduce the transplant organ gap and increase the yields of food for an expanding population. Lastly, the ability to modify stem cells of hematopoietic, embryonic, or somatic origin will allow for more relevant disease modeling, and more targeted and effective therapies. Collectively, the efficiency of gene editing nucleases and the ability to apply them across cells of multiple species allows for new research opportunities, more flexibility, and greater accuracy in choosing the model best suited for genome manipulation.

Keywords *Clustered* regularly-interspaced short palindromic repeats (CRISPR)/Cas9 • Embryonic stem cell (ESC) • Hematopoietic stem cell (HSC) • Homologous recombination (HR) • Insertions/deletions (indels) • Inducible pluripotent stem cell (iPSC) • Meganuclease (MN) • Non-homologous end joining (NHEJ) • Oligonucleotide donor (ODN) • Reprogramming • Somatic cell nuclear transfer (SCNT) • Transcription activator-like effector nuclease (TALEN) • Zinc finger nuclease (ZFN)

M.J. Osborn, Ph.D.
Department of Pediatrics, Division of Blood and Marrow Transplantation, Masonic Cancer Center, Stem Cell Institute, and Center for Genome Engineering, University of Minnesota, Minneapolis, MN, USA
e-mail: osbor026@umn.edu

J. Tolar, M.D., Ph.D. (✉)
Department of Pediatrics, Division of Blood and Marrow Transplantation, Stem Cell Institute, University of Minnesota, MMC 366, 420 Delaware Street SE, Minneapolis, MN 55455, USA
e-mail: tolar003@umn.edu

T. Cathomen et al. (eds.), *Genome Editing*, Advances in Experimental Medicine and Biology 895, DOI 10.1007/978-1-4939-3509-3_12

By definition, stem cells are capable both of self-renewal and of differentiation into cell types of multiple lineages. Cell types include embryonic stem cells (ESCs), inducible pluripotent stem cells (iPSCs), and hematopoietic stem cells (HSCs). ESCs were first described in 1998 [1] and are derived from the inner cell mass of a blastocyst-stage embryo before the separation of the germ and somatic lineages occurs [2]. This property confers upon them a putative selective advantage under normal circumstances according to the Germ Plasm Theory of August Weismann [3]. This theory states that the germline gives rise to somatic cells and that mutations acquired in germ cells will affect the soma but not vice versa [4]. As such, the genome of the germline is guarded and guided by selection against the accumulation of deleterious mutations. While the potential for these cells as tools of regenerative and transplant medicine is promising, the nature in which they are derived is the subject of intense legal and ethical debate that greatly restricts their widespread use.

In the last 60 years one portion of the Weismann theory was disproved while another was validated. In the 1950s and 1960s, Robert Briggs, Thomas King and John Gurdon showed that when a nucleus from a differentiated cell is transferred to an enucleated oocyte, a complete organism (a frog) can develop [5, 6]. In 1997 Ian Wilmut and colleagues utilized this concept of somatic cell nuclear transfer (SCNT) to create Dolly the sheep [7]. This procedure relies on the transfer of the nucleus from a somatic cell into an enucleated oocyte. Moreover, work from the Yamanaka laboratory in 2006 showed that by the addition of defined transcriptional factors, somatic cells could be reprogrammed to a pluripotent state; these cells are termed inducible pluripotent stem cells (iPSCs) [8]. These studies refuted the Weismann theory in regards to the irreversible 'stemness' of somatic cells, but validated it in that somatic genome aberrations can impact the phenotype of the new cell type [9].

A second technology has merged with the stem cell field and enables users the freedom to make precise alterations to the genome. This technology uses the genome editing class of nucleases that exist in multiple formats and architectures that have the unifying characteristic of recognizing and contacting a unique sequence of DNA. The majority of studies have tethered these proteins to nuclease domains or used their inherent ability to cut DNA. Once the DNA is broken, two predominant repair pathways have been exploited for genome engineering: non-homologous end-joining (NHEJ) and homologous recombination (HR). NHEJ is an error-prone pathway that, in the absence of a donor template, repairs the DNA break in a fashion that is associated with small insertions or deletions ('indels') that can permanently disrupt coding DNA sequences. Gene repair relies on the error-free HR pathway. When both nucleases and a donor template are provided, DNA breaks are repaired using the donor as a template as part of a "cut and replace" strategy that allows for precise insertion of user-defined sequences. The ability to employ nucleases for NHEJ and HR allows for great flexibility and accuracy in gene knockout, knock-in and repair strategies, and allows investigators to tailor cells for research in disease modeling, in drug discovery, and for gene correction. The nuclease molecules that have been employed for genome modification are meganucleases (MNs), zinc finger nucleases (ZFNs), transcription activator-like effector nucleases (TALENs), and *clustered* regularly-interspaced short palindromic repeats (CRISPRs)/Cas9 system.

Meganucleases

Multiple MN family members exist, and one class, containing the LAGLIDADG protein motif, appears to be the most amenable MN for reengineering for unique target sites [10]. MNs have been employed in human cells for modification, but not directly in pluripotent stem cells [11, 12]. In rodents, proof of principle for MN activity was established by knocking-in an I-*Sce*I target site into the villin gene locus, followed by transient expression of the I-*Sce*I MN that resulted in a 100-fold stimulation of HR [13]. Precision targeting was described by Ménoret et al., who utilized the I-CreI homing endonuclease as a template for rational sequence-specific reengineering for the *Rag1* gene [14]. Injection of plasmid DNA into zygotes resulted in 0.6 % of rats and 3.4 % of mice born with mutation at this locus, and with a significant decrease of T- and B-cells and a normal NK-cell component [14]. Despite their ability to disrupt endogenous genes, MN use for widespread genome editing has been restricted by their costly and lengthy generation procedure; however, their monomeric, highly active nature may make them uniquely suited for future therapeutic genome engineering.

Zinc Finger Nucleases

The Zif268 transcription factor that binds to and regulates the expression of multiple cellular genes serves as the foundational template for the generation of custom-engineered ZFNs. By altering the specificity of the DNA binding residues of Zif268, a multitude of gene-specific ZFNs have been produced that recognize user-defined sequences. Typically ZFNs employed for sequence-specific binding are organized into an array that binds a particular sequence of DNA. Each individual unit, or finger, of the array is comprised of approximately 30 amino acids and contacts three base pairs of DNA [15]. Multiple fingers are connected to one another, comprising a ZFN monomeric array, such that each array recognizes and binds to a specific sequence termed a 'half-site.' Thus, the fully formed complex recognizes a target site comprised of the two half-sites that are separated from one another by a 'spacer sequence,' in which the double-stranded break (DSB) is mediated by the nuclease component of the complex (Fig. 1) [15, 16]. To date, ZFNs have been widely employed in stem cells in mammals.

Genetically engineered rodents have been the core tool employed by researchers for genome modification, with the mouse established as the most commonly used animal model system [17]. As functional ESC lines exist for the mouse but not for many other species, intrazygotic injections can be performed to augment the direct use of rodent ESCs.

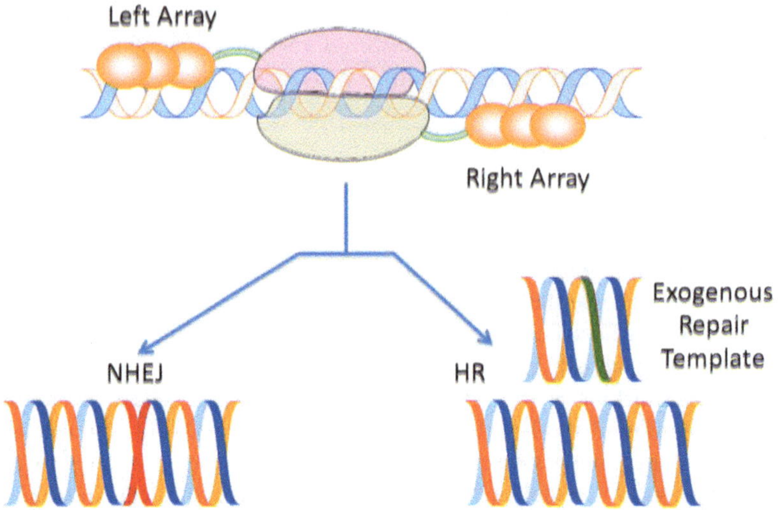

Fig. 1 ZFN Nuclease Architecture. At top is shown is a three finger ZFN with the fully formed left and right array heterodimeric complex. Each individual finger contacts three bp of DNA and the arrays are tethered to the FokI nuclease domain (pink and tan ovals) that mediates a DSB in the spacer region. DNA breaks can be resolved by error prone mutagenic non-homologous endjoining (NHEJ) or homologous recombination (HR). Red indicates NHEJ-mediated insertion/deletion and HDR occurs from an exogenous donor template (right). Image generated using software from Motifolio Inc. Sykesville, MD

Rodents

In 2009, the first documented modification of a rodent genome was reported by Geurts et al., who utilized ZFNs in rat embryos. They targeted an exogenous integrated reporter and the endogenous *IgM* and Rab38 genes, and showed rates of disruption of 25–100 % with stable germline transmission [18]. A second study disrupted the rat *Il2rg* locus at rates of ~20 % [19]. As such, these studies developed rat models of hypertension [18] and immunodeficiency [19], and a platform for monoclonal antibody production [18]. Targeted integration into the rat genome was also shown by Cui et al., who introduced donor-derived sequences into the *Mdr1a* and *PXR* genes by HR [20]. These studies are significant due to the facts that the rat genome had previously proved to be largely intractable for targeting [18] and that the rat may be a better model than the mouse in addiction and other neurobehavioral studies [21].

ZFNs were successfully used for knockout or knock-in of target genes in mice in 2010 [22, 23]. For the knockout, the *Mdr1a*, *Jag1*, and *Notch3* genes were targeted in the C57BL/6 or FVB/N strains of mice [22]. Rates of editing in live-birthed animals ranged from 20 to 75 % and the disrupted alleles were efficiently transmitted through the germline [22]. By successfully targeting the *ROSA26* locus with ZFNs and an exogenous donor template, Meyer et al. documented a tenfold or greater increase in HR compared to an earlier report [23, 24]. These studies were transformative due to the rapidity (as little as 4 months) with which genome engineering could now be performed in rodents of any strain [22]. This is an important consideration that will facilitate the generation of more relevant and penetrant disease models. For example, non-obese diabetic (NOD) mice have been an important tool for the study of type 1 diabetes. Using ZFNs in NOD embryos, Chen et al. were able to knockout the *Tnfrsf9* gene that encodes CD137 [25]. Their report showed that CD137 was not required for the development of insulitis, but was a factor in promoting overt diabetes progression [25]. Thus ZFNs and strain-specific embryos allowed for disease modeling in a pure NOD background and removed any potential confounding and/or disease-masking factors that might result from contamination of genetic material obtained from ESCs of one strain and implanted into a second. More expansive disease modeling in mice is possible because of ZFNs as well. Meyer et al. showed the efficacy of introducing ZFNs and single-stranded oligonucleotides into mouse embryos in order to introduce missense mutations into a specific gene [26]. This approach simplifies the manner in which relevant animal models can be created and should facilitate more widespread in vivo development and modeling of diseases, genes, and therapeutic interventions. A second study that reinforces this concept was performed by Osiak et al., who devised a selection-free methodology for generating knockout mouse ESCs [27]. The ability to precisely target a specified locus in the genome without the addition of selectable markers has streamlined the generation process and minimized the potential for ectopic transgene/exogenous sequence interference in disease modeling.

Larger animal models have also been employed for ZFN-mediated engineering with the hopes of increasing the yield/mass of animals for food production or making them safer by reducing allergenicity or unwanted traits. These studies serve as general models for eventual human ex vivo therapies in that the animal studies largely center on modification of skin cells with reprogramming into pluripotent stem cells. In humans this will rely likely on iPSC-derived technologies. In animals it will most likely center on SCNT, a cloning technique used to produce an animal from a single cell nucleus placed in a surrogate oocyte from which the nucleus has been removed [28, 29].

Swine

In 2011, Whyte et al. showed the ability of ZFNs to knockout a GFP reporter gene in porcine fibroblasts, and Hauschild et al. successfully disrupted both alleles of the porcine α1,3-galactosyltransferase (*GGTA1*) gene [30, 31]. *GGTA1* encodes for the enzyme required to generate Gal epitopes [31], which are highly immunogenic and result in a hyperacute rejection response in xenotransplantation models [32]. The ability to mitigate or remove the antigens responsible for organ rejection is significant because it raises the possibility of engineering xenogeneic organs to help address the gap of approximately 70,000 persons per year between people receiving organ transplants and those on the waiting list [33, 34].

Cattle

The modification of the bovine genome holds promise in efforts to improve the quantity and quality of dairy output. Yu et al. used ZFNs to induce NHEJ-mediated disruption of the beta-lactoglobulin (*BLG*) gene in the bovine genome that encodes a protein that is a major antigen in cow's milk [35]. Liu et al. employed the 'nickase' version of ZFNs whereby one ZFN monomer is not capable of cleaving the DNA strand. This results in a single-stranded DNA break, a modification that greatly improves the accuracy of targeted gene knock-in [36]. Using ZFN nickases designed for the second intron of the bovine *CSN2* locus, they knocked-in a copy of the *Staphylococcus simulans* lysostaphin gene. Mastitis is a common disease in dairy cows and prophylactic antibiotic regimens are instituted in approximately 90 % of herds [37], as healthier dairy cows have a higher yield and quality of milk production [37]. The ability, documented by Yu and colleagues, to engineer dairy cows with ZFNs and SCNT to express lysostaphin or other similar genes may beneficially impact cattle farming due to the ability of *Staphylococcus simulans*-derived lysostaphin to control *Staphylococcus aureus*-caused mastitis [38]. A major detriment to this strategy is the presence of antibiotics in milk consumed by humans that may contribute to antibiotic resistance [39]. Antimicrobial resistance is a critical threat to public health, as antibiotic use in humans is thought to result in a gain of 2–10 years of life expectancy [40, 41].

Humans

Stem cell candidates that are actively being pursued for gene editing in humans are HSCs, ESCs, and iPSCs. HSCs are a highly desirable target cell population due to their pluripotent ability to differentiate into all blood lineages and their use in transplant centers around the globe for dozens of malignant and non-malignant diseases (Fig. 2). For humans, there are a limited number of ESC lines for unrestricted use. Multiple wild-type and disease-model iPSCs, however, are able to circumvent the potentially restrictive constraints associated with ESCs.

Approaches for HSC genome modification in human cells have predominantly centered on the use of ZFNs and have used both the NHEJ and HR arms of the DNA repair pathway. HR represents the gold standard for future gene therapies whereby a mutation can be specifically and seamlessly corrected. Lombardo et al. reached maximal rates of 0.11 % gene targeting at the *CCR5* locus utilizing ZFNs and a donor containing GFP or a puromycin resistance gene [42]. In 2014, a second groundbreaking study detailed the use of ZFNs for the insertion of a gene-targeting construct into the *AAVS1* or the *IL2RG* locus [43]. This study maximized recent advances in HSC culture and expansion to achieve higher rates of HDR [44–46]. Antagonism of the aryl hydrocarbon receptor with StemRegenin 1 (SR1) drives relative CD34+ HSC expansion (as a result of diminished rate of spontaneous

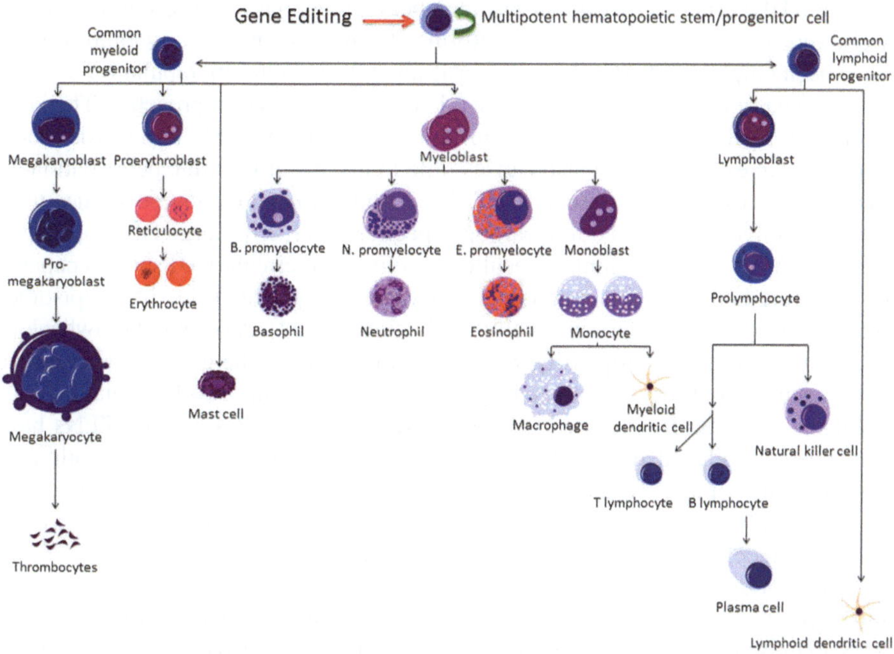

Fig. 2 Human Hematopoiesis. The hematopoietic stem/progenitor cell possesses self renewal capacity (green arrow) and the ability to form the common lymphoid (CLP) and common myeloid (CMP) progenitors. CLPs give rise to the cells of the lymphoid compartment: T-, B-, and NK Cells (right). The CMP generates lineage restricted progenitors (myeloblasts, proerythroblasts, and megakaryoblasts) that themselves differentiate into the terminal hematopoietic myeloid and erythroid effector cells. Red arrow indicates the cell target for genome engineering. Image generated using software from Motifolio Inc. Sykesville, MD

differentiation of HSCs ex vivo) with a corresponding ability to engraft at higher rates in xenotransplanted human-murine chimeric mice [47]. 16,16 Dimethyl prostaglandin E2 (dmPGE2), a stable analogue of PGE2, has been shown to regulate hematopoiesis and enhance engraftment [48, 49]. The use of both SR1 and dsPGE2 preserved the hematopoietic stem cell phenotype, and delivery of ZFN mRNA and donor template in an integrase-deficient lentivirus (IDLV) resulted in an increased proportion of mice showing engraftment of gene-edited human cells [43]. Further, the engrafted cells showed preferential outgrowth and functionality when challenged with an allogeneic tumor. The studies were performed in a manner to support rapid translation to the clinic: GMP-level mRNA and IDLV procedures are in place and the donor construct for the *IL2RG* gene was generated in a manner to splice endogenous exons 1–4 with donor-derived exons 5–8; this would therefore be broadly applicable in SCID. Further, by demonstrating *AAVS1* HDR, the procedure is transferrable to other disorders and represents a strategy for broad application to the many disorders currently treated by hematopoietic cell transplantation.

The NHEJ arm of DNA repair has therapeutic potential for HSC engineering as well, and multiple gene delivery platforms for maximizing gene disruption rates have been employed. HSC gene transfer is a key concept for manipulation of these cells and has encompassed viral, non-viral, and protein-based approaches. The ideal strategy is to deliver the genetic cargo in a format that not only allows for robust gene/protein expression but also carries little to no risk of genomic perturbation by the nuclease expression cassette itself. The Lombardo and Genovese studies used IDLV to deliver ZFNs and donor cargo, respectively [42, 43]. Cargo DNA packaged with the D64V integrase mutant resulted in a linear or circularized species capable of transiting both the cellular and nuclear membranes where they exist as episomes that are lost through progressive cellular division [50–54]. A potential drawback of IDLVs is their known propensity to integrate into areas of the genome where there have been DNA breaks generated by a nuclease or as the result of endogenous breaks at genomic fragile sites [55, 56]. Adenoviral cassettes bearing ZFNs have also been introduced into HSCs and have resulted in *CCR5* gene disruption by NHEJ at rates of approximately 25 % with potentially offsetting toxicity [57]. This comparatively high rate was due to modulation of the mitogen-activated protein kinase/extracellular signal-regulated kinase pathway [57]. Other studies have shown the value of drug-induced enhancement of nuclease gene transfer and represent an intriguing avenue of research to further maximize nuclease activity [58]. Because ZFN expression in 293 adenoviral producer cells may be toxic and impact titers, Saydaminova et al. devised a strategy to 'detarget' ZFN expression in 293 cells by including microRNA (MiRNA) sequences that repressed gene expression in producer but not hematopoietic cells [59]. As such, they observed gene modification rates of ~12 % in engrafted HSCs. Both the Li and Saydaminova studies observed that the gene-modified cells resulted in a reduced ability to engraft in a murine model; however, they achieved clinically meaningful *CCR5* disruption rates that show the usefulness of adenoviral gene delivery in HSCs [57, 59–61].

Non-viral approaches for nuclease delivery include protein- and messenger RNA (mRNA)-based platforms that are highly desirable because their transient expression has no potential for genome integration. Protein-based delivery strategies have employed ZFN-transferrin conjugates to deliver ZFNs via transferrin receptor-mediated endocytosis [62]. These complexes were delivered to human HSCs and mediated both NHEJ and HR in reporter assays in transformed cell lines [62]. Using an optimized electroporation-based procedure with the Amaxa Nucleofector instrument, mRNA delivery has resulted in *CCR5* gene disruption in HSCs at a rate of approximately 15 % [63]. Furthermore, this group employed the HSCs in a pre-clinical humanized mouse model to show that the modified cells are resistant to HIV-1 infection [63, 64]. The ability to disrupt HIV entry portals in HSCs with ZFNs is significant due to the recent treatment of patients using hematopoietic cell transplantation of grafts with homozygous CCR5Δ32 mutations that disrupt cellular entry of the HIV particles [65]. This treatment protocol was initiated for an individual with HIV/AIDS who developed acute myeloid leukemia (the so-called 'Berlin patient') in an attempt to cure both the malignancy and the HIV infection [65, 66]. Because of the paucity of CCR5Δ32/CCR5Δ32 donors, ZFN-modified

HSCs are thought to be an ideal strategy for widespread implementation of this regimen. A further consideration for nuclease-based gene disruption therapies for HIV includes the *CXCR4* gene that is also a co-receptor for HIV cellular entry, and a significant number of HIV patients harbor the *CXCR4* variant HIV strain [67]. As such, a recent combinatorial ZFN approach has been investigated in the laboratory. Didigu et al., using *CXCR4* and *CCR5* adenoviral-borne ZFNs, showed the ability to remove both HIV co-receptors simultaneously in human T-cells [68]. A potential clinical limitation of this approach is the fact that *CXCR4* is a critical homing molecule for HSCs [68] and its disruption may perturb normal HSC homeostasis.

Their blood lineage plasticity and expansive clinical application makes HSCs a desirable cell type for genome engineering with designer nucleases; however, their limited ability to form extrahematopoietic tissue restricts their wide use in comprehensive disease modeling and in regenerative medicine outside the lymphohematopoietic system. ESCs and iPSCs represent powerful tools for filling this void and performing nuclease genome modification of multi-lineage stem cells (Figs. 3 and 4).

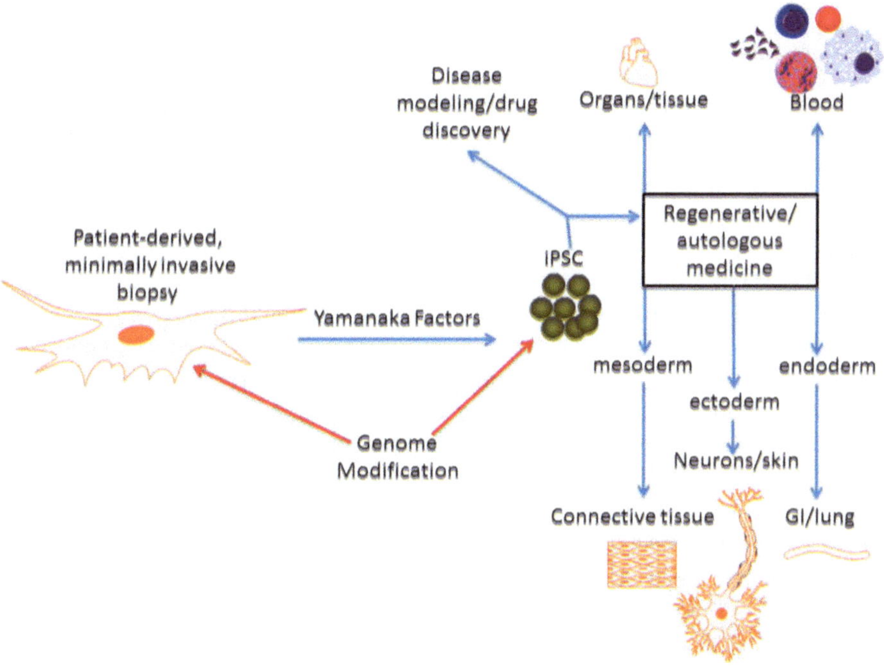

Fig. 3 Inducible pluripotent stem cell engineering. Gene editing nucleases can be introduced into somatic cells prior to reprogramming or directly to iPSCs. Somatic cells can be obtained from multiple sources as part of a minimally invasive biopsy from a patient. The Yamanaka factors (Oct3/4, Sox2, Klf4, c-Myc) are added to the template cell to induce pluripotency. These cells serve as a platform for disease modeling, drug discovery, and autologous therapies/regenerative medicine due to their broad differentiation potential. Image generated using software from Motifolio Inc. Sykesville, MD

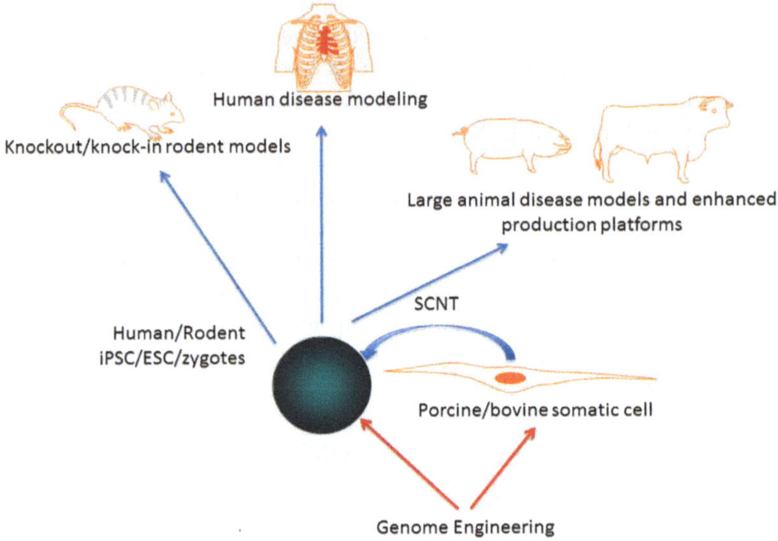

Fig. 4 Multi-species embryonic stem cell engineering. For livestock models, genome engineering can be performed in somatic cells followed by somatic cell nuclear transfer (SCNT) to an enucleated oocyte. Small animal model gene editing can be initiated in primary ESCs or early stage embryos followed by implantation. Human stem cell gene editing occurs in primary cells in vitro. Genome engineering provides whole live birth animal models for discovery, and human ESCs (or iPSCs) serve as a platform for in vitro gene/drug discovery and lineage commitment studies. Image generated using software from Motifolio Inc. Sykesville, MD

Human ESC engineering with ZFNs was first described by Lombardo et al., who targeted the HUES-3 and HUES-1 cell lines with ZFNs designed for the *CCR5* gene [42]. They introduced ZFNs and a donor sequence containing GFP into exon 3 of the *CCR5* gene and observed up to ~5 % rates of targeted integration. Importantly, the cells maintained their pluripotency and ability to self-renew [42]. This study established a precedent for inserting genes of interest into a specified spot in the ESC genome via HR. Hockemeyer et al. extended this to allow for placement of an inducible expression cassette at the so-called 'safe harbor' locus *AAVS1*. Using ZFNs for the first exon of the *PPP1R12C* gene on chromosome 19, they introduced a 'stand alone' expression cassette containing a promoter, a puromycin gene, and a polyadenylation signal (or gene trap targeting vector) containing a splice acceptor-2A-puromycin gene that relied on proper targeting and splicing with the first exon of the *PPP1R12C* gene [69]. As such, gene targeting at the AAV locus allows for placement of a gene with a promoter that drives the desired level of expression or is controlled by the native *PPP1R12C* promoter that is constitutively active [70]. Employing this strategy does not appear to alter the pluripotent nature of iPSCs or ESCs [69–71]. Lombardo and colleagues showed that iPSCs stably expressed the *AAVS1* localized GFP gene as well as the endogenous *TRA-1, Nanog, Sox2*, and *Oct4* markers of pluripotentiality [71]. Wang et al. utilized ZFNs to mediate HR-directed insertion of 2A-puromycin resistance gene designed to splice into exon

1 of *PPP1R12C* followed by an ubiquitin promoter-driven tricistronic construct containing red fluorescent protein, firefly luciferase, and *herpes simplex* virus thymidine kinase genes [72]. This strategy allowed for imaging using fluorescence, bioluminescence, and positron emission tomography of ESCs, iPSCs, and endothelial cells and cardiomyocytes derived from the stem cell progeny [72]. While the HR-directed integration of transgenes into the *AAVS1* site does not appear to disrupt the transcriptional profile of the *PPP1R12CI* gene or those in the immediate vicinity, there is still debate as to whether this locus is a true 'safe harbor' [71, 73]. More gene-specific targeting is preferable, and possible, using ZFNs in stem cells. In the same study as their *AAVS1* targeting approach, Hockemeyer et al. also established that ZFNs designed for the *OCT4* (*POU5F1*) locus could mediate insertion of a novel OCT4-EGFP reporter that allowed the pluripotent state of hESCs to be monitored with real-time imaging [69]. Additionally, they showed that the *PITX3* gene, normally silenced in ESCs and iPSCs, could be targeted, thus indicating that sequences could be targeted in ESCs/iPSCs regardless of their transcriptional status [69].

The ability to modify genes in ESCs and iPSCs is important to disease modeling in vitro, which has become a new foundation for the acceleration of translational research. To establish ZFNs as a tool for gene alterations in order to imitate human disease, Zou et al. targeted the phosphatidylinositol N-acetylglucosaminyltransferase subunit A (*PIG-A*) gene in ESCs and iPSCs, which generated cells that phenotypically mimic cells of patients with the severe blood disorder paroxysmal nocturnal hemoglobinuria (PNH) [74]. Their strategy relied on NHEJ-mediated deletions in H1 ESCs as well as HR-induced mutagenesis. By disrupting exon 6 with ZFNs and a targeting donor construct, they were able to derive clonal isolates that maintained pluripotency in iPSCs and ESCs [74]. These studies established ZFN-mediated genome editing as a powerful tool for stem cell disease modeling. The promise of such an approach is inherently reliant on uniform tools for study that will recapitulate the disease phenotype, will not introduce confounding variables, and will account for the significant differences that are present at multiple loci between any two individuals (or populations) that serve as the study materials.

A potentially significant source of variability in stem cell engineering is the cell generation/propagation process. ESC lines approved for widespread study can exhibit variability in regards to their ability and susceptibility to differentiate, as well as in their epigenetic and genetic stability [75]. These factors contribute to genetic heterogeneity and can preclude accurate comparisons in vitro [75]. For iPSCs, there is an additional layer of complexity due to the method by which the cells are generated. Currently, chemical, protein, viral, and non-viral platforms exist for reprogramming cells into iPSCs [8, 76–78]. Variegation effects, ectopic transgene expression when integrating vectors are used, and copy number variants that are present at the time of reprogramming all contribute to potentially confounding factors in disease modeling [79, 80]. As a solution to this, Soldner et al. used ZFNs to generate isogenic control and Parkinson's disease (PD) cell lines [79]. This work centered on engineering the A53T or E46K PD mutations into disease-free ESCs or repairing the mutation in PD patient-derived iPSCs [79]. In this way they devised an elegant solution to the numerous genetic differences and modifiers that exist between individuals and ESC and iPSC clones. A recently described, straightforward, and uniform procedure for generating isogenic cell lines using established ZFNs will further facilitate the implementation of this strategy [81].

Further work also showed the ability to revert a disease mutation back to wild-type status using genome engineering. Using ZFNs to correct the mutation that causes sickle cell anemia has been documented [82–84], with Zou et al. showing the ability of corrected iPSCs to differentiate into cells of the erythroid lineage [83]. Yusa et al. further expanded the ZFN correction repertoire to include the Glu342Lys mutation in the α_1-antitrypsin gene that is responsible for α_1-antitrypsin deficiency, with corrected hepatocyte-like cells, showing proper enzymatic activity in vitro [85]. Rahman et al. employed ZFNs in fibroblasts and iPSCs to correct the DNA-dependent protein kinase defect that causes severe combined immunodeficiency and successfully generate phenotypically rescued T-lymphocytes [86].

ZFNs have also been used for whole chromosomal gene editing by Jiang et al., who employed a novel approach by reprogramming Down syndrome patient fibroblasts into iPSCs and then using ZFNs for the *DYRK1A* locus on chromosome 21 for targeted insertion of an inducible X-inactivation (*XIST*) transgene [87]. The transactivator for *XIST* was integrated into the *AAVS1* locus using ZFNs, and the cells were screened for a single chromosomal copy of *XIST* at the *DYRK1A* locus [87]. Six clones were tested. Induction of the system with doxycycline *XIST* transgene expression on chromosome 21 caused stable heterochromatin modifications and gene silencing to form a 'chromosome 21 Barr body' [87]. Genome-wide transcriptional profiling showed silencing of 95 % of the expressed genes on the targeted chromosome [87]. These studies have expanded the power of ZFNs to chromosomal gene therapy for trisomy 21, as well as allowed for Down syndrome modeling in vitro to identify trisomy 21-induced genome/transcriptome dysregulation [87].

Despite numerous reports in the literature documenting the ability and efficiency of ZFN-mediated genome editing in multiple organisms, ZFN technology has several limitations. The construction of highly active ZFNs is not efficient enough to allow for widespread use. While the modularity of the platform has been improved upon, the most efficient ways to generate ZFNs that take into consideration the context dependency of adjacent fingers still require industry affiliations or the acquisition of specialized and lengthy procedures [88, 89]. These drawbacks, combined with the advent of the TALEN platform, which can be constructed from freely available modules, have caused a shift in the landscape of stem cell and genome engineering.

TALEN

The TAL effector proteins are derived from plant pathogenic bacteria of the genus *Xanthomonas*. DNA recognition is conferred by a central repeat region comprised of 14–24 tandem 34 amino acid repeats and is governed by a simple code allowing

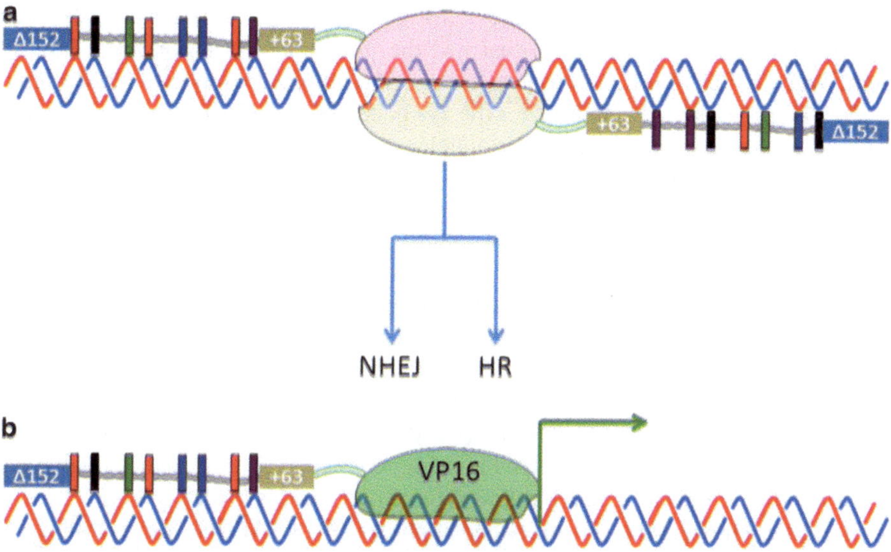

Fig. 5 TALEN nuclease and transcriptional activator architecture. (A) TALE nuclease. A TAL is comprised of an N-terminal deletion of 152 residues and 63 wild-type TAL sequences at the C-terminus and repeat regions containing two repeat variable di-residues (colored boxes) that show a simple recognition code for each of the 4 DNA bases. This code can be utilized to make a user-defined TALEN that is fused to the nuclease domain of the FokI enzyme (pink and tan ovals). TALE nucleases are heterodimeric and contain a left and right array that co-localize at the target site to mediate a double-stranded break that can be resolved by NHEJ or HR. (B) TALE activator. The FokI nuclease domain is replaced by multiple copies of the VP16 domain derived from the herpes simplex virus encoded (or other component derived) transcriptional activator(s). TALEN monomers fused to activators can be targeted to target sequences for transcriptional upregulation. Image generated using software from Motifolio Inc. Sykesville, MD

for rapid assembly [90–92]. Similar to ZFNs, TALENs function as heterodimers (Fig. 5a). In contrast to ZFNs, however, each TALEN monomer can be engineered to recognize between 12 and 32 bp of DNA. The TALEN sequence specificity is therefore expected to be exquisite, as the chances of this length of sequence being repeated in a given genome is low. TALENs have been employed in a similar fashion as ZFNs in multiple models, except for HSCs.

Rodents

In 2011 Tesson et al. employed TALENs to target the rat *IgM* locus in order to eliminate *IgM* function [93]. They observed higher rates of editing with DNA-encoding TALENs but no bi-allelic modifications. In contrast, mRNA delivery, while lower in overall frequency of cutting, resulted in 50 % bi-allelic modification with higher doses of mRNA [93]. TALENs were also applied to rats for the knockout of the *TLR4* gene that is involved in ethanol-induced behavioral effects. Ferguson et al. introduced

TALENs designed for exon 1 of the *TLR4* gene and injected them into one-cell embryos [21]. One rat out of 13 that had the mutation was successfully bred to homozygosity [21]. Further advances in rates of editing were reported by Mashimo et al., who co-delivered TALENs for the *Tyr* gene with mRNA for the Exonuclease 1 (*Exo1*) gene that possesses 5′–3′exonuclease activity and observed a substantial increase in the number of knockout rats [94]. Importantly, the tandem delivery of TAL and *Exo1* mRNA did not result in toxicity, which showed that genome editing with ectopic manipulation of the DNA repair pathways can further improve efficacy [95].

In 2013 Sung et al. described the first application of TALENs in mouse embryos [96]. They targeted the *Pibf1* and *Sepw1* genes, and observed that higher doses of TALEN mRNA resulted in more bi-allelic modifications, while lower doses resulted in a greater number of founder animals. This latter effect was associated with off-target toxicity for the *Pibf1* TALEN and, in contrast to other studies that suggested higher doses of nucleases mediated higher rates of editing, in their study they observed higher rates of editing at lower doses of TALENs [96].

In studies showing the versatility and applicability of TALENs in generating different mouse strain knockout models, Davies et al. employed TALENs in oocytes from three different strains to inactivate the *Zic2* gene in order to create a murine model of holoprosencephaly, one of the most common human congenital anomalies [97, 98]. By injecting TALEN mRNA, they observed editing rates of 25 % on the C3H/HeH background and 10 % in C57BL/6J animals [92]. In similar studies, Qiu et al. showed equivalent editing rates in both C57BL/6 and FVB/N zygotes [99]. Collectively these studies allow for greater flexibility and purity in introducing desired mutations in various strains of mice. It is thus conceivable that the differential genetic backgrounds of mice could be employed to mimic similar scenarios observed in human population studies where constitutional genetic modifiers can influence physiological responses [100].

Wang et al. targeted the *Sry* and *Uty* genes on the Y chromosome [101] in a study that represents a further advance in mouse genetics afforded by gene editing and TALENs in particular. Previously, there was a near-complete lack of knockout mice with deletions of genes on the Y chromosome due to the difficulty in targeting it by conventional methods [101]. The use of TALENs allowed for both *Sry* and *Uty* targeting with high efficiency [101]. This will allow for more detailed studies of Y-chromosome gene function to be performed. MiRNA gene targeting of miR-10a and miR-10b has also been reported by Takada et al. This versatility, ease of generation, and efficiency of use will facilitate strain-specific gene discovery studies [102].

Targeted knock-ins have also allowed for the generation of humanized mouse mutation models of disease. Wefers et al. modeled Hermansky–Pudlak syndrome by introducing TALEN mRNA and a single-stranded oligonucleotide donor (ODN) containing a missense mutation in the *RAB38* gene [103]. In a second set of studies, this group further optimized the procedure by utilizing a novel TALEN scaffold that included a plasmid-coded poly(A) tail, thereby removing the in vitro polyadenylation step [104]. In their first study they had observed <2 % rates of ODN-mediated HR, but the optimized TALEN mRNA species mediated the introduction of amyotrophic lateral sclerosis ODN-derived mutations into the *Fus* gene at a rate of 6.8 % [104]. They also observed that multiple founders showed a mixture of HR and NHEJ. Similar findings were also observed in the Low et al. study using TALENs and ODN donors to correct the *Crb1rd8* mutation, which is present in many mouse strains and complicates studies of retinal disease [105]. The use of ODN donors is highly appealing as they are easily synthesized and can be delivered at high concentration in a low volume. These reports and others show their ability to be integrated into the genome for the generation of novel disease polymorphisms [106, 107]. However, the mix of truncated ODN NHEJ and HR events may make them suboptimal and inefficient in conditions where it is necessary to screen numerous clones without the benefit of a drug resistance marker.

Classic gene targeting has relied on the use of double-stranded DNA donors, typically borne on a plasmid to the targeted locus, that contain arms of homology of variable lengths and commonly carry a drug resistance gene to allow for selection. The disadvantage of this approach is that the size of the plasmid can limit the dose that can be delivered. The advantages are that the lengthiness of the donor arms mitigates truncations due to incomplete or short crossover events, and the presence of a selectable marker (e.g., drug resistance gene) can greatly aid in the recovery of properly targeted genes. A landmark study, using a donor such as this, established a precedent for one of the most desirable aspects of genome engineering: ex vivo gene correction of pluripotent stem cells in a clinically relevant model. In 2007, Hanna et al. derived a fibroblast cell line from a humanized mouse model of sickle cell anemia, reprogrammed these cells into iPSCs, performed gene correction using a plasmid donor, differentiated the cells into hematopoietic progenitor cells, and transplanted them into sickle cell mice to reconstitute normal erythropoiesis [108]. The ability to employ this approach without using engineered nucleases but still with sufficient efficiency provides an even greater rationale for combining gene targeting with designer nucleases to make such an approach possible in humans.

Swine

TALENs have been used for modification of the porcine genome for knocking out genes for large animal disease models. Carlson et al. used TALENs to generate a live model of hypercholesterolemia by TALEN-mediated indels at the low-density lipoprotein receptor locus in fibroblasts that were then utilized for SCNT [107]. This group has also used a similar approach to generate porcine models of infertility and colon cancer. TALEN-edited fibroblasts were used for SCNT and production of founder animals containing knockout alleles in the *DAZL* and *APC* genes [109].

Injection of TALEN mRNA targeting the *GGTA1* locus into porcine embryos resulted in rates of editing of >70 % [110]. Follow-up studies to disrupt the gene in fibroblasts using a donor containing neomycin revealed >85 % targeting following drug selection, with >25 % of the cells containing homozygous gene editing events [110]. A corresponding study using ZFNs observed ~1 % bi-allelic disruption [31].

This latter study did not use a drug selection procedure, so a direct comparison is difficult; however, the ability to recover higher rates of edited cells is an important consideration for genome modification.

Cattle

In 2012, the ability of TALENs to function efficiently in large animal model embryos was documented. These studies showed modification of the *ACAN11* or *ACAN12* genes using an early generation TALEN scaffold that contained more native bacterial sequences [91, 111]. The injections that produced the highest indel frequency were associated with developmental delay, prompting them to reformulate TALENs using the N- and C-terminal truncations that appear to be optimal for TALEN activity with minimal toxicity in mammalian cells [112, 113]. When using this architecture with TALENs designed for the *GT-GDF83.1* gene, they saw greater frequency of indels without a significant impact on development rate in vitro [107].

Non-Human Primates (NHP)

In 2014, Liu et al. employed TALENs for genome modification in rhesus and cynomolgus monkey cells and tissue [114]. By injecting TALEN plasmids targeting the *MECP2* gene along with a third plasmid encoding *RAD51*, included to promote NHEJ, they observed miscarriages of male fetuses in accordance with Rett syndrome-associated male lethality. With this procedure they were able to bring to term a single female with *MECP2* indels showing the ability to edit the genome of NHPs and showed that plasmid DNA mediated higher rates of modification than did mRNA [114].

Humans

In 2011, Hockemeyer et al. generated TALENs, tested them for the same gene targets that they had previously described for ZFNs, and observed a comparable efficiency [69, 115]. On-target donor integration was observed at all of the five target sites: *OCT4* (5′ and 3′ prime targets), *AAVS1*, and *PITX3* (start and stop codon targets) in human ESCs and iPSCs, each of which retained their pluripotent properties [65]. These key findings—combined with the drastic improvements in availability of TALEN modules, how quickly they can be generated, the enhanced targeting repertoire, and efficiency with which they operate—allow for unparalleled access and opportunities to the stem cell and gene editing fields. Multiple investigators have capitalized on this by using stem cell engineering for disease modeling or correction. Ding et al. demonstrated the synergistic power of stem cells and TALENs by targeting and creating mutations in 15 genes [116]. They went on to phenotype three disease genes in stem cells: *SORT1*, *AKT2*, and *PLIN1* [116]. *SORT1* has a role in the regulation of blood insulin, cholesterol levels, and neuronal viability. They used TALENs to disrupt exon 2 or 3 of *SORT1* in two different ESC lines. Differentiation into hepatocyte like-cells revealed that *SORT1* reduces apoB-levels in the blood, thereby lowering cholesterol and suggesting protection from atherosclerosis [116]. Furthermore, these studies showed that *SORT1* appears to be essential for insulin-responsive glucose uptake, suggesting a role for *SORT1* in human insulin sensitivity [116]. Lastly, neuronal differentiation of *SORT1* null cells validated previous data showing the necessity of *SORT1* for brain-derived neurotrophic factor (proBDNF)-mediated apoptosis in neurons [116]. These data agree with the reported requirement of *SORT1* for proBDNF-induced programmed cell death in human motor neurons [116]. These authors also performed in vitro disease modeling in *AKT2* and *PLIN1* TALEN-edited cells. The E17K missense mutation in the *AKT2* gene has been associated with insulin resistance and increased body fat; however, protein studies of this variant were lacking due to an inability to perform studies in physiologically relevant cell types [116, 117]. To address this, TALENs were employed to knockout exon 2 of *AKT2* in HUES 9 cells with a frequency of 17/192 clones possessing indels, followed by a second round of gene targeting to obtain 2/96 clones possessing a compound heterozygote mutation [116]. To more precisely model the E17K mutation, TALENs and a 67 nucleotide ODN were introduced into HUES 9 cells to derive an *AKT2*^{E17K} heterozygous cell line for study [116]. Differentiation of each cell population into hepatocyte-like cells and adipocytes revealed a dominant function for the E17K mutation that causes diabetic-like symptoms in patients [116, 117]. Using a similar strategy, these authors used TALENs to generate frameshift mutations in exon 8 of the *PLIN1* gene in HUES 9 cells. They obtained 70/293 clones with indels, one of which closely mimicked a natural, elongated variant (Val398fs mutation) and a second resulted in a truncated form of *PLIN1* [116]. By differentiating the cells into adipocytes, they observed that the elongated variant protein altered storage and droplet formation in adipocytes in a fashion similar to lipodystrophy patients [116].

These studies highlight the power of gene targeting in stem cells for in vitro disease modeling and discovery. By performing their analyses in a single cell line, they mitigated patient-to-patient or clonal variations that might impact disease gene manifestation similar to a previous study of isogenic disease modeling [79]. Further, although their results were consistent with a previous human study [118], they were different from a murine knockout model [118], highlighting the value of species-specific disease modeling [116]. Moreover, the relative ease with which TALENs can be generated greatly expands their broad applicability, as evidenced by the 15 gene targeting strategy with multiple lineage differentiation and phenotypic disease modeling performed on three of the genes in this study [116].

The ability of maintaining consistency for disease gene modeling and discovery in stem cells was further enhanced by the DICE (dual integrase cassette exchange) system [119]. This system relies on the insertion of a 'landing pad' cassette containing phiC31 and Bxb1 attP sites at the *H11* locus in ESCs or iPSCs [119]. These attP sites are non-overlapping in regards to recognition by the phiC31 and Bxb1 integrases, and result in the highly precise placement of a desired transgene, in a single copy, in the same orientation. This eliminates copy number and expression variation as a result of random genomic integration sites with differential epigenetic landscapes [119]. Utilizing the landing pad sequence insertion using TALENs resulted in a ~8-fold increase in efficiency in ESCs, two iPSC lines, and the 1754 iPSC line derived from an individual with Parkinson's disease [119]. A similar approach relying on loxp-Cre transgene exchange at the *AAVS1* locus has also been described [120]. As such, disease-causing or therapeutic genes can be introduced into multiple stem cell platforms and in various loci (e.g., *AAVS1*, *H11*), in a manner that reduces the unpredictability associated with classical transgenesis.

TALEN gene discovery has also extended to miRNA genes. TALENs designed to delete the miR-302/367 cluster in human fibroblasts showed that these miRNA are required for reprogramming to iPSCs [121]. This study also expanded the TALEN functionality to modifications to the epigenome by fusing TALEN arrays to a Kruppel-associated box (KRAB) transcriptional repressor [121]. Utilizing the TAL-KRAB repressor fusion, these authors observed a ~4-fold decrease in reprogramming efficiency when knocking down this miR cluster. TALEN transcriptional activators have also been described with TAL DNA binding domains fused to the activation domain of VP16 (Fig. 5b). This complex could be targeted to the *Oct4* locus and moderately upregulated transcription in ESCs [122]. Fusion of TALE repeats to the TET1 hydroxylase catalytic domain has also allowed for modification of methylated CpG residues with concomitant gene expression upregulation [123]. As such, the application of TALENs as genome editing nucleases that result in permanent modifications, as well as genome 'rheostats' that can temporarily activate or repress gene function, will allow for greater application of this technology.

TALEN gene correction has also been documented directly in iPSCs. Using a plasmid donor with alpha-1 antitrypsin (*AAT*) gene arms of homology flanking a fusion-resistance gene comprised of a puromycin gene fused to a truncated thymidine kinase gene, Choi et al. achieved correction of the *AAT* gene [124]. Using puromycin selection, they recovered 66/66 properly targeted iPSCs with 25–33 % demonstrating bi-allelic correction [124]. Finally, using a *piggyBac* transposon system, they subsequently removed the resistance genes, leaving behind only donor-derived sequences and showing functional correction of iPSC-derived hepatocyte-like cells [124]. Sun et al. used a similar donor strategy for the correction of the E6V mutation in the hemoglobin beta (*HBB*) gene that causes sickle cell disease [125]. More than 60 % of their puromycin-resistant clones showed on-target gene correction, and the iPSCs remained karyotypically normal and retained pluripotency [125]. Osborn et al. applied TALENs for gene correction in primary fibroblasts derived from a patient with a severe blistering disorder, recessive dystrophic epidermolysis bullosa (RDEB) [56]. The cells were subsequently reprogrammed into iPSCs that, when injected into immune-deficient mice, formed teratomas with skin organoids showing the proper deposition of the type VII collagen protein that is missing in RDEB patients [56]. Collectively, these studies form the basis for future ex vivo therapies where patient cells are corrected either pre- or post-reprogramming and provide individualized genomics-based therapeutic and regenerative medicine.

TALENs targeting capacity surpasses that of ZFNs, and they are easier to generate [111, 126]. TALENs can also be fused to differential domains (e.g., transcriptional activator and repressor) to confer new activities in a site-specific manner. Despite the great enthusiasm for TALENs, the technology is still relatively new and some issues remain unresolved. First, the ability to efficiently deliver TALENs is compromised by their size, making adenoviral vectors well suited for delivery while other delivery platforms (retro-, lenti-, and adeno-associated viral vehicles) can be refractory to this technology [127]. Second, the ability to multiplex TALENs currently requires generating and introducing multiple site-specific proteins that have to be generated individually and then delivered efficiently to achieve the desired events. For these reasons, there is currently great interest in the CRISPR/Cas9 system, which affords the user an even greater ease and flexibility for stem cell engineering.

CRISPR/Cas9

CRISPR/Cas9 exists naturally in archaea and bacteria as part of an adaptive immunity mechanism for bacteriophage defense [128]. The system is comprised of the Cas9 nuclease and of a guide RNA (gRNA) species (Fig. 6) that possesses sequence specificity in relation to a protospacer adjacent motif (PAM) that can differ between different Cas9 orthologs [129, 130]. The system has been adapted for mammalian use in multiple platforms and is highly user friendly.

Rodents

Mashiko et al. were able to target either the murine *Cetn1* or *Prm1* genes by using a simplified delivery method where the Cas9 nuclease and gRNA were contained on a single plasmid injected into mouse zygotes [131]. Higher doses of plasmid were associated with more efficient knockout rates, and only 2/46 modified animals contained a random integrant of the Cas9 expression cassette [131]. This relatively low integration rate simplifies the delivery/injection of this platform as it obviates the need for RNA generation. However, RNA delivery has been associated with higher overall rates of editing. Li et al. were unable to modify the *Th* gene with low doses of DNA in FVB/N mice, and the rate of editing only increased to 0.3 % when using 2.5 times more DNA [132]. In contrast, the use of RNA Cas9 and gRNA increased gene targeting in C57Bl/6 mice at the *Th* locus to 6.7 % [132]. Fujii et al. showed that CRISPR/Cas9 with two independently targetable gRNAs of 80

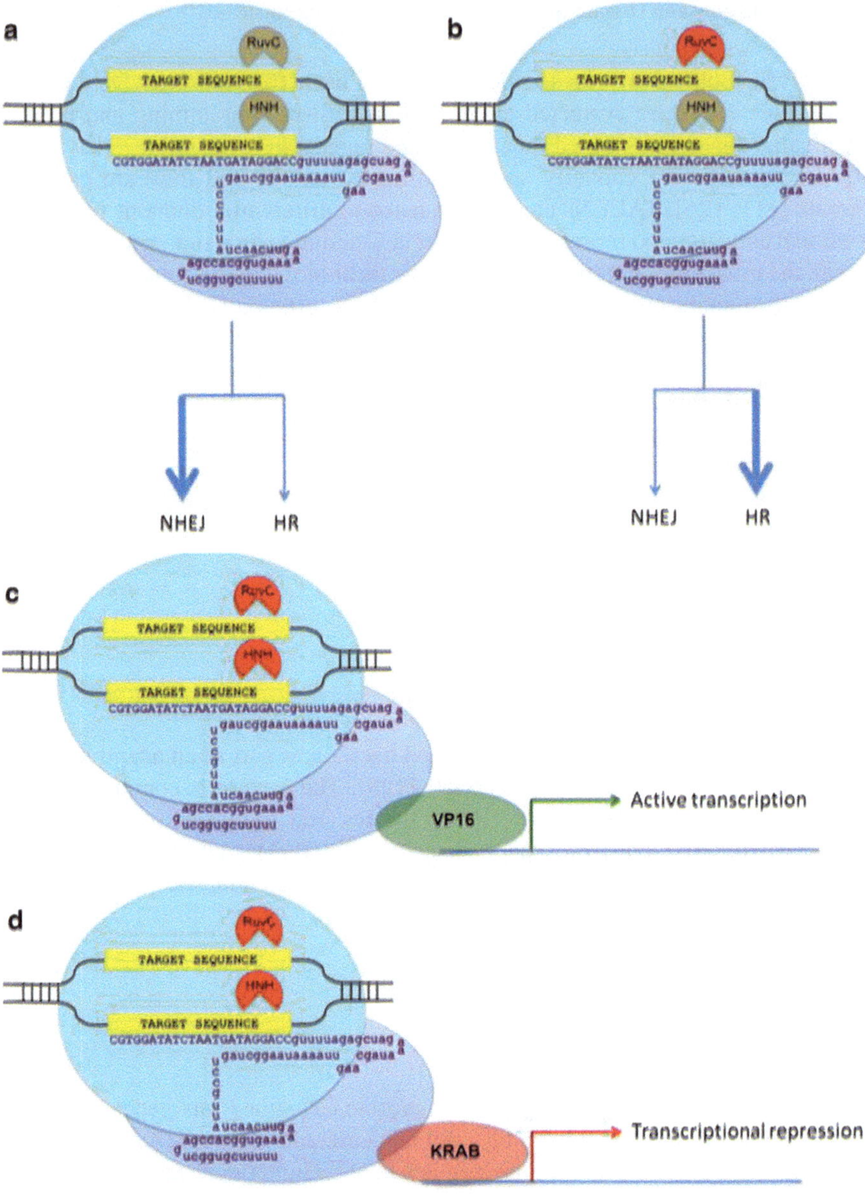

Fig. 6 CRISPR nuclease and nickase. The CRISPR/Cas9 system is comprised of a guide RNA (purple sequence) that possesses secondary structure that interacts with the Cas9 nuclease. The complex binds a target sequence and two domains in Cas9 mediate sense and antisense DNA cutting: HNH cleaves the complementary strand and RuvC cleaves the noncomplementary strand. (A) CRISPR/Cas9 nuclease. The gRNA and Cas9 co-localize at the target site, and both strands of DNA are cut promoting NHEJ but allowing for HR. (B) CRISPR/Cas9 nickase. The D10A mutation inactivates the RuvC domain (indicated by red) and results in only one strand of the DNA helix being cleaved with preferential promotion of HR over NHEJ. (C) CRISPR/Cas9 activator and (D) CRISPR/Cas9 repressor. Inactivation of the HNH and RuvC domains removes the nuclease activity of Cas9. Fusing Cas9 with transcriptional activating (e.g., VP16) or repressive domains (e.g., Krueppel-associated box (KRAB)) converts the complex into a gene expression modulatory platform

nucleotides (compared to the typical ~40 nucleotide length) could cause large scale deletions (~10 kb) that were transmittable to progeny mice [133]. These data showed the ability of multiple gRNAs to function simultaneously in murine zygotes. Expanding this to different genes, Zhou et al. attempted to target the murine *B2m, IL2rg, Prf1, Prkdc*, and *Rag1* genes in order to generate immunodeficient animals [134]. While the Cas9 and gRNAs were delivered as RNA species, the researchers observed a dose-dependent associated increase in efficiency similar to the Mashiko study. This study, in one step, recapitulated three different immunodeficient mouse models (*IL2rg* and *Prkdc*-deficient NSG model; *Rag1* and *IL2rg* null BRG model; and NSG B2m−/− with *B2m, IL2rg*, and *Prkdc*) [134]. Furthermore, they documented three animals that showed disruption of all five target genes [134]. Wang et al. also targeted 5 genes (*Tet1, 2, 3, Uty*, and *Sry*), and 10 % of the ES clones screened showed mutations in all eight alleles of the five genes [135]. In this and a second report authored by the group, they also demonstrated the ability to introduce double- or single-stranded donor templates to insert small alterations into the targeted sequence [135, 136]. These alterations included point mutations, the insertion of a loxP site, or a V5 epitope tag [135, 136]. In this way the one-step generation of mice carrying designed point mutations, floxed alleles for conditional mutations, and fluorescent marker- or V5 epitope-tagged genes was efficiently achieved. These strategies offer a path forward for gene discovery, as well as a method for protein-based studies using candidates tagged with a sequence (e.g., V5 epitope) where protein-specific antibodies may not exist.

The immense power of the CRISPR/Cas9 system multiplexing ability is being fully realized in genome-wide gene discovery studies [137]. Koike-Yusa and colleagues introduced 87,897 guide RNAs targeting 19,150 mouse protein-coding genes into murine ESCs that constitutively expressed Cas9 [137]. They then screened them for resistance to 6-thioguanine or *Clostridium septicum* alpha-toxin and identified 31 genes involved in these phenotypes [137].

Higher order mammalian embryonic stem cells have also been targeted with CRISPR gene editing. Rat zygotes injected with Cas9 and gRNA RNA species targeting the *Mc3r* or *Mc4r* genes showed disruption frequencies, with germline transmission, of 0.8 % and 10.6 %, respectively [132]. Conditional gene targeting has also been utilized in rats with CRISPR/Cas9, allowing for temporal and tissue-specific control of gene inactivation. Ma et al., using a single circular donor, targeted three genes——*Dnmt1, Dnmt3a*, and *Dnmt3b*——and generated offspring with *floxed* alleles [138]. An especially powerful tool is the generation of haploid rat ESCs that can be efficiently targeted with CRISPR/Cas9 to achieve homozygous gene disruption or insertion that produces fertilized oocytes and generates fertile offspring [139]. Thus, the efficient ability to modulate the rat genome opens new avenues of investigation due to its superiority to the murine model, particularly for toxicology and pharmacology studies [138, 140].

Non-Human Primates (NHP)

As a model system the NHP is closer to humans than any other organism. Generating suitable disease models has been hampered; however, by the lengthy time to reproductive maturation and gestation period that produces few offspring [141]. As a solution to this, Wan and colleagues employed CRISPR/Cas9 to achieve biallelic *p53* gene knockout animals. Their study found that high doses of gRNA and Cas9 negatively impacted development to the morula/blastocyst stage [141]. However, when a lower dose of Cas9 and gRNA was injected into 108 zygotes, 62 morphologically normal embryos developed that were implanted into 13 surrogates, four of whom became pregnant [141]. Two pregnancies were carried to term with three offspring, two of which carried *p53* gene modification events [141]. This study also showed that CRISPR/Cas9-mediated HDR was possible in embryos, opening a new avenue of approach for generating NHP models that closely mimic human disorders.

Humans

The first two reports describing the CRISPR/Cas9 system for use in mammalian cells revealed important considerations and opportunities for use in human stem cells [129, 130]. Mali et al. first showed the ability of CRISPR/Cas9 to mediate NHEJ-induced indels at the *AAVS1* locus on chromosome 19 in the PGP1 iPSC line [129]. The Zhang laboratory first showed the multiplexing ability of this system [130] and maximized this potential in the HUES62 ESC line by targeting 18,080 genes with 64,751 unique gRNAs to enable both negative and positive selection screening of genes involved in viability [142]. This approach represents an advance over RNA interference (RNAi)-based approaches that can only target the transcriptome, whereas the CRISPR technology allows for targeting of the entire genetic landscape (e.g., promoters and enhancers) [142].

Despite being able to design gRNAs for over 50 % of the genome, the wild-type *Streptococcus pyogenes* Cas9 can only target sequences possessing a $G(N_{20})GG$ motif [129, 130]. Differential Cas9 variants may enhance targeting and represent an opportunity to simultaneously and independently target sequences using orthogonal variants. The Cas9/gRNA platform from *Neisseria meningitidis* recognizes a 5′-NNNNGATT-3′ PAM, thus giving it a different targeting profile from *S. pyogenes* [143, 144]. Using *Neisseria*, Hou et al. were able to disrupt a reporter gene knocked into the DNMT3b locus in H9 human ESCs [143]. Further, they were able to increase rates of gene targeting in the H1 and H9 ESCs and the iPS005 [145] cell lines using a puromycin selection strategy that showed ~60 % were correctly targeted with single insertion events [143]. Cas9 from *Staphylococcus aureus* can target sequences containing NNAGAAW, and its main benefit is that it is ~1 kb smaller than other described Cas9 cDNAs, thus making it able to be packaged in

AAV virions to facilitate robust in vitro and in vivo gene editing in a highly specific manner [146]. Rational engineering has also facilitated the expansion of the targeting capacity of Cas9. Kleinstiver et al. utilized combinatorial design, structural data, and a bacterial selection-based evolutionary system to modify *S. pyogenes*, *S. thermophilis*, and *S. aureus* Cas9 proteins to facilitate new PAM recognition variants that showed functionality in zebrafish and human cells [147]. Thus, the directed evolution of existing and application of new Cas9 variants will further facilitate enhanced applications for this platform.

To that end, direct CRISPR gene repair has also been achieved in murine and human stem cells. Schwank et al. described CRISPR/Cas9 efficacy in small and large intestine stem cells derived from cystic fibrosis patients with a homozygous F508 deletion in exon 11 of the *CFTR* gene [148]. The clonally expanded organoids displayed full functionality and provided proof of principle for use of CRISPR/Cas9 in human disease correction [148]. In mice with a mutation in the cataract-causing *Crygc* gene, Wu et al. showed that gRNA/Cas9 RNA injection into zygotes could result in correction using an exogenous donor oligonucleotide or the endogenous normal allele [149]. Human iPSC disease correction has been reported by multiple investigators for several disorders (beta-thalassemia, muscular dystrophy, chronic granulomatous disease) [150–155] many of which rely on the introduction of selectable markers to force preferential outgrowth of modified cells. Often this necessitates a second step using cre-lox or a transposable element to remove the selectable marker. Grobarczyk et al. described a manner in which selection-free procedures used mechanical picking and enzymatic dissociation of cells to obtain ~2 % of cells with HDR or ~15 % with NHEJ [156]. Miyaoka et al. detail a strategy whereby pools of iPSCs treated with a nuclease and an oligonucleotide donor can be screened *en masse* with droplet digital PCR to detect the events followed by fractionation and subcloning of the pools to isolate corrected clones [157]. As such, the pairing of nuclease mediated modification with selection- free techniques will foster the generation of minimally invasive procedures to create stem cells for downstream applications. Further, because the Cas9 protein contains RuvC and HNH domains, each responsible for generating single-strand DNA breaks ('nicks') on opposite strands of the DNA helix, inactivation of one of these domains converts Cas9 into a DNA nickase capable of cutting only one strand (Fig. 6b) [129, 130, 158]. A nick to a single strand has been shown to preferentially promote error- free HDR, while nucleases often have competing HDR and NHEJ events [159, 160]. Furthermore, paired nickases have been shown to promote greater specificity [161–163]. As such, the extensive engineering of Cas9 allows for maximal efficiencies for differential purposes (e.g., HDR vs. NHEJ). As an extension of this, Cas9 is capable of functioning as a fusion protein that allows for subtler, transient gene regulatory alterations. Inactivation of each of the RuvC and HNH domains results in a catalytically inactive protein that retains binding ability to the gRNA and can acquire new functions by fusing it with other domains. This strategy has been used in stem cells by fusing Cas9 to the VP48 (three copies of a minimal VP16 sequence) activation domain (Fig. 6c) [164]. The *SOX2*, *OCT4*, and *IL1RN* genes were able to be simultaneously activated in a highly specific manner that was documented by genome-wide microarray expression analysis [164].

The modulation of genes in human stem cells employed a *SOX17* gene Cas9-VP16 activator or a *OCT4A* Cas9-KRAB repressor to interrogate the regulatory governors of differentiation and show the expansive ability of CRISPRs to mediate differential effects (Fig. 6d) [165]. This system provides a platform for the interrogation of the underlying regulators governing specific differentiation decisions, which can then be employed to direct cellular differentiation down desired pathways. This approach has been further expanded upon to achieve higher levels of transcriptional activation in iPSCs using a 'tripartite activator' comprised of VP64-p65-Rta fused to inactive Cas9 [166]. The gRNA itself can also be modified, resulting in enhanced efficacy as shown by Konermann et al., who included aptamer sequences to the gRNA that facilitated recruitment of non-Cas9 bound effector domains [167]. Larger scale epigenome modifications by targeting enhancers (proximal and distal) provide a further layer of targeting capability [168].

The high degree to which the CRISPR/Cas9 system can accommodate both iterative and directed design alterations that confer new functionality represents an incredibly agile platform. Its application in stem cells for disease modeling, generation, and correction, as well as its ability operate as an epigenetic modulator, makes it a tool capable of permanent and transient effects in a dynamic manner.

Summary

The future of genome editing and stem cell biology is bright, having been founded on molecular and cellular studies that have changed the manner in which biomedical research is performed. The first description of iPSCs [8] has primary importance, as it represents a powerful tool for multiple areas of study including developmental and stem cell biology. In addition, iPSCs represent a platform for in vitro disease modeling and drug discovery, and are potential sources of autologous cells for transplantation and regenerative medicine. The fact that they are derived from postnatal cells essentially removes the ethical hurdles that limit ESC applications. Further investigations into variations that exist in iPSCs and ESCs are required in order to identify the best platform for the next generation of therapies. Major applications of ESCs in the future will be for livestock manipulations and in creating rodent models of human disease.

A second transformative tool has been gene-specific nucleases and their application in stem cells. Within the past decade the field has expanded at a high rate of speed, from ZFNs that were available in few laboratories to TALENs and CRISPRs that are available to any researcher with basic molecular biology skills. Importantly, each iteration of nuclease platform is additive to the field and does not resign the previous versions to obscurity. The design and application of each class is mutually reinforcing to the entire field. ZFNs, TALENs, and CRISPRs are all being actively pursued in stem cells, and whether one will dominate is unknown. Driving factors in this decision are presented in Table 1.

Table 1 Nuclease consideration matrix

Platform	Cost	Specificity	Availability	Applicability	Ability	Multiplex	Vectorability
Meganuclease	$50,000	Low–high	low	R,M	Nuclease	No	Multiple
ZFN	$5000	Low–high	Low–medium	H,M,R,B,P	Nuclease, nickase	No	Multiple
				S, ESC, iPSC			
TALEN	$500	Low–high	High	H,M,R,B,P, NHP	Nuclease, nickase, activator, epigenome	No	Few
				S, ESC, iPSC			
CRISPR	$5	Low–high	Highest	H,M,R,B,P, NHP	Nuclease, nickase, activator, repressor, epigenome	Expansive	Multiple
				S, ESC, iPSC			

Approximate cost of gene-specific nuclease generation is shown at left. The specificity relies primarily on the complexity of the target site in regards to sequence heterogeneity. Nucleases designed for sequences of low complexity are expected to manifest with more off-target activity. A greater degree of stringency results in more specificity. Availability is dictated by cost, the generation process complexity, and the availability of the core reagents required for candidate generation. Applicability to reported animal and human cells: H=human, M=mouse, R=rat, B=bovine, P=porcine, S=somatic cell, ESC=embryonic stem cell, iPSC=inducible pluripotent stem cell. Ability indicates the properties of the classes of proteins to function with expanded properties of nickases, activators, or repressors. Multiplex and vectorability refer to multigene targeting simultaneously, as well as the core platform effector protein to be packaged and delivered in a broad range of viral or non-viral expression cassettes. MN, ZFN, and TALEN can target a small number of genes simultaneously, while CRISPR/Cas9 possess a genome-level multiplex ability. TALENs have the most restricted vectorability profile

Both nucleases and pluripotent stem cells have potentially deleterious aspects that could limit their effectiveness. For pluripotent stem cells this relates to the presence or accumulation of genetic and epigenetic modifications prior to or during reprogramming. Both ESCs and iPSCs are subject to these modifications in vitro, which may manifest in the same line or even within the same culture vessel during propagation [9]. Aneuploidy has been reported in iPSCs, their parental cellular precursors, and ESCs. Studies by the International Stem Cell Initiative suggest that karyotypic abnormalities may occur in as many as one out of every three cell lines [169]. Trisomy 12 is the most common abnormality in human ESCs and iPSCs, and chromosome 17 trisomy occurs frequently in murine ESCs [169, 170]. Interestingly, the *NANOG* gene, the master regulator of induced pluripotency, resides on chromosome 12, so the gene dosage advantage of cells trisomic for chromosome 12 may represent a positive selection in vitro [171, 172]. Copy number and single nucleotide variations appear to be derived primarily from the source cell and do not appear to play a role in preferential reprogramming or amplification of deleterious sequences [173]. These findings suggest that rigorous screening of template cells could minimize downstream adverse events, and that reprogramming by the addition of transcriptional activators is not likely to be mutagenic [9].

However, reprogramming itself results in significant alterations to the epigenetic landscape. During this process there is a global resetting of the epigenetic profile on the X chromosome as well as at multiple discrete loci [174]. For example, in human female cells a copy of the X chromosome is inactivated at random and this pattern is retained in daughter cells both prior to and, apparently, during reprogramming [175]. Prolonged culture, however, is associated with erosion of X inactivation in both ESCs and iPSCs, and appears to amplify and can even take over the culture population [176, 177]. Local epigenetic modifications may occur as well. Moreover, maintenance of the methylation profile of the parental cell type (such as epidermal or hematopoietic cell) may persist through reprogramming ('epigenetic memory'). In support of this, cell line-specific DNA methylation patterns have been reported that may significantly impact cellular phenotype when they are differentiated into the desired lineage [173, 178, 179].

Lineage plasticity is the hallmark of pluripotent cells. The most common assay to establish this plasticity is the teratoma assay in immune-deficient animals. The propensity for teratoma formation of undifferentiated cells is a concern for clinical application. Methods for removing contaminating iPSCs from the pool of induced lineage-committed cells have been described and may lower or even remove this significant hurdle to clinical use [180, 181]. Aberrant methylation status and subtle genetic changes may also impact clinical use of iPSC-derived cells by compromising their physiological function or by causing immunogenicity [182]. Rigorous quality assurance and control must therefore be performed before, during, and after reprogramming to insure safety and efficacy. As costs for whole genome, exome, and epigenome sequencing continue to decrease, these methodologies may allow for tandem quality assessments of stem cell manipulation and nuclease-induced deleterious genetic changes.

The choice and efficacy of a particular class of engineered nuclease depends on many factors: cost, ability to be designed by individual users (i.e., availability) for desired genomic targets, broad applicability, functionality (e.g., nuclease, nickases etc.), and ability to be delivered efficiently (factors shown in Table 1). Perhaps the most critical parameter is the consideration of safety. By definition, engineered nucleases are designed to recognize a specific DNA sequence; however, they may also exhibit off-target (OT) effects due to overlapping or low-complexity sequence recognition between the primary target and the OT site. Current methods for predicting and identifying OT sites include in vitro modeling, systematic evolution of ligands by exponential enrichment (SELEX), and unbiased genome-wide screens. SELEX has been performed with both ZFNs [183] and TALENs [115], but the correlation to actual in vitro or in vivo target sites has not been fully validated [184]. A newer methodology has shown better predictability for ZFNs [185], and a database for predicting CRISPR OT sites is also available [186, 187].

The unbiased approaches include integrase-deficient lentiviral gene trapping (IDLV) and mapping using linear amplification mediated (LAM) PCR [55, 188], GUIDE-seq [189], Digenome-seq [190], LAM PCR high-throughput, genome-wide, translocation sequencing (HTGTS) [191], and direct in situ breaks labeling, enrichment on streptavidin and next generation sequencing (BLESS) [146, 192]. These highly sensitive methodologies are effective at discovering and mapping OT sites; however, OT effects are not unprecedented and do not necessarily preclude clinical application [55, 193]. Moreover, there does not appear to be an emergent and dominant nuclease class that is devoid of OT potential, and each target site and nuclease candidate must be considered individually. At present, the overriding factor dictating specificity appears to be related primarily to the sequence being targeted and its heterogeneity/complexity. Toward achieving maximal efficiency, direct reengineering of the nuclease to an obligate heterodimer [194] or nickase/paired nickase versions [162, 163] can greatly reduce OT effects. Further, truncation of the gRNA target sequence appears to increase specificity [195]. The ability to assess OT effects at the genome level, target sequences of sufficient complexity (i.e., minimal overlap to secondary/OT loci), and engineer components of the nuclease architecture to achieve more stringency will allow for the safest reagents for clinical use.

Together stem cells and nucleases have made a tremendous impact in mammalian models of genetic disease modeling, gene discovery, and functional gene analysis. Translational applications hold great promise as well by virtue of precision nuclease-mediated modifications in pluripotent stem cells and their subsequent directed differentiation into terminal effector cells. As proof of principle, non-genome edited iPSCs have, for the first time, have been employed in a patient with age-related macular degeneration [196]. This foundational study provides a path forward for gene-edited stem cells. However, to fully realize this potential, safety concerns must be rigorously assessed at the stages surrounding reprogramming and after nuclease delivery. Further advances in high throughput sequencing technologies at the genome, exome, and epigenome levels and assessment of large data sets utilizing bioinformatics analysis [197] will facilitate the entrance of nuclease-modified cells for clinical use.

These analytical parameters must be considered in the context of societal and ethical issues that will guide the very choice of stem cell that is acceptable for use. *Primum non nocere* (First, do no harm) [196] is the core tenant of medical intervention and ethics and iPSCs, due to their derivation from adult tissues, are largely accepted in this context while ESCs may not be. The ability to modify genomes adds a layer of complexity to the ethical debate and, despite the call for a moratorium on germ line genome modification [198], a 2015 report by Liang et al. utilized CRISPR/Cas9 to modify the hemoglobin B locus in human zygotes at a frequency that was highly inefficient [199]. Therefore, the complicated biological development, optimization, assessment, and use of stem cell and gene editing models and technologies must exist and evolve in the complex societal and organismal arenas that themselves may have significant potential for 'off-target' effects.

Acknowledgements M.J.O. is supported by funding from the Lindahl Family and the Corrigan Family. J.T. is supported by grants from the National Institutes of Health (R01 AR063070 and R01 AR059947), the Department of Defense (USAMRAA/DOD Dept of the Army W81XWH-12-1-0609 and USAMRAA/DOD W81XWH-10-1-0874), DebRA International, Sohana Research Fund, Richard M. Schulze Family Foundation, Jackson Gabriel Silver Fund, Epidermolysis Bullosa Medical Research Fund, and University of Pennsylvania grant MPS I-11-009-01. M.J.O. and J.T. are also supported by the Children's Cancer Research Fund, Minnesota. Research reported in this publication was supported by the National Center for Advancing Translational Sciences of the National Institutes of Health Award Number UL1TR000114 (M.J.O.). The content is solely the responsibility of the authors and does not necessarily represent the official views of the National Institutes of Health.

References

1. Thomson JA, et al. Embryonic stem cell lines derived from human blastocysts. Science. 1998;282:1145–7.
2. Puri MC, Nagy A. Concise review: embryonic stem cells versus induced pluripotent stem cells: the game is on. Stem Cells. 2012;30:10–4.
3. Weismann A, Poulton EB, Schönland S, Shipley AE. Essays upon heredity and kindred biological problems. Oxford: Clarendon; 1891.
4. Solana J. Closing the circle of germline and stem cells: the primordial stem cell hypothesis. Evodevo. 2013;4:2.
5. Gurdon JB. The developmental capacity of nuclei taken from intestinal epithelium cells of feeding tadpoles. J Embryol Exp Morphol. 1962;10:622–40.
6. Briggs R, King TJ. Transplantation of living nuclei from blastula cells into enucleated frogs' eggs. Proc Natl Acad Sci U S A. 1952;38:455–63.
7. Wilmut I, Schnieke AE, McWhir J, Kind AJ, Campbell KH. Viable offspring derived from fetal and adult mammalian cells. Nature. 1997;385:810–3.
8. Takahashi K, et al. Induction of pluripotent stem cells from adult human fibroblasts by defined factors. Cell. 2007;131:861–72.
9. Liang G, Zhang Y. Genetic and epigenetic variations in iPSCs: potential causes and implications for application. Cell Stem Cell. 2013;13:149–59.

10. Silva G, et al. Meganucleases and other tools for targeted genome engineering: perspectives and challenges for gene therapy. Curr Gene Ther. 2011;11:11–27.
11. Popplewell L, et al. Gene correction of a duchenne muscular dystrophy mutation by meganuclease-enhanced exon knock-in. Hum Gene Ther. 2013;24:692–701.
12. Dupuy A, et al. Targeted gene therapy of xeroderma pigmentosum cells using meganuclease and TALEN. PLoS One. 2013;8, e78678.
13. Cohen-Tannoudji M, et al. I-SceI-induced gene replacement at a natural locus in embryonic stem cells. Mol Cell Biol. 1998;18:1444–8.
14. Menoret S, et al. Generation of Rag1-knockout immunodeficient rats and mice using engineered meganucleases. FASEB J. 2013;27:703–11.
15. Durai S, et al. Zinc finger nucleases: custom-designed molecular scissors for genome engineering of plant and mammalian cells. Nucleic Acids Res. 2005;33:5978–90.
16. Porteus MH, Carroll D. Gene targeting using zinc finger nucleases. Nat Biotechnol. 2005;23:967–73.
17. Capecchi MR. Gene targeting in mice: functional analysis of the mammalian genome for the twenty-first century. Nat Rev Genet. 2005;6:507–12.
18. Geurts AM, et al. Knockout rats via embryo microinjection of zinc-finger nucleases. Science. 2009;325:433.
19. Mashimo T, et al. Generation of knockout rats with X-linked severe combined immunodeficiency (X-SCID) using zinc-finger nucleases. PLoS One. 2010;5, e8870.
20. Cui X, et al. Targeted integration in rat and mouse embryos with zinc-finger nucleases. Nat Biotechnol. 2011;29:64–7.
21. Ferguson C, McKay M, Harris RA, Homanics GE. Toll-like receptor 4 (Tlr4) knockout rats produced by transcriptional activator-like effector nuclease (TALEN)-mediated gene inactivation. Alcohol. 2013;47:595–9.
22. Carbery ID, et al. Targeted genome modification in mice using zinc-finger nucleases. Genetics. 2010;186:451–9.
23. Meyer M, de Angelis MH, Wurst W, Kuhn R. Gene targeting by homologous recombination in mouse zygotes mediated by zinc-finger nucleases. Proc Natl Acad Sci U S A. 2010;107:15022–6.
24. Brinster RL, et al. Targeted correction of a major histocompatibility class II E alpha gene by DNA microinjected into mouse eggs. Proc Natl Acad Sci U S A. 1989;86:7087–91.
25. Chen YG, et al. Gene Targeting in NOD mouse embryos using zinc-finger nucleases. Diabetes. 2014;63:68–74.
26. Meyer M, Ortiz O, Hrabe de Angelis M, Wurst W, Kuhn R. Modeling disease mutations by gene targeting in one-cell mouse embryos. Proc Natl Acad Sci U S A. 2012;109:9354–9.
27. Osiak A, et al. Selection-independent generation of gene knockout mouse embryonic stem cells using zinc-finger nucleases. PLoS One. 2011;6, e28911.
28. Campbell KH, McWhir J, Ritchie WA, Wilmut I. Sheep cloned by nuclear transfer from a cultured cell line. Nature. 1996;380:64–6.
29. Ogura A, Inoue K, Wakayama T. Recent advancements in cloning by somatic cell nuclear transfer. Philos Trans R Soc Lond B Biol Sci. 2013;368:20110329.
30. Whyte JJ, et al. Gene targeting with zinc finger nucleases to produce cloned eGFP knockout pigs. Mol Reprod Dev. 2011;78:2.
31. Hauschild J, et al. Efficient generation of a biallelic knockout in pigs using zinc-finger nucleases. Proc Natl Acad Sci U S A. 2011;108:12013–7.
32. Platt JL, Lin SS. The future promises of xenotransplantation. Ann N Y Acad Sci. 1998;862:5–18.
33. DuBose J, Salim A. Aggressive organ donor management protocol. J Intensive Care Med. 2008;23:367–75.
34. Salim A, et al. The effect of a protocol of aggressive donor management: implications for the national organ donor shortage. J Trauma. 2006;61:429–33; discussion 433-5.
35. Yu S, et al. Highly efficient modification of beta-lactoglobulin (BLG) gene via zinc-finger nucleases in cattle. Cell Res. 2011;21:1638–40.

36. Liu X, et al. Zinc-finger nickase-mediated insertion of the lysostaphin gene into the beta-casein locus in cloned cows. Nat Commun. 2013;4:2565.
37. Oliver SP, Murinda SE. Antimicrobial resistance of mastitis pathogens. Vet Clin North Am Food Anim Pract. 2012;28:165–85.
38. Schuhardt VT, Schindler CA. Lysostaphin therapy in mice infected with Staphylococcus aureus. J Bacteriol. 1964;88:815–6.
39. van den Bogaard AE, Stobberingh EE. Epidemiology of resistance to antibiotics. Links between animals and humans. Int J Antimicrob Agents. 2000;14:327–35.
40. Hollis A, Ahmed Z. Preserving antibiotics, rationally. N Engl J Med. 2013;369:2474–6.
41. McDermott W, Rogers DE. Social ramifications of control of microbial disease. Johns Hopkins Med J. 1982;151:302–12.
42. Lombardo A, et al. Gene editing in human stem cells using zinc finger nucleases and integrase-defective lentiviral vector delivery. Nat Biotechnol. 2007;25:1298–306.
43. Genovese P, et al. Targeted genome editing in human repopulating haematopoietic stem cells. Nature. 2014;510:235–40.
44. Watts KL, et al. CD34(+) expansion with Delta-1 and HOXB4 promotes rapid engraftment and transfusion independence in a Macaca nemestrina cord blood transplant model. Mol Ther. 2013;21:1270–8.
45. Boitano AE, et al. Aryl hydrocarbon receptor antagonists promote the expansion of human hematopoietic stem cells. Science. 2010;329:1345–8.
46. Delaney C, Varnum-Finney B, Aoyama K, Brashem-Stein C, Bernstein ID. Dose-dependent effects of the notch ligand Delta1 on ex vivo differentiation and in vivo marrow repopulating ability of cord blood cells. Blood. 2005;106:2693–9.
47. Mohrin M, et al. Hematopoietic stem cell quiescence promotes error-prone DNA repair and mutagenesis. Cell Stem Cell. 2010;7:174–85.
48. North TE, et al. Prostaglandin E2 regulates vertebrate haematopoietic stem cell homeostasis. Nature. 2007;447:1007–11.
49. Goessling W, et al. Prostaglandin E2 enhances human cord blood stem cell xenotransplants and shows long-term safety in preclinical nonhuman primate transplant models. Cell Stem Cell. 2011;8:445–58.
50. Vargas Jr J, Gusella GL, Najfeld V, Klotman ME, Cara A. Novel integrase-defective lentiviral episomal vectors for gene transfer. Hum Gene Ther. 2004;15:361–72.
51. Nightingale SJ, et al. Transient gene expression by nonintegrating lentiviral vectors. Mol Ther. 2006;13:1121–32.
52. Yanez-Munoz RJ, et al. Effective gene therapy with nonintegrating lentiviral vectors. Nat Med. 2006;12:348–53.
53. Philippe S, et al. Lentiviral vectors with a defective integrase allow efficient and sustained transgene expression in vitro and in vivo. Proc Natl Acad Sci U S A. 2006;103:17684–9.
54. Leavitt AD, Robles G, Alesandro N, Varmus HE. Human immunodeficiency virus type 1 integrase mutants retain in vitro integrase activity yet fail to integrate viral DNA efficiently during infection. J Virol. 1996;70:721–8.
55. Gabriel R, et al. An unbiased genome-wide analysis of zinc-finger nuclease specificity. Nat Biotechnol. 2011;29:816–23.
56. Osborn MJ, et al. TALEN-based gene correction for epidermolysis bullosa. Mol Ther. 2013;21(6):1151–9.
57. Li L, et al. Genomic editing of the HIV-1 coreceptor CCR5 in adult hematopoietic stem and progenitor cells using zinc finger nucleases. Mol Ther. 2013;21:1259–69.
58. Rahman SH, et al. The nontoxic cell cycle modulator indirubin augments transduction of adeno-associated viral vectors and zinc-finger nuclease-mediated gene targeting. Hum Gene Ther. 2013;24:67–77.
59. Saydaminova K, et al. Efficient genome editing in hematopoietic stem cells with helper-dependent Ad5/35 vectors expressing site-specific endonucleases under microRNA regulation. Mol Ther Methods Clin Dev. 2015;1:14057.
60. Stephen SL, et al. Chromosomal integration of adenoviral vector DNA in vivo. J Virol. 2010;84:9987–94.

61. Harui A, Suzuki S, Kochanek S, Mitani K. Frequency and stability of chromosomal integration of adenovirus vectors. J Virol. 1999;73:6141–6.
62. Chen Z, et al. Receptor-mediated delivery of engineered nucleases for genome modification. Nucleic Acids Res. 2013;41, e182.
63. Holt N, et al. Human hematopoietic stem/progenitor cells modified by zinc-finger nucleases targeted to CCR5 control HIV-1 in vivo. Nat Biotechnol. 2010;28:839–47.
64. Hofer U, et al. Pre-clinical modeling of CCR5 knockout in human hematopoietic stem cells by zinc finger nucleases using humanized mice. J Infect Dis. 2013;208 Suppl 2:S160–4.
65. Allers K, et al. Evidence for the cure of HIV infection by CCR5Delta32/Delta32 stem cell transplantation. Blood. 2011;117:2791–9.
66. Hutter G, et al. Long-term control of HIV by CCR5 Delta32/Delta32 stem-cell transplantation. N Engl J Med. 2009;360:692–8.
67. de Mendoza C, et al. Prevalence of X4 tropic viruses in patients recently infected with HIV-1 and lack of association with transmission of drug resistance. J Antimicrob Chemother. 2007;59:698–704.
68. Didigu CA, et al. Simultaneous zinc-finger nuclease editing of the HIV coreceptors ccr5 and cxcr4 protects CD4+ T cells from HIV-1 infection. Blood. 2013.
69. Hockemeyer D, et al. Efficient targeting of expressed and silent genes in human ESCs and iPSCs using zinc-finger nucleases. Nat Biotechnol. 2009;27:851–7.
70. DeKelver RC, et al. Functional genomics, proteomics, and regulatory DNA analysis in isogenic settings using zinc finger nuclease-driven transgenesis into a safe harbor locus in the human genome. Genome Res. 2010;20:1133–42.
71. Lombardo A, et al. Site-specific integration and tailoring of cassette design for sustainable gene transfer. Nat Methods. 2011;8:861–9.
72. Wang Y, et al. Genome editing of human embryonic stem cells and induced pluripotent stem cells with zinc finger nucleases for cellular imaging. Circ Res. 2012;111:1494–503.
73. Sadelain M, Papapetrou EP, Bushman FD. Safe harbours for the integration of new DNA in the human genome. Nat Rev Cancer. 2012;12:51–8.
74. Zou J, et al. Gene targeting of a disease-related gene in human induced pluripotent stem and embryonic stem cells. Cell Stem Cell. 2009;5:97–110.
75. Lengner CJ, et al. Derivation of pre-X inactivation human embryonic stem cells under physiological oxygen concentrations. Cell. 2010;141:872–83.
76. Narsinh KH, et al. Generation of adult human induced pluripotent stem cells using nonviral minicircle DNA vectors. Nat Protoc. 2011;6:78–88.
77. Lin T, et al. A chemical platform for improved induction of human iPSCs. Nat Methods. 2009;6:805–8.
78. Kim D, et al. Generation of human induced pluripotent stem cells by direct delivery of reprogramming proteins. Cell Stem Cell. 2009;4:472–6.
79. Soldner F, et al. Generation of isogenic pluripotent stem cells differing exclusively at two early onset Parkinson point mutations. Cell. 2011;146:318–31.
80. Hussein SM, et al. Copy number variation and selection during reprogramming to pluripotency. Nature. 2011;471:58–62.
81. Dreyer AK, Cathomen T. Zinc-finger nucleases-based genome engineering to generate isogenic human cell lines. Methods Mol Biol. 2012;813:145–56.
82. Sebastiano V, et al. In situ genetic correction of the sickle cell anemia mutation in human induced pluripotent stem cells using engineered zinc finger nucleases. Stem Cells. 2011;29:1717–26.
83. Zou J, Mali P, Huang X, Dowey SN, Cheng L. Site-specific gene correction of a point mutation in human iPS cells derived from an adult patient with sickle cell disease. Blood. 2011;118:4599–608.
84. Voit RA, Hendel A, Pruett-Miller SM, Porteus MH. Nuclease-mediated gene editing by homologous recombination of the human globin locus. Nucleic Acids Res. 2013.
85. Yusa K, et al. Targeted gene correction of alpha1-antitrypsin deficiency in induced pluripotent stem cells. Nature. 2011;478:391–4.

86. Rahman SH, et al. Rescue of DNA-PK signaling and T-cell differentiation by targeted genome editing in a prkdc deficient iPSC disease model. PLoS Genet. 2015;11, e1005239.
87. Jiang J, et al. Translating dosage compensation to trisomy 21. Nature. 2013;500:296–300.
88. Maeder ML, et al. Rapid "open-source" engineering of customized zinc-finger nucleases for highly efficient gene modification. Mol Cell. 2008;31:294–301.
89. Maeder ML, Thibodeau-Beganny S, Sander JD, Voytas DF, Joung JK. Oligomerized pool engineering (OPEN): an 'open-source' protocol for making customized zinc-finger arrays. Nat Protoc. 2009;4:1471–501.
90. Christian ML, et al. Targeting G with TAL effectors: a comparison of activities of TALENs constructed with NN and NK repeat variable di-residues. PLoS One. 2012;7, e45383.
91. Cermak T, et al. Efficient design and assembly of custom TALEN and other TAL effector-based constructs for DNA targeting. Nucleic Acids Res. 2011;39, e82.
92. Moscou MJ, Bogdanove AJ. A simple cipher governs DNA recognition by TAL effectors. Science. 2009;326:1501.
93. Tesson L, et al. Knockout rats generated by embryo microinjection of TALENs. Nat Biotechnol. 2011;29:695–6.
94. Mashimo T, et al. Efficient gene targeting by TAL effector nucleases coinjected with exonucleases in zygotes. Sci Rep. 2013;3:1253.
95. Certo MT, et al. Coupling endonucleases with DNA end-processing enzymes to drive gene disruption. Nat Methods. 2012;9:973–5.
96. Sung YH, et al. Knockout mice created by TALEN-mediated gene targeting. Nat Biotechnol. 2013;31:23–4.
97. Davies B, et al. Site specific mutation of the Zic2 locus by microinjection of TALEN mRNA in mouse CD1, C3H and C57BL/6J oocytes. PLoS One. 2013;8, e60216.
98. Brown SA, et al. Holoprosencephaly due to mutations in ZIC2, a homologue of Drosophila odd-paired. Nat Genet. 1998;20:180–3.
99. Qiu Z, et al. High-efficiency and heritable gene targeting in mouse by transcription activator-like effector nucleases. Nucleic Acids Res. 2013;41, e120.
100. Kahle M, et al. Phenotypic comparison of common mouse strains developing high-fat diet-induced hepatosteatosis. Mol Metab. 2013;2:435–46.
101. Wang H, et al. TALEN-mediated editing of the mouse Y chromosome. Nat Biotechnol. 2013;31:530–2.
102. Takada S, et al. Targeted gene deletion of miRNAs in mice by TALEN system. PLoS One. 2013;8, e76004.
103. Wefers B, et al. Direct production of mouse disease models by embryo microinjection of TALENs and oligodeoxynucleotides. Proc Natl Acad Sci U S A. 2013;110:3782–7.
104. Panda SK, et al. Highly efficient targeted mutagenesis in mice using TALENs. Genetics. 2013;195:703–13.
105. Low BE, et al. Correction of Crb1rd8 allele and retinal phenotype in C57BL/6N mice via TALEN-mediated homology-directed repair. Invest Ophthalmol Vis Sci. 2013.
106. Orlando SJ, et al. Zinc-finger nuclease-driven targeted integration into mammalian genomes using donors with limited chromosomal homology. Nucleic Acids Res. 2010;38, e152.
107. Carlson DF, et al. Efficient TALEN-mediated gene knockout in livestock. Proc Natl Acad Sci U S A. 2012;109:17382–7.
108. Hanna J, et al. Treatment of sickle cell anemia mouse model with iPS cells generated from autologous skin. Science. 2007;318:1920–3.
109. Tan W, et al. Efficient nonmeiotic allele introgression in livestock using custom endonucleases. Proc Natl Acad Sci U S A. 2013;110:16526–31.
110. Xin J, et al. Highly efficient generation of GGTA1 biallelic knockout inbred mini-pigs with TALENs. PLoS One. 2013;8, e84250.
111. Christian M, et al. Targeting DNA double-strand breaks with TAL effector nucleases. Genetics. 2010;186:757–61.
112. Miller JC, et al. A TALE nuclease architecture for efficient genome editing. Nat Biotechnol. 2011;29:143–8.

113. Mussolino C, et al. A novel TALE nuclease scaffold enables high genome editing activity in combination with low toxicity. Nucleic Acids Res. 2011;39:9283–93.
114. Liu H, et al. TALEN-mediated gene mutagenesis in rhesus and cynomolgus monkeys. Cell Stem Cell. 2014;14:323–8.
115. Hockemeyer D, et al. Genetic engineering of human pluripotent cells using TALE nucleases. Nat Biotechnol. 2011;29:731–4.
116. Ding Q, et al. A TALEN genome-editing system for generating human stem cell-based disease models. Cell Stem Cell. 2013;12:238–51.
117. Hussain K, et al. An activating mutation of AKT2 and human hypoglycemia. Science. 2011;334:474.
118. Musunuru K, et al. From noncoding variant to phenotype via SORT1 at the 1p13 cholesterol locus. Nature. 2010;466:714–9.
119. Zhu F, et al. DICE, an efficient system for iterative genomic editing in human pluripotent stem cells. Nucleic Acids Res. 2013.
120. Zhu H, et al. Baculoviral transduction facilitates TALEN-mediated targeted transgene integration and Cre/LoxP cassette exchange in human-induced pluripotent stem cells. Nucleic Acids Res. 2013;41, e180.
121. Zhang Z, et al. Dissecting the roles of miR-302/367 cluster in cellular reprogramming using TALE-based repressor and TALEN. Stem Cell Reports. 2013;1:218–25.
122. Bultmann S, et al. Targeted transcriptional activation of silent oct4 pluripotency gene by combining designer TALEs and inhibition of epigenetic modifiers. Nucleic Acids Res. 2012;40:5368–77.
123. Maeder ML, et al. Targeted DNA demethylation and activation of endogenous genes using programmable TALE-TET1 fusion proteins. Nat Biotechnol. 2013;31:1137–42.
124. Choi SM, et al. Efficient drug screening and gene correction for treating liver disease using patient-specific stem cells. Hepatology. 2013;57:2458–68.
125. Sun N, Zhao H. Seamless correction of the sickle cell disease mutation of the HBB gene in human induced pluripotent stem cells using TALENs. Biotechnol Bioeng. 2013.
126. Cermak T, et al. Efficient design and assembly of custom TALEN and other TAL effector-based constructs for DNA targeting. Nucleic Acids Res. 2011;39(17):7879.
127. Holkers M, et al. Differential integrity of TALE nuclease genes following adenoviral and lentiviral vector gene transfer into human cells. Nucleic Acids Res. 2013;41, e63.
128. Jinek M, et al. A programmable dual-RNA-guided DNA endonuclease in adaptive bacterial immunity. Science. 2012;337:816–21.
129. Mali P, et al. RNA-guided human genome engineering via Cas9. Science. 2013;339(6121):823–6.
130. Cong L, et al. Multiplex genome engineering using CRISPR/Cas systems. Science. 2013;339(6121):819–23.
131. Mashiko D, et al. Generation of mutant mice by pronuclear injection of circular plasmid expressing Cas9 and single guided RNA. Sci Rep. 2013;3:3355.
132. Li D, et al. Heritable gene targeting in the mouse and rat using a CRISPR-Cas system. Nat Biotechnol. 2013;31:681–3.
133. Fujii W, Kawasaki K, Sugiura K, Naito K. Efficient generation of large-scale genome-modified mice using gRNA and CAS9 endonuclease. Nucleic Acids Res. 2013;41, e187.
134. Zhou J, et al. One-step generation of different immunodeficient mice with multiple gene modifications by CRISPR/Cas9 mediated genome engineering. Int J Biochem Cell Biol. 2014;46:49–55.
135. Wang H, et al. One-step generation of mice carrying mutations in multiple genes by CRISPR/Cas-mediated genome engineering. Cell. 2013;153:910–8.
136. Yang H, et al. One-step generation of mice carrying reporter and conditional alleles by CRISPR/Cas-mediated genome engineering. Cell. 2013;154:1370–9.
137. Koike-Yusa H, Li Y, Tan E-P, Velasco-Herrera MDC, Yusa K. Genome-wide recessive genetic screening in mammalian cells with a lentiviral CRISPR-guide RNA library. Nat Biotechnol. 2014;32(3):267–73.

138. Ma Y, et al. Generating rats with conditional alleles using CRISPR/Cas9. Cell Res. 2014;24:122–5.
139. Li W, et al. Genetic modification and screening in Rat using haploid embryonic stem cells. Cell Stem Cell. 2013.
140. Jacob HJ, Kwitek AE. Rat genetics: attaching physiology and pharmacology to the genome. Nat Rev Genet. 2002;3:33–42.
141. Wan H, et al. One-step generation of p53 gene biallelic mutant Cynomolgus monkey via the CRISPR/Cas system. Cell Res. 2015;25:258–61.
142. Shalem O, et al. Genome-scale CRISPR-Cas9 knockout screening in human cells. Science. 2014;343:84–7.
143. Hou Z, et al. Efficient genome engineering in human pluripotent stem cells using Cas9 from Neisseria meningitidis. Proc Natl Acad Sci U S A. 2013;110:15644–9.
144. Esvelt KM, et al. Orthogonal Cas9 proteins for RNA-guided gene regulation and editing. Nat Methods. 2013;10:1116–21.
145. Chen G, et al. Chemically defined conditions for human iPSC derivation and culture. Nat Methods. 2011;8:424–9.
146. Ran FA, et al. In vivo genome editing using Staphylococcus aureus Cas9. Nature. 2015;520:186–91.
147. Kleinstiver BP, et al. Engineered CRISPR-Cas9 nucleases with altered PAM specificities. Nature. 2015.
148. Schwank G, et al. Functional repair of CFTR by CRISPR/Cas9 in intestinal stem cell organoids of cystic fibrosis patients. Cell Stem Cell. 2013;13:653–8.
149. Wu Y, et al. Correction of a genetic disease in mouse via use of CRISPR-Cas9. Cell Stem Cell. 2013;13:659–62.
150. Sim X, Cardenas-Diaz FL, French DL, Gadue P. A doxycycline-inducible system for genetic correction of iPSC disease models. Methods Mol Biol. 2016;1353:13–23.
151. Song B, et al. Improved hematopoietic differentiation efficiency of gene-corrected beta-thalassemia induced pluripotent stem cells by CRISPR/Cas9 system. Stem Cells Dev. 2015;24:1053–65.
152. Li HL, et al. Precise correction of the dystrophin gene in duchenne muscular dystrophy patient induced pluripotent stem cells by TALEN and CRISPR-Cas9. Stem Cell Reports. 2015;4:143–54.
153. Xie F, et al. Seamless gene correction of beta-thalassemia mutations in patient-specific iPSCs using CRISPR/Cas9 and piggyBac. Genome Res. 2014;24:1526–33.
154. Flynn R, et al. CRISPR-mediated genotypic and phenotypic correction of a chronic granulomatous disease mutation in human iPS cells. Exp Hematol. 2015;43(10):838–48.e3.
155. Huang X, et al. Production of gene-corrected adult beta globin protein in human erythrocytes differentiated from patient iPSCs after genome editing of the sickle point mutation. Stem Cells. 2015;33:1470–9.
156. Grobarczyk B, Franco B, Hanon K, Malgrange B. Generation of isogenic human iPS cell line precisely corrected by genome editing using the CRISPR/Cas9 system. Stem Cell Rev. 2015;11(5):774–87.
157. Miyaoka Y, et al. Isolation of single-base genome-edited human iPS cells without antibiotic selection. Nat Methods. 2014;11:291–3.
158. Gasiunas G, Barrangou R, Horvath P, Siksnys V. Cas9-crRNA ribonucleoprotein complex mediates specific DNA cleavage for adaptive immunity in bacteria. Proc Natl Acad Sci U S A. 2012;109(39):E2579–86.
159. Davis L, Maizels N. Homology-directed repair of DNA nicks via pathways distinct from canonical double-strand break repair. Proc Natl Acad Sci U S A. 2014;111:E924–32.
160. Osborn M, Gabriel R, Webber BR, DeFeo AP, McElroy AN, Jarjour J, Starker CG, Wagner JE, Joung JK, Voytas DF, von Kalle C, Schmidt M, Blazar BR, Tolar J. Fanconi anemia gene editing by the CRISPR/Cas9 system. Hum Gene Ther. 2014;26(2):114–26.
161. Guilinger JP, Thompson DB, Liu DR. Fusion of catalytically inactive Cas9 to FokI nuclease improves the specificity of genome modification. Nat Biotechnol. 2014;32:577–82.

162. Wyvekens N, Topkar VV, Khayter C, Joung JK, Tsai SQ. Dimeric CRISPR RNA-guided FokI-dCas9 nucleases (RFNs) directed by truncated gRNAs for highly specific genome editing. Hum Gene Ther. 2015.
163. Tsai SQ, et al. Dimeric CRISPR RNA-guided FokI nucleases for highly specific genome editing. Nat Biotechnol. 2014;32:569–76.
164. Cheng AW, et al. Multiplexed activation of endogenous genes by CRISPR-on, an RNA-guided transcriptional activator system. Cell Res. 2013;23:1163–71.
165. Kearns NA, et al. Cas9 effector-mediated regulation of transcription and differentiation in human pluripotent stem cells. Development. 2014;141:219–23.
166. Chavez A, et al. Highly efficient Cas9-mediated transcriptional programming. Nat Methods. 2015;12:326–8.
167. Konermann S, et al. Genome-scale transcriptional activation by an engineered CRISPR-Cas9 complex. Nature. 2015;517:583–8.
168. Hilton IB, et al. Epigenome editing by a CRISPR-Cas9-based acetyltransferase activates genes from promoters and enhancers. Nat Biotechnol. 2015;33:510–7.
169. Amps K, et al. Screening ethnically diverse human embryonic stem cells identifies a chromosome 20 minimal amplicon conferring growth advantage. Nat Biotechnol. 2011;29:1132–44.
170. Ben-David U, Benvenisty N. High prevalence of evolutionarily conserved and species-specific genomic aberrations in mouse pluripotent stem cells. Stem Cells. 2012;30:612–22.
171. Mayshar Y, et al. Identification and classification of chromosomal aberrations in human induced pluripotent stem cells. Cell Stem Cell. 2010;7:521–31.
172. Mayshar Y, Yanuka O, Benvenisty N. Teratogen screening using transcriptome profiling of differentiating human embryonic stem cells. J Cell Mol Med. 2011;15:1393–401.
173. Ruiz S, et al. Analysis of protein-coding mutations in hiPSCs and their possible role during somatic cell reprogramming. Nat Commun. 2013;4:1382.
174. Liang G, Zhang Y. Embryonic stem cell and induced pluripotent stem cell: an epigenetic perspective. Cell Res. 2013;23:49–69.
175. Tchieu J, et al. Female human iPSCs retain an inactive X chromosome. Cell Stem Cell. 2010;7:329–42.
176. Mekhoubad S, et al. Erosion of dosage compensation impacts human iPSC disease modeling. Cell Stem Cell. 2012;10:595–609.
177. Anguera MC, et al. Molecular signatures of human induced pluripotent stem cells highlight sex differences and cancer genes. Cell Stem Cell. 2012;11:75–90.
178. Lister R, et al. Hotspots of aberrant epigenomic reprogramming in human induced pluripotent stem cells. Nature. 2011;471:68–73.
179. Ruiz S, et al. Identification of a specific reprogramming-associated epigenetic signature in human induced pluripotent stem cells. Proc Natl Acad Sci U S A. 2012;109:16196–201.
180. Lee MO, et al. Inhibition of pluripotent stem cell-derived teratoma formation by small molecules. Proc Natl Acad Sci U S A. 2013;110:E3281–90.
181. Tang C, et al. An antibody against SSEA-5 glycan on human pluripotent stem cells enables removal of teratoma-forming cells. Nat Biotechnol. 2011;29:829–34.
182. Okita K, Nagata N, Yamanaka S. Immunogenicity of induced pluripotent stem cells. Circ Res. 2011;109:720–1.
183. Pattanayak V, Ramirez CL, Joung JK, Liu DR. Revealing off-target cleavage specificities of zinc-finger nucleases by in vitro selection. Nat Methods. 2011;8(9):765–70.
184. Cheng L, Blazar B, High K, Porteus M. Zinc fingers hit off target. Nat Med. 2011;17(10):1192–3.
185. Sander JD, et al. In silico abstraction of zinc finger nuclease cleavage profiles reveals an expanded landscape of off-target sites. Nucleic Acids Res. 2013;41, e181.
186. Ran FA, et al. Double nicking by RNA-guided CRISPR Cas9 for enhanced genome editing specificity. Cell. 2013;154:1380–9.
187. Ran FA, et al. Genome engineering using the CRISPR-Cas9 system. Nat Protoc. 2013;8:2281–308.

188. Paruzynski A, et al. Genome-wide high-throughput integrome analyses by nrLAM-PCR and next-generation sequencing. Nat Protoc. 2010;5(8):1379–95.
189. Tsai SQ, et al. GUIDE-seq enables genome-wide profiling of off-target cleavage by CRISPR-Cas nucleases. Nat Biotechnol. 2015;33:187–97.
190. Kim D, et al. Digenome-seq: genome-wide profiling of CRISPR-Cas9 off-target effects in human cells. Nat Methods. 2015;12:237–43, 1 p following 243.
191. Frock RL, et al. Genome-wide detection of DNA double-stranded breaks induced by engineered nucleases. Nat Biotechnol. 2015;33:179–86.
192. Crosetto N, et al. Nucleotide-resolution DNA double-strand break mapping by next-generation sequencing. Nat Methods. 2013;10:361–5.
193. Tebas P, et al. Gene editing of CCR5 in autologous CD4 T cells of persons infected with HIV. N Engl J Med. 2014;370:901–10.
194. Doyon Y, et al. Enhancing zinc-finger-nuclease activity with improved obligate heterodimeric architectures. Nat Methods. 2011;8(1):74–9.
195. Fu Y, Sander JD, Reyon D, Cascio VM, Joung JK. Improving CRISPR-Cas nuclease specificity using truncated guide RNAs. Nat Biotechnol. 2014;32:279–84.
196. Nakano-Okuno M, Borah BR, Nakano I. Ethics of iPSC-based clinical research for age-related macular degeneration: patient-centered risk-benefit analysis. Stem Cell Rev. 2014; 10:743–52.
197. Mardis ER. The $1,000 genome, the $100,000 analysis? Genome Med. 2010;2:84.
198. Lanphier E, Urnov F, Haecker SE, Werner M, Smolenski J. Don't edit the human germ line. Nature. 2015;519:410–1.
199. Liang P, et al. CRISPR/Cas9-mediated gene editing in human tripronuclear zygotes. Protein Cell. 2015;6:363–72.

Index

T. Cathomen et al. (eds.), *Genome Editing*, Advances in Experimental Medicine and Biology 895, DOI 10.1007/978-1-4939-3509-3

MIX
Papier aus verantwortungsvollen Quellen
Paper from responsible sources
FSC® C105338

If you have any concerns about our products,
you can contact us on
ProductSafety@springernature.com

In case Publisher is established outside the EU,
the EU authorized representative is:
Springer Nature Customer Service Center GmbH
Europaplatz 3, 69115 Heidelberg, Germany

Printed by Libri Plureos GmbH
in Hamburg, Germany